UNDERSTANDING COMPLEX ECOSYSTEM DYNAMICS

A Systems and Engineering Perspective

UNDERSTANDING COMPLEX ECOSYSTEM DYNAMICS

A Systems and Engineering Perspective

WILLIAM S. YACKINOUS

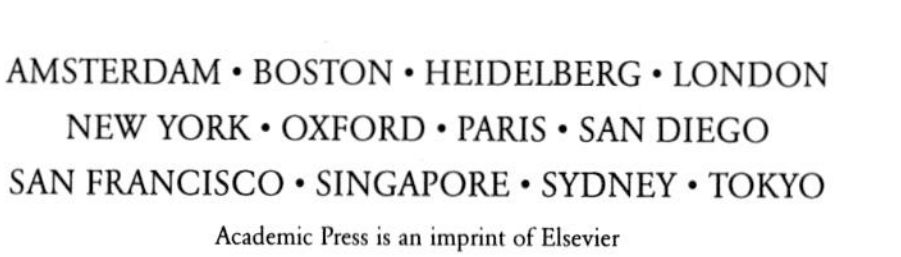
AMSTERDAM • BOSTON • HEIDELBERG • LONDON
NEW YORK • OXFORD • PARIS • SAN DIEGO
SAN FRANCISCO • SINGAPORE • SYDNEY • TOKYO
Academic Press is an imprint of Elsevier

Academic Press is an imprint of Elsevier
125 London Wall, London, EC2Y 5AS, UK
525 B Street, Suite 1800, San Diego, CA 92101-4495, USA
225 Wyman Street, Waltham, MA 02451, USA
The Boulevard, Langford Lane, Kidlington, Oxford OX5 1GB, UK

ISBN: 978-0-12-802031-9

Library of Congress Cataloging-in-Publication Data
A catalog record for this book is available from the Library of Congress

British Library Cataloguing in Publication Data
A catalogue record for this book is available from the British Library

For information on all Academic Press publications
visit our website at http://store.elsevier.com/

Cover Credit: Joyce Yackinous

CONTENTS

PREFACE

"Look deep, deep into nature, and then you will understand everything better."
Albert Einstein[1]

I have had a 34-year professional career as a systems engineer at Bell Laboratories. Originally, The Bell Telephone Laboratories was the research and development arm of the Bell System in the United States. From the 1920s to about 2000, Bell Labs was acknowledged by many to be the premier R&D labs in the world. The term "systems engineering" was coined at Bell Laboratories in the 1940s. Throughout my career, I worked to solve systems problems for Bell Labs and for its clients across the United States and around the world. During that time, I acquired a very significant set of skills and perspectives with respect to the practice of systems engineering and the systems approach. In my Bell Labs work, I focused on *building* human-made systems (some would say "artificial" systems). I have always thought, however, that this same set of skills and perspectives could be applied very beneficially to *understanding* natural systems—specifically natural ecological systems. After taking early retirement from Bell Labs in 2001, I began to do just that. I began to follow Einstein's admonition (above) in earnest.

My first task was to explicitly define and describe the systems and systems engineering skill set that I had acquired during my career. The science and engineering working environment at Bell Labs was outstanding and full of opportunities. Over the years, I was fortunate[2] to be in a position to earn every major Bell Labs technical award and honorarium including the highest and most prestigious—Bell Labs Fellow. As a result, I frequently had the opportunity to give talks and seminars on systems engineering both inside Bell Labs and at major universities. I gave presentations, on behalf of Bell Laboratories, to graduate engineering and computer science students and faculty at Stanford University, UC Berkeley, Cal Tech, UCLA, University of Arizona, University of Texas, Purdue University, and Oxford University. My objective in these various talks was to describe my view of the required skills and perspectives necessary for the successful practice of systems

[1] Albert Einstein, in *The Quotable Einstein*, Princeton University Press, 1996, p. 32.

[2] I have found that such opportunity and good fortune is directly related to having very good bosses. They were plentiful at Bell Labs. I've had some of the best.

engineering. Beginning with the material comprising the talks, I have developed "A Systems Engineering Skills Framework." We will discuss that in Chapter 1.

My next task—and my primary ongoing work objective—was the application of those systems and engineering skills to increase understanding of complex natural systems. Given the state of our planet's environment, I decided that the most important and relevant natural systems for me to consider were ecological systems. I needed to engage, therefore, in very substantial ecological system study and research. To that end, I pursued a PhD in Ecology at the University of Georgia's Eugene P. Odum School of Ecology. Dr. Eugene Odum was a very influential ecologist, a pioneer in the area of systems ecology, and a founder of the ecology program at the University of Georgia. Dr. Bernard C. Patten was my PhD major professor and my doctoral committee chairperson. He is a systems ecologist and is recognized as one of the premier ecological system modelers and ecological network analysts in the field. I studied with Dr. Patten and the Systems and Engineering Ecology group—an interdisciplinary group of professors and graduate students at the University of Georgia. I received my PhD degree in Ecology in December 2010. My doctoral research and dissertation on "Emerging Principles of Ecological Network Dynamics" are prominently reflected in this book.

As I have explained, the systems and engineering skills and perspectives that I acquired at Bell Labs are absolutely essential to my ecosystem research. My related systems and engineering academic background also plays an important role. My early degrees—bachelors and masters—are in electrical engineering. My master's degree work (and early Bell Labs work) included the areas of communications theory, signal processing theory, information theory, control theory, and network theory. My more recent PhD academic work has enhanced my understanding of the various important subsets of complex systems theory. All of these scientific disciplines and their associated methods and tools contribute to my current complex system dynamics work. You will notice that throughout the book.

Highly complex systems (including ecosystems) take the form of networks. My work addresses challenging and open issues of complex system network dynamics. It pushes boundaries and explores frontiers. Thus far, mainstream network science has focused primarily on understanding network structure. Network dynamics is a much tougher issue. Melanie Mitchell, in her book on complexity, has said, "To understand the dynamics . . . network science will have to characterize networks in which the nodes and links continually change in both time and space. This will be a major challenge, to say the least" (Mitchell, 2009). Duncan Watts has said, "Next to the mysteries of dynamics on a network ... the problems of networks that we have encountered up to now are just pebbles on the seashore" (Watts, 2003). In this book, we take on the challenge of understanding complex system network dynamics—specifically complex ecosystem network dynamics.

The complex system dynamics research described in the book is innovative and unique in many respects. The full set of experiences, approaches, methods, and tools employed here have never before been applied to ecological systems (or to complex systems in general). The work represents a "fresh look" at ecological network dynamics. I have begun with substantial systems and engineering experience and knowledge; combined that with PhD-level study of ecology; and further, have supplemented those resources with additional study of complex systems theory in the areas of networks, nonlinear dynamics, cellular automata, and roughness (fractals). I have assimilated, extended, and combined all of these resources in new ways to create a fresh view of complex system dynamics. That process is the working definition of innovation and creativity. Creativity is "the ability to discover new relationships, to look at subjects from new perspectives, and to form new combinations from two or more concepts already in the mind" (Evans, 1991). Scientific discovery often arises from "picking up the stick from the other end" (Butterfield, 1960). The other end, in this case, is the systems and engineering perspective.

This book clearly takes an interdisciplinary perspective and contains many new ideas that sometimes do *not* represent merely small, incremental changes to existing scientific paradigms. Although such new perspectives and ideas are often necessary to push research boundaries, they are not always easily accepted by natural science investigators who prefer more traditional approaches and paradigms. (Thomas Kuhn (1996), in his classic book *The Structure of Scientific Revolutions*, comprehensively describes these effects.) Some of the ideas and hypotheses in the book, therefore, may be controversial. That circumstance may generate an increased focus on important complex system issues and motivate scientists across disciplines to ponder and explore these issues. In my view, that would be a very desirable outcome of the book. It would help to further our collective understanding of complex system dynamics.

Throughout this book, I reference and quote a lot of very smart people. Except for a few, I know these people only by their work. Their work has been exceedingly helpful. I quite agree with the sentiment that Thoreau[3] expressed long ago in his *Journals*: "He who speaks with most authority on a given subject is not ignorant of what has been said by his predecessors. He will take his place in a regular order, and substantially add his own knowledge to the knowledge of previous generations." I certainly do not claim that I speak with "most authority"—but I am pleased to have the opportunity to express my views in this book.

We began this preface with some advice from Albert Einstein. I have attempted to look deep into nature. I think that now I really do understand everything better. I'm still working on it . . ..

Bill Yackinous
November 2014

[3] Henry David Thoreau, *Journals*, December 31, 1859.

INTRODUCTION

The primary purpose of my work in this book is to increase the understanding of complex system dynamics—in particular, complex ecological system dynamics. My systems and engineering perspective is foundational to this effort. I begin by defining and describing pertinent systems and engineering skills and practices, including an explanation of the systems approach and its major elements. Consistent with the systems approach, I then formulate an ecosystem dynamics functionality-based framework to guide my investigations. Complex systems theory, across many subject matter areas, is crucial to the work of this book. I cover relevant network theory, nonlinear dynamics theory, cellular automata theory, and roughness (fractal) theory in some detail. This material serves as an important resource as we proceed in the book. Next, in the context of all of the foregoing investigation, I construct a view of the characteristics of ecological network dynamics. This view, in turn, is the basis for the central hypothesis of the book, i.e., ecological networks are ever-changing networks with propagation dynamics that are punctuated, local-to-global, and perhaps most importantly *fractal*. To analyze and fully test this hypothesis I define, design, and develop an innovative ecological network dynamics model. The modeling approach seeks to emulate features of real-world ecological networks. The approach does not make *a priori* assumptions about ecological network dynamics, but rather lets the dynamics develop as the model simulation runs. Model analysis results corroborate the hypothesis. Additional important insights and principles are suggested by the model analysis results and by the other supporting investigations of this book—and may serve as a basis for going-forward complex system dynamics research, not only for ecological systems but also for complex systems in general.

"MAP" OF THE BOOK

The book has six major parts comprised of nineteen chapters. There is also an appendix. An overview of each of the book's components follows.

Part I The Systems and Engineering Perspective

Part I (Chapters 1 through 4) provides a comprehensive look at the systems and engineering perspective that is foundational to the work of this book.

Chapter 1 describes my view of systems engineering skills, the systems approach, and the associated systems perspectives that can be beneficially applied to understanding highly complex natural systems. Chapter 2 discusses additional views on systems thinking from the scientific community—as well as more of my own views. In Chapter 3, I detail three important concepts that I consider to be major elements of the systems approach. They are: a blend of synthesis and analysis; network thinking; and the systems triad. Note that traditional scientific research (including ecological research) is most often conducted using a reductionist approach rather than a systems approach. Chapter 4 addresses a significant potential problem with the indiscriminate use of reductionism, i.e., reductionism can isolate the target of investigation from the larger system in which it resides and thereby cause information loss.

Part II A Function-Structure-Process Framework for Ecological System Dynamics

In Part II (Chapters 5 through 7), I construct and describe a functionality-based framework that provides a unifying context for exploring principles of ecosystem dynamics. In systems engineering, I have found that such a framework is essential for specifying and guiding the design and development of artificial (human-made) systems. In systems ecology, such a framework is equally essential for understanding natural systems. Chapter 5 provides an overview of this ecosystem dynamics framework—which consists of operational, developmental, and core functional tiers. In Chapter 6, one of the core ecological system functions, regulation/adaptation, is discussed in detail. Chapter 7 addresses the developmental tier. I make the case that the species evolution function provides the basis for a universal development model. The operational tier becomes our focus in Parts IV and V of the book. (All of the elements of the framework are discussed, to varying degrees, throughout the book.)

Part III Complex Systems Theory: Networks, Nonlinear Dynamics, Cellular Automata, and Fractals (Roughness)

In Part III (Chapters 8 through 12), we conduct an extensive review of the pertinent extant complex systems theory. Chapters 8 and 9 cover network theory. Those two chapters address the structure aspects and the dynamics aspects, respectively, of complex networks. Chapter 10 reviews nonlinear dynamics theory. Chapter 11 is about cellular automata investigations and associated emerging complex system principles. Chapter 12 addresses fractals (roughness theory). In some areas, I provide additional commentary based on my systems, engineering, and ecological perspectives. The material of Part III serves as a valuable and necessary resource for our work. Application and, in some cases, extensions of the theory contribute to a "synthesis of ideas" that is pursued in the subsequent parts of the book.

Part IV A View of the Characteristics of Ecological Network Dynamics

Based on knowledge of the systems approach (Part I), the ecosystem dynamics framework (Part II), and applicable complex systems theory (Part III), a view of the characteristics of ecological network dynamics is constructed in Part IV (Chapters 13–15). First, we do a bit more investigation to properly set the stage. Chapter 13 addresses the human perceptual context in which we are working, especially the human tendency to see smoothness, stability, and continuity in the natural world—even when they are absent. Chapter 14 considers the nature of order and complexity in ecological systems—and their relationships—to gain additional insights into the behavior of highly complex systems. Now we are ready to proceed with the dynamics characteristics. Chapter 15 describes a comprehensive view of the behavioral characteristics of ecological network dynamics, which is the basis for the central hypothesis of the book: ecological networks are ever-changing, "flickering" networks with propagation dynamics that are punctuated, fractal, local-to-global, and enabled by indirect effects.

Part V Modeling Ecological Network Dynamics and the Generation and Analysis of Results

In Part V (Chapters 16–18), I describe the development of an innovative ecological network dynamics model, the generation of results, and the analysis of those results in order to test our central characteristics hypothesis. Model requirements are the subject of Chapter 16. Model software design and development are covered in Chapter 17. The software implements the ecological network operational model, the required analysis activities, and the needed graphics capabilities. Chapter 18 is all about results. Ecological network dynamics results are generated, displayed, and analyzed. The specific dynamics results categories are: operational propagation flow; network propagation events; propagation path length; indirect effects; and network connectivity. The characteristics hypothesis is fully tested—and corroborated.

Part VI Pulling It All Together

In Part VI (Chapter 19), we pull everything together. The chapter begins with a brief summary of the key aspects of the work covered in the book. We then take a broader and more interpretative look at the work with respect to the perspective taken, what we have found, what it means, and its potential influence on work in this area going forward. We see that, although our focus has been ecological systems, there are implications for all complex systems.

Appendix

Complex system dynamics modeling is an important part of my work and an important part of this book. I have developed the software that implements my complex ecosystem dynamics model using the MATLAB[1] programming environment and language. I want readers to have access to the model and the complete MATLAB programming code. Readers are invited to explore, run, and experiment with the model software in order to enhance their understanding of complex system network dynamics—as well as to develop and test their own ideas. The full complex ecosystem dynamics model code is available on the book's companion website.[2] The programming code is heavily commented to explain and describe the software.

In the appendix to this book, I provide selected excerpts of the model programming code. I refer to this material as I discuss the model in the main body of the book. These excerpts provide the reader with easy access to examples of the software without having to navigate through the full set of MATLAB m-files while reading the book. There are nine code excerpts ranging from code that establishes the model network structure and relationships, to code that describes and implements propagation process flow, to code that describes and implements ecosystem dynamics analysis activities.

A "FRESH" LOOK AT COMPLEX SYSTEM DYNAMICS

The book takes a fresh, interdisciplinary look at complex system dynamics and contains many new ideas, perspectives, and areas of emphasis. Here are a few examples. (1) In complexity theory, the emphasis is often on chaotic systems. While chaotic systems are complex, they are not the most complex. They do not exhibit coherent process-over-structure flow that delivers

[1] MATLAB release R2009a, The MathWorks, Inc., February 12, 2009.
[2] To access the companion site, go to booksite.elsevier.com/9780128020319.

meaningful function. I focus on the most-complex systems, e.g., highly complex ecological systems. (2) Extant work on fractals typically emphasizes static spatial fractals (also called structure fractals). Even Benoit Mandelbrot (the originator of fractal geometry) dealt primarily with *structure fractals*. I, on the other hand, focus on dynamic *process fractals*—which have both spatial and temporal aspects. (3) Although most of the extant research efforts on networks have dealt with network structure rather than network dynamics, I have taken on the challenge of complex system network dynamics in this book. (4) I have found that traditional ecosystem models do not adequately represent the real-time ecological system dynamics that I wish to represent. Therefore, I have developed a new innovative modeling approach that does so. (5) Finally, much of the writing on systems thinking is limited to discussion of broad generalizations. My view of the systems approach focuses on tested and proven elements that I have applied successfully in my work for years—and have applied throughout this book as well.

PART I

The Systems and Engineering Perspective

This part provides a comprehensive look at the systems and engineering perspective that is foundational to the work of this book. My views as well as other views from the scientific community are addressed. The systems approach and the reductionist approach are compared and contrasted.

CHAPTER 1

A Systems Engineer's Perspective

This chapter describes my view of the systems engineering skill set, the systems approach, and the associated systems perspectives that can be very beneficially applied to understanding highly complex natural systems. The vehicle for accomplishing this is my systems engineering skills framework. The framework defines the systems engineering domain, and then addresses the four key skill areas: engineering technical methods, system development skills, systems thinking, and communication skills. Engineering technical methods include mathematical methods, empirical methods, and computer simulation. System development skills include system design, system modeling, and system architecting. The major elements of systems thinking are iterative synthesis and analysis, network thinking, and the systems triad. Necessary communication skills are also discussed. Application of these skills to ecological systems is not a giant leap. In the spirit of the systems approach, it is a reasonable and prudent pursuit.

1.1 INTRODUCTION

As I discussed in the Preface to this book, I had a 34-year career as a systems engineer at Bell Laboratories. During that time, I acquired a very substantial set of skills and perspectives with respect to the practice of systems engineering and the systems approach. I frequently had the opportunity to give talks and seminars on systems engineering both inside Bell Labs and at major universities. The university audiences were composed of graduate engineering and computer science students and faculty. My objective in these various talks was to describe my view of the required skill set necessary for the successful practice of systems engineering. It is these same skills and perspectives that I endeavor to apply to ecological systems in this book.

I have explicitly defined and described the systems and systems engineering skill set that I acquired during my career. Starting with the material comprising my Bell Labs and university talks, I have developed a systems engineering skills framework. In this chapter, I'll provide an overview of the skills framework and discuss my systems engineering perspectives. Areas

Understanding Complex Ecosystem Dynamics
http://dx.doi.org/10.1016/B978-0-12-802031-9.00001-2

requiring further elaboration regarding their application to ecological systems are discussed in more detail in subsequent chapters.

1.2 DEFINITIONS

Let's begin with a few definitions and then get right to the framework. The Merriam-Webster online[1] definition of *system* includes 1: a regularly interacting or interdependent group of items forming a unified whole ... as (a): a group of interacting bodies under the influence of related forces ... (b): a group of body organs that together perform one or more vital functions ... (c): a group of related natural objects or forces ... (d): a group of devices or artificial objects or an organization forming a network. The etymology is given as late Latin and Greek with origins from *synistanai*, to combine. The Next Generation Education Project[2] provides a concise and useful definition of system: "A group of interacting, interrelated, or interdependent elements that together form a complex whole."

Systems engineering can be defined as an interdisciplinary approach and means:

1. To enable the realization of successful human-made systems.
2. To increase understanding of natural systems.

The first part of this definition is from INCOSE (The International Council on Systems Engineering).[3] I added the second part.

The term *system(s) engineer* has been used rather loosely by the technical community over the years, and has been given very different meanings in different contexts. For example, the title Microsoft Certified Systems Engineer[4] referred to someone who knows something about Microsoft computer operating systems and has passed some related exams. That usage of the term systems engineer has little to do with what I am talking about here. I consider a much broader context.

1.3 SYSTEMS ENGINEERING SKILLS FRAMEWORK

The systems engineering skills framework is depicted in Figure 1.1.

[1] See http://www.merriam-webster.com/. (Most dictionary definitions in this book are from Merriam-Webster online, unless indicated otherwise.)
[2] See http://www.mdrc.org/, the website of the nonprofit, nonpartisan education and social policy research organization.
[3] See http://www.incose.org/.
[4] Microsoft "retired" this certification several years ago.

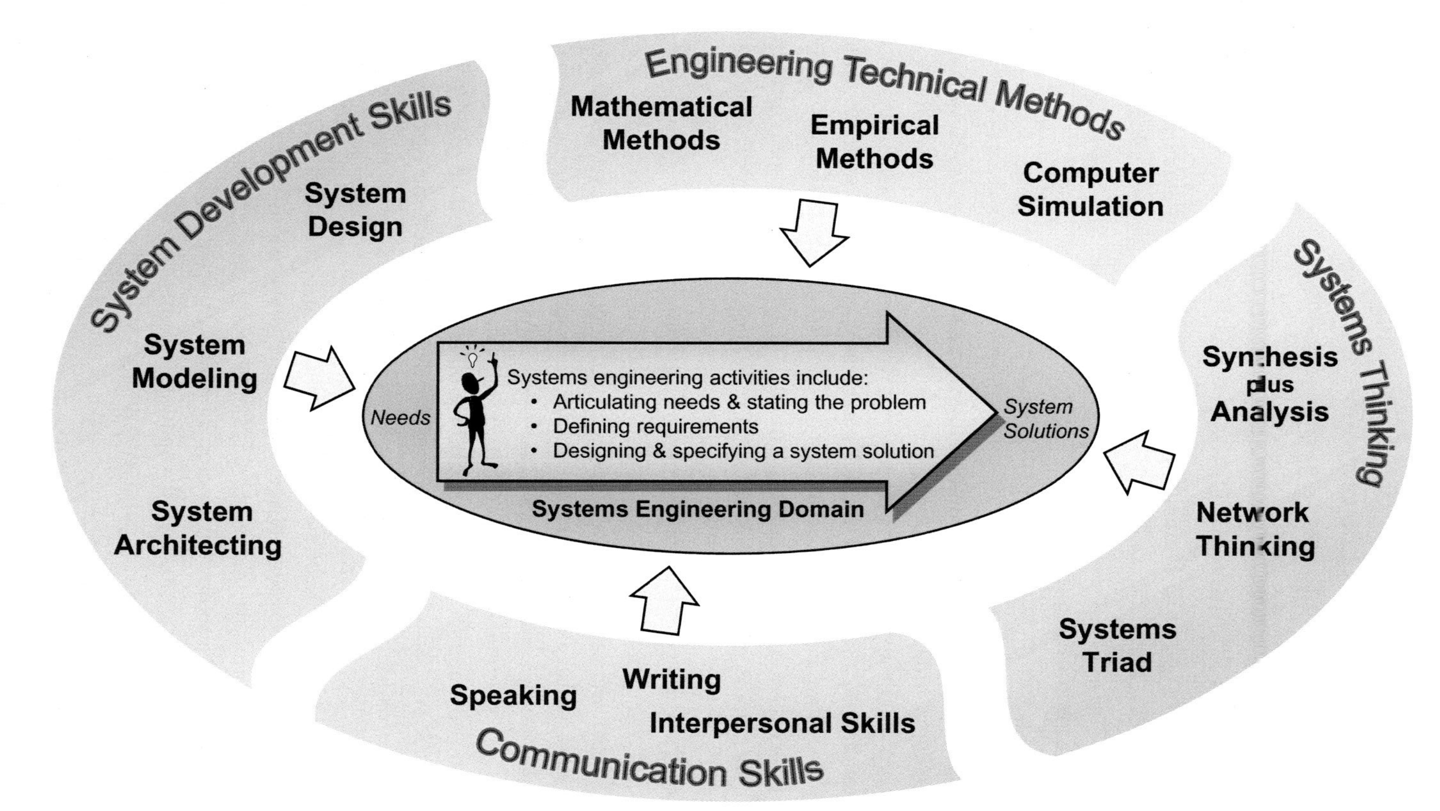

Figure 1.1 The systems engineering skills framework.

Application of these skills to ecological systems is not a giant leap. In the spirit of the systems approach and systems thinking, it is a reasonable and prudent pursuit. The five framework components—systems engineering domain, engineering technical methods, system development skills, systems thinking, and communication skills—are described in Sections 1.3.1–1.3.5, respectively.

1.3.1 Systems Engineering Domain

As shown in the center of Figure 1.1, traditional systems engineering activities include

- Articulating needs and stating the problem.
- Defining requirements.
- Designing and specifying a system solution.

All of these more traditional activities pertain to *building human-made systems*. I want to expand the systems engineering domain and activities to include *understanding natural systems*.

So we want to apply systems engineering skills and techniques to increase understanding of natural systems. Which natural systems in which natural science disciplines can best be addressed by the systems approach and systems engineering? The answer is the most complex systems in the most complex natural science disciplines. Ecosystems and the ecological sciences are in this group.

Consider the complexity spectrum of the natural sciences. Proceeding from least complex to most complex, we find physics and then chemistry at the lower end, and ecological sciences and social sciences at the higher end. Scientific disciplines in the physics realm focus on systems that are (relatively) less complex and more well defined. The system problems that are addressed are known to or expected to have solutions (if only we are clever enough to find them). The "many-body problem," for example, seems too complex. Herbert Simon (1996) points out that since the time of Newton, a comprehensive solution to the many-body problem has never been found. Simon states, "Since Newton, astronomers have been able to compute the motion of a system of two bodies that exercise mutual gravitational attraction on each other. With three or more bodies, they have never obtained more than approximations to the motion." The n-body problem with $n \geq 3$ is too complex for the methods of physics. The physical sciences, of course, have made tremendous progress in understanding systems that do reside in their limited domain. Progress is much more difficult in the ecological sciences.

The ecological sciences do not have the luxury of addressing less-complex systems. All of the most important ecological systems are highly complex. In my view, the systems approach and systems engineering are *essential* for understanding complex ecological systems. These are among the most complex systems in the most complex natural science disciplines. That's where the most help is needed. Ecology is an especially fertile application area for systems and engineering approaches.

1.3.2 Engineering Technical Methods

Traditionally, we have usually thought of two general categories of technical methods: theoretical and empirical. Now, of course, computers are an important part of the technical methods across all of the sciences and engineering. Melanie Mitchell (2009) explains the situation this way: "The more mathematically oriented sciences such as physics, chemistry, and mathematical biology have traditionally concentrated on studying simple, idealized systems that are more tractable via mathematics. However, more recently, the existence of fast, inexpensive computers has made it possible to construct and experiment with models of systems that are too complex to be understood with mathematics alone. . . . The traditional division of science into theory and experiment has been complemented by an additional category: computer simulation."

My categorization of engineering technical methods is shown in Figure 1.1. There are three broad categories:

Mathematical methods
Empirical methods
Computer simulation

The effective systems engineer is proficient in all three.

The three categories are not mutually exclusive and, in my view, all three can have theoretical aspects. Additionally, the first two categories very often employ the third. For example, regarding mathematical methods, in some cases we can write mathematical equations for a system (e.g., a differential equation description of a simplified dynamical system), but cannot solve the equations in closed form. We need to use a computer-based numerical approach (computer simulation) to obtain a solution. In other cases, except perhaps for some specific aspects of a given highly complex system, we cannot even write (let alone solve) mathematical equations that describe the system. Computer simulation of an appropriate system model can provide a means for us to make progress in understanding the complex system. All three of these technical method categories are used extensively in the work described in this book. For the moment, I'll comment further on mathematical methods and then empirical methods. Computer simulation of a cellular automata model of complex ecosystem dynamics is the subject of Part V of the book.

1.3.2.1 Mathematical Methods

Mathematics is very important in all of the natural sciences. Mathematics-based methods are rigorous, objective, and definite—and appealing for those reasons. They work well for less-complex systems, and can also work well for certain aspects of highly complex systems. In my systems engineering work, I use mathematical methods whenever they are helpful and appropriate. For complex systems, I find that mathematics is necessary but not sufficient.

To better understand my view of mathematical methods, let's consider a specific engineering example, that is, mathematics-based optimization theory, which is often used in engineering system design. As we proceed, I think you'll see both the benefits (for simpler systems) and the significant limitations (for complex systems) of this approach.

We'll use Jasbir Arora's (1989) description of optimization theory in system design.[5] Perhaps the first thing that should be mentioned is that Arora does not discuss how to create a system design. (I'll touch on that topic later in this chapter and in Part II of the book.) Rather, Arora addresses the details of optimizing an already-proposed design: "It is assumed that various preliminary analyses have been completed and a detailed design of a concept or a subproblem needs to be carried out." Arora's stated objectives are to provide mathematical foundations for engineering design ("It is my sincere belief that methods of optimum design will form a core for the engineering design process") and to provide a means of systematizing the engineering design practice ("Optimization is viewed as a tool for the systematic design of engineering systems"). Arora's key areas of coverage are

- Mathematical formulation of design problems

  "A proper mathematical statement of the problem is critical for designing workable systems."
- Solution methods (usually computer-based)

  "Many optimization problems must be solved using numerical algorithms because they are difficult to solve by analytical procedures."

  "Optimum design ... requires sophisticated computer programs."
- Dealing with nonlinearity

  "Solutions of many optimization problems are the roots of nonlinear equations."

[5] Unless indicated otherwise, the quotes in this optimum design description are from Arora (1989).

Let's discuss the first of Arora's key areas—design problem formulation—in more detail. The goal is to transcribe the design problem into a well-defined mathematical formulation "where a measure of performance is to be optimized while satisfying all the constraints." This approach "forces the designer to identify explicitly a set of design variables, a cost function to be minimized [or an objective function to be maximized], and the constraint functions for the system. This rigorous formulation of the design problem helps the designer to gain a better understanding of the problem. Proper mathematical formulation of the design problem is a key to good solutions." The steps in the procedure are

- Analyze the design problem.
  (Generate pertinent mathematical expressions that describe the design problem.)
- Identify and define the design variables.
  (Design variables are parameters chosen to represent the design.)
- Specify an objective function or cost function.
  (This is a function of some or all of the design variables that provides a criterion to measure the "performance" of candidate designs. Either an objective function or a cost function can be used. An objective function is to be maximized while a cost function is to be minimized.)
- Identify the system constraints.
  (The constraints specify limits on the values that design variables can take. Express the constraints in terms of some or all of the design variables and identify a resulting feasible region. This region specifies the set of feasible designs, i.e., designs that satisfy all of the constraints.)

Figure 1.2 provides a very simple *design problem formulation* example: optimizing the design of a beverage can.

After design problem formulation, how do you actually find the optimum solution? Arora discusses several solution methods, with a focus on *search methods.* (Search methods work for the beverage can example in Figure 1.2.) The search procedure generally involves starting with an estimate of the solution and then iteratively improving the estimate until the optimum point is reached. This is a search in a design space. The procedure can get extremely complicated even for a relatively simple system. (We won't cover the details here; they are provided in Arora's (1989) book.)

Arora considers a variety of design optimization problems, in addition to the beverage can problem, in his book. "Throughout the text, simple design

A. Analyze the design problem.

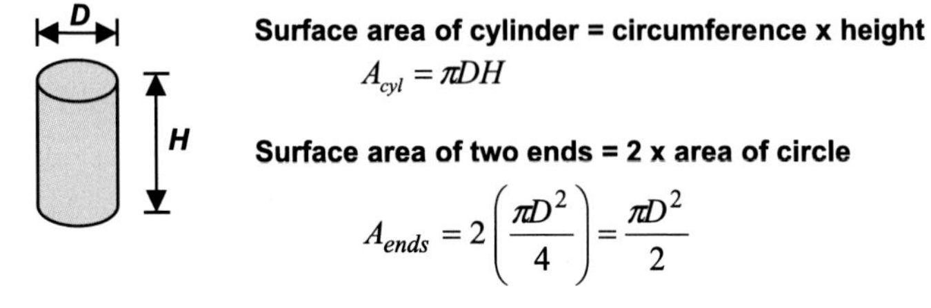

Surface area of cylinder = circumference x height

$$A_{cyl} = \pi DH$$

Surface area of two ends = 2 x area of circle

$$A_{ends} = 2\left(\frac{\pi D^2}{4}\right) = \frac{\pi D^2}{2}$$

B. Identify and define the design variables.

D **= diameter of the can.**

H **= height of the can.**

C. Specify a cost function.

It is desirable to minimize the cost of manufacturing the cans. Cost is related directly to the surface area of the sheet metal, so the design objective is to minimize the total surface area of sheet metal required for each can.

Cost function to be minimized:

$$\text{Total surface area} = A_{cyl} + A_{ends}$$

$$f(D,H) = \pi DH + \frac{\pi D^2}{2} \qquad \text{(nonlinear)}$$

D. Identify the system constraints.

The can must hold at least 400 ml (400 cm³) of fluid.

$$\text{Volume of can} = \frac{\pi D^2 H}{4}$$

$$\frac{\pi D^2 H}{4} \geq 400\ cm^3 \qquad \text{(nonlinear)}$$

There are additional constraints due to fabrication, handling, aesthetic, and shipping considerations.

$$3.5\ cm \leq D \leq 8\ cm$$

$$8\ cm \leq H \leq 18\ cm$$

Design problem summary:

Find the design variables D and H that minimize the cost function (surface area function) subject to the volume and dimension constraints.

Figure 1.2 Design problem formulation—beverage can example.

problems involving two to three design variables and three to four constraints are solved in detail . . . ." The specific problems include

A two-bar bracket to support a given force without structural failure.

A two-member frame with out-of-plane loads.

A three-bar structure with multiple constraint requirements.

A cantilever truss.

All of these represent small-scale and low-complexity mechanical systems. Arora admits that there are numerous "complex applications for which explicit dependence of the problem functions on design variables is not known." Such cases cannot be handled by mathematical optimization theory. Perhaps mathematics-based optimization methods can be useful tools for treating small subsets of real-world complex systems. For highly complex natural systems (e.g., ecological systems), much more is needed.

Herbert Simon (1996) has recognized that a comprehensive search through a design space is often not practical and, for many real-world systems, not possible. Simon coined the term "satisficing" to mean finding a solution that is "good enough." He says, "We must trade off *satisficing* in a nearly-realistic model against optimizing in a greatly simplified [and mathematically rigorous] model." "In the face of real-world complexity, [we turn] to procedures that find good enough answers to questions whose best answers are unknowable."

We see here that mathematical methods can be useful, but they are not sufficient. Stephen Wolfram (2002) said, "Like most other fields of human inquiry mathematics has tended to define itself to be concerned with just those questions that its methods can successfully address."

1.3.2.2 Empirical Methods

Empirical methods can address highly complex systems. Empirical methods (based on observation, experience, and experiments) include *prototyping* and *explicit experimentation*.

For building human-made complex systems, prototyping has traditionally been an important means of system experimentation. To help ensure that the actual system will work as intended, one builds an approximate and simplified representation of the system. The objective is to produce, as quickly and inexpensively as possible, a system "mock-up" or model that captures the essential features and characteristics of the actual system. We run the prototype and see what happens. When done well, the results can be an extremely valuable indicator of actual system behavior.

For highly complex natural systems, a similar approach—which can be called explicit experimentation—can be used. The goal in this case is not to build artificial systems, but rather to understand the behavior of natural systems. Stephen Wolfram (2002) utilizes explicit experimentation in his book, *A New Kind of Science*. (In fact, he may have originated the term "explicit experimentation.") Wolfram realized that cellular automata models could capture essential features and characteristics of natural systems. In a sense, certain cellular automata could act as natural system prototypes. Wolfram set up a series of cellular automata programs, ran them, and observed their behavior in order to understand a range of natural system phenomena. I do that kind of explicit experimentation in Part V of this book. I define and implement a cellular automata model in order to understand the dynamical behavior of complex ecosystems. (We'll discuss Wolfram's work further in Chapter 11.)

1.3.3 System Development Skills

In this subsection, system development skills (i.e., system design, system modeling, and system architecting) are discussed. Refer again to Figure 1.1. This skill set is needed for developing artificial (human-made) systems, and is very much applicable to understanding natural systems. Interestingly, *system design* approaches for artificial systems often tend to mimic natural system "design" approaches. (See Chapter 7 on evolution and development for more details on this.) *System modeling* is widely used for understanding both artificial and natural systems. (We will encounter natural system modeling throughout the book and especially in Part V.) An explicit *system architecture* is a required part of the development of a human-made complex system. The concept of system architecture can also be applied to complex natural systems, and can significantly enhance our understanding of these natural systems. (We'll develop and discuss a functionality-driven system architecture for ecological systems in Part II of the book.)

1.3.3.1 System Design

Design is an important systems engineering skill. K. Preston White Jr., an engineering professor at the University of Virginia, says that design is a central skill for building human-made systems. White (1998) explains, "Design is the essence of engineering." "As a profession . . . engineering is concerned primarily with design—the design of processes, structures, machines, circuits, and software—and with the combinations of these elements we call systems." "Design is central to the practice of systems engineering and systems engineers understand that design is a creative, iterative, decision-making process." Gerald M. Weinberg also sees design as a creative and iterative process. Weinberg (1975) says, "learning to design is learning to generate and evaluate models" in an iterative fashion. (This iterative generation and evaluation essentially mimics natural system "design" approaches. In my view, it mimics the *choice generation* and *selection* process of evolution. See Chapter 7.)

A high-level, conceptual engineering design model, therefore, can be depicted as shown in Figure 1.3.

The engineer generates design candidates and then performs a comparative evaluation of the candidates. In the next iteration, some candidates may be eliminated, others may be modified, and new ones may be added. The iterative process continues until it converges on an "acceptable" design, i.e., a design that is well suited to and can work effectively in the particular engineering environment under consideration.

Note that Herbert Simon (1996) uses the term *design* to mean a strictly human activity that applies only to artificial systems. Simon defines *artificial systems* as human-made systems that are intended to satisfy human goals and purposes. *Artifacts* are representations of this human artifice. Simon states, "Historically and traditionally, it has been the task of the science disciplines to teach about natural things: how they are and how they work. It has been the task of engineering schools to teach about artificial things: how to make artifacts that have desired properties and how to design." As long as the

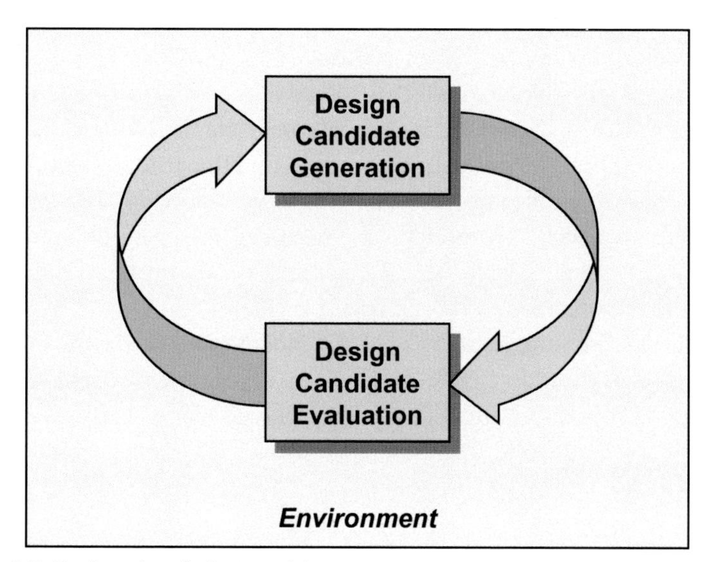

Figure 1.3 Engineering design model.

context is clear, that view is fine and has merit. I, however, obviously think that design concepts are applicable to both artificial and natural systems.[6]

I see two broad classes of system design investigation:

- Artificial system development—seeking to develop the engineering design of a new human-made system or to understand the design of an existing human-made system (with unknown design specifics).
- Natural system inquiry—seeking to understand the "design" of an existing natural system (e.g., an ecological system).

In this book, I am mostly interested in the natural system inquiry class.

1.3.3.2 System Modeling

System modeling is a key system development skill. Beyond that, however, modeling is pervasive across all of systems engineering and shows up to some degree throughout the systems engineering skills framework. We've already seen that modeling is a very important part of engineering technical methods and of system design.

What is the definition of a *model*? According to Webster's Third International Dictionary, a model is a description . . . to help us visualize, often in a simplified way, something that cannot readily be observed. Many systems investigators have expressed their views on models. Here are some useful samples. "Models, abstractions of reality, are critical in the engineering of systems" (Buede, 2000). "A model is an abstraction of something for the purpose of understanding it" . . . "Engineers, artists, and craftsmen have built models for thousands of years" (Rumbaugh et al., 1990). "A model is simply a representation of relevant characteristics of a reality" (Echenique, 1963). The objective of a "good model" is to provide an abstract representation of effects that are important in determining the behavior of a system (Wolfram, 2002). "The problem . . . is not to find the 'true anatomy' of anything, but the level of [anatomy] that will serve our purposes" (Weinberg and Weinberg, 1988). "A model is a conceptual surrogate for some other thing and may range from a set of equations describing a machine to an artist's concept of a landscape" (Warfield, 1976).

The last quote touches on the area of model types or classes. Here's an often-used classification scheme (see, e.g., Flood & Carson, 1993 or Hazelrigg, 1996) that I have updated. There are four broad classes of system models: iconic, symbolic, computer-based, and analogical. Iconic models

[6] By the way, the creationist term "intelligent design" is an entirely different matter and has no application here.

are physical models that are representative in form, but not necessarily size or detail. Examples are vehicle scale models or "an artist's concept of a landscape." Symbolic models represent a system by means of symbols. Mathematical models are in this class. Computer-based models are very useful when it is not possible to solve (or even write) mathematical equations for a system or otherwise describe the system of interest. Analogical models include any type of model from one domain that can be used to represent a system of another domain. Henry David Thoreau[7] has said, "All perception of truth is the detection of an analogy; we reason from our hands to our head."

Modeling is widely used for understanding both artificial and natural systems. In the natural systems research that is the focus of this book, we use mathematical models (throughout the book) and an analogical computer-based cellular automata model (in Part V).

1.3.3.3 System Architecting

An explicit system architecture is required when developing a human-made complex system, and also has applicability for understanding natural systems. In my view (based on decades of systems engineering experience), a system architecture consists of a functional architecture that describes what the system does and a physical architecture that describes how the functional architecture is implemented. As I will discuss in Chapter 3, function is preeminent. It drives the system.

In Part II of the book, we will develop and discuss what could be called a functionality-driven system architecture for ecological systems. To emphasize the three architecture aspects that matter most, however, I call it a function-structure-process framework. The framework can provide a unifying context for exploring principles of complex ecosystem dynamics. That's what this book is all about. The framework can be the vehicle for comprehensive and disciplined identification of ecological system functions, the structure upon which the system functions are implemented, and the processes that operate within that structure to deliver the functions.

1.3.4 Systems Thinking

Many of us have heard or read some of the usual pronouncements about systems thinking. For example, (a) General systems thinking provides

[7] Henry David Thoreau, Journals, September 5, 1851.

cross-disciplinary tools for solving systems problems. Instead of seeing separate fields of knowledge, systems thinking lets you see the commonality among the disciplines. Or (b) General systems approaches are simply effective approaches that work across disciplines. The well-respected Gerald Weinberg has said (Weinberg, 1975), "By elevating particular disciplinary insights to a general framework and language, we make some ideas of each discipline available for the use of all." While all of these statements may be true and admirable, they are very general and therefore, by themselves, are not very useful.

Most of us are familiar with the so-called systems anthem: the whole is greater than the sum of the parts. I suppose that's true, as long as one does not consider connections or interactions as "parts." We are also told that systems have somewhat mysterious *emergent properties*, i.e., properties above and beyond the properties of the parts that comprise the systems. We are sometimes informed that this has something to do with *synergy*. It seems to me that these gems are statements of the lack of knowledge about systems, rather than something useful that might help us understand.

I try to focus on useful things. Please refer back to the "systems thinking" component of Figure 1.1. Note the items displayed there:

- A blend of synthesis and analysis
- Network thinking
- The systems triad

I have found these concepts extremely useful in my systems work—both my earlier work on human-made systems and my current work on natural systems. I consider these concepts major elements of systems thinking and the systems approach. We will discuss these three elements and their application to ecological systems in Chapter 3, with additional detail in Chapters 4 and 5. The first two elements have sufficiently descriptive labels. To allay any mystery about the third, the vertices of the systems triad are function, structure, and process.

1.3.5 Communication Skills

Doing good systems and engineering work—whether for human-made systems or natural systems—is only half the job. Communicating that good work to others is the remaining half. There is wide agreement on this point. "Although systems engineering is seen as a technical activity . . . it is, in fact, a sociotechnical activity" (Harris, 2001). "Clearly, technical competence is a necessary condition for success as a systems engineer, but it is not sufficient.

Systems engineers must also exhibit formidable skills in their human relations" (Reilly, 1993).

This book is not the proper venue for a detailed discussion of communication skills, so I will simply note a few of my views on clarity in communications:

- If you can't communicate it clearly, you probably don't understand it yourself.
- To communicate, you have to know what you're talking about. Grasp the thing; words will follow.
- When writing or speaking, your primary objective and your challenge is to make your audience understand you.
- Be careful with technical jargon. It may make insiders feel special, but it's a barrier to understanding for everyone else.

Philosopher Karl Popper (1983) stated, "What can be said can and should always be said more and more simply and clearly." John Ruskin, Victorian art critic, said in 1888, "The greatest thing a human soul ever does in this world is to see something, and tell what it saw in a plain way."

1.4 TECHNICAL RATIONALITY AND REFLECTIVE RATIONALITY

Earlier in this chapter, when discussing engineering technical methods, I said that mathematical methods are useful in systems work, but they are not sufficient. Other technical methods are also required. In the previous subsection we said that, although technical competence is required for success in systems work, technical methods are not sufficient either. (That point was noted in the context of communication skills, but in my view it has much broader application than just communication.) The systems approach is fundamentally multidisciplinary. The "disciplines" we are talking about, however, are not limited to the technical disciplines. The more human-oriented disciplines have an important role as well.

What, then, is the full set of required competencies and perspectives needed for success in the systems approach? I say that the full set can be viewed as consisting of two important pieces: *technical rationality* and *reflective rationality*.[8] Technical rationality includes the "harder" (more formal) science/engineering competencies, and is more akin to *analysis* (taking apart)

[8] Donald A. Schön also uses the terms *technical rationality* and *reflective rationality* and discusses these concepts in his books *The Reflective Practitioner* (1983) and *Educating the Reflective Practitioner* (1990). Some of Schön's ideas are incorporated here in Section 1.4.

abilities. Refer again to the systems engineering skills framework displayed in Figure 1.1. Technical rationality is perhaps most evident in the engineering technical methods skill set, but is present in all of the skill sets. Reflective rationality includes human-oriented competencies, and is more akin to *synthesis* (putting together) abilities. In my view, reflective rationality is also present in all of the Figure 1.1 skill sets.

The following few paragraphs provide some brief supporting background and discussion regarding the tradition of technical rationality, its potential shortcomings, and why we need more for dealing with highly complex systems.

Technical rationality is based fundamentally on Western materialist philosophy, reductionism, and the scientific method. The technical rationality view is very widely held and its advocates can be said to range from the Socratics in ancient Greece, to Galileo, Bacon, Descartes, and Newton during the Enlightenment in seventeenth-century Europe, to current scientists and engineers around the world. The approach has been extremely successful, but is not sufficient for all systems—particularly not for many highly complex systems.

Technical rationality emphasizes instrumental problem solving (applying well-defined solution techniques to well-formed problems). If the problem is complex and doesn't fit the available techniques or is not well formed, we often adjust the problem until it does fit. We then solve the wrong problem—and we solve it very efficiently! "The problem adjustment is often not stated explicitly and, as a consequence, an impression is created that the original problem was solved while, in fact, it was not" (Klir, 1991). "We often fail not because we fail to solve the problem we face, but because we fail to face the right problem" (Gharajedaghi, 1999). Yet we really like our well-defined formal technical methods because they do not require too much thinking about difficult complex issues. "By leaning on correctness, it [is] possible to alleviate the burden of decision" (Alexander, 1964).

Albert Einstein understood what was going on. Einstein (1950) said, "Perfection of means and confusion of goals seem—in my opinion—to characterize our age." We are very good at means, i.e., developing and executing techniques and methods. We are not so good at goals, i.e., understanding how or even whether to apply those means. "We know how to teach people how to build ships but not how to figure out what ships to build."[9]

[9] This 1974 quote appears in Schön, *Educating the Reflective Practitioner* (1990) and is attributed to Alfred Kyle, an engineering school dean.

Stephen J. Gould (2003) has argued that science has overly favored technical rationality—this one mode of knowing with "quantitative and experimental techniques so brilliantly suited to the resolution of relatively simple systems." We need additional modes of knowing. We need technical rationality *and* reflective rationality. Donald Schön (1983) says we need to "develop an epistemology of practice which places technical problem solving within a broader context of reflective inquiry . . . ." Sam Alessi (2000) also supports this view: "It is likely the basic paradigm [for systems engineering] must be a balance of both a humanistic/subjective and a materialist/objective problem-solving framework."

So, for successful complex systems work, we need multiple modes of knowing—both technical rationality and reflective rationality. Those ideas are pursued in more detail in the latter part of Chapter 2. Furthermore, we need systems thinking and the systems approach. We need to employ a blend of analysis and synthesis. Not only do we have to know how to break things apart (reductionism), we also have to know how to put things together in order to understand the larger picture. We need network thinking so that we not only focus on system parts, but also focus on the even more important interactions among the parts. We need what I have defined as the systems triad to understand the roles of function, structure, and process in complex system dynamics. I discuss these major elements of the systems approach in detail in Chapter 3, and I apply them to my work throughout the book.

The *summary message* of this chapter is this: the systems approach, systems engineering skills, and the associated systems perspectives can be very beneficially applied to understanding highly complex natural systems. Ecological systems are an especially fertile application area.

CHAPTER 2

More Views on Systems Thinking

Chapter 1 addressed my perspectives on systems engineering skills, the systems approach, and systems thinking. Here in this chapter, we'll cover other views on systems thinking from the scientific community, as well as more of my own views. Section 2.1 of this chapter provides background on the origin and evolution of various systems thinking movements. Section 2.2 describes Warren Weaver's ranges of system complexity. My opinion is that Weaver gave us (in 1948) a very useful and helpful way of thinking about system complexity. Sections 2.3 and 2.4 take a philosophical turn. Section 2.3 addresses "the many ways of knowing" and the importance to systems work of the corresponding spectrum of disciplines. Section 2.4 describes and discusses a debate between E. O. Wilson and S. J. Gould about the needed integration of these disciplines across the sciences and the humanities domains.

2.1 ORIGIN AND EVOLUTION OF SYSTEMS THINKING MOVEMENTS

Systems thinking has had a fragmented history. Some aspects have been around for centuries, yet it is still an immature field. In this section, we'll discuss cybernetics, general systems theory, some related efforts, and finally an assessment of the work.

2.1.1 Cybernetics

Cybernetics, often regarded as a systems thinking discipline, has been in existence in some form for a very long time. The word *cybernetics* comes from the Greek *kybernetes*, which means steersman. Plato used the term *kybernetics* when discussing self-governance of people. Physicist André-Marie Ampère used the word cybernetics in 1834 to refer to the science of governing in the context of social science. In the latter part of the eighteenth century, James Watt developed the steam engine and included a cybernetics-related governor mechanism to control the engine's speed. Later, in 1868, James Clerk Maxwell published a theoretical article on cybernetics that described the principles of feedback in mechanical self-regulating governor devices. Norbert

http://dx.doi.org/10.1016/B978-0-12-802031-9.00002-4

Weiner has been a leader and central figure in modern cybernetics. In his 1948 book (Wiener, 1948), he defined cybernetics as the study of communication and control in the animal and in the machine. The field of modern cybernetics focuses on how systems function regardless of the system context—social, mechanical, or biological. It tantalizes with an ambitious promise: to unite different disciplines by showing that the same basic system principles can be applied in all of them. Weiner talked about the opportunity for important interdisciplinary systems work and how the work should be conducted: "A proper exploration . . . made by a team of scientists, each a specialist in his own field but each possessing a thoroughly sound and trained acquaintance with the fields of his neighbors; all in the habit of working together, of knowing one another's intellectual customs, and of recognizing the significance of a colleague's new formal expression . . . to understand the region as a whole, and to lend one another the strength of that understanding."[1]

The basic model of cybernetics is the feedback control model, shown in Figure 2.1. This, of course, is a fundamental system regulation mechanism. The system measures the difference between what is and what should be (some defined reference state or goal), and then generates an error that is fed back for the purpose of moving the system closer to the goal.

Modern cybernetics and general systems theory developed side by side in the post-World War II period.

2.1.2 General Systems Theory

Ludwig von Bertalanffy began an effort to establish general systems theory in the 1940s. "Bertalanffy envisaged a framework of concepts and theory that would be equally applicable to many fields of inquiry" (Flood and Carson, 1993). It would be valid for systems in general. The original context was the field of biology and the early emphasis was on mathematical approaches. The premise was that an efficient science could be developed that spans disciplines—and unifies the sciences.

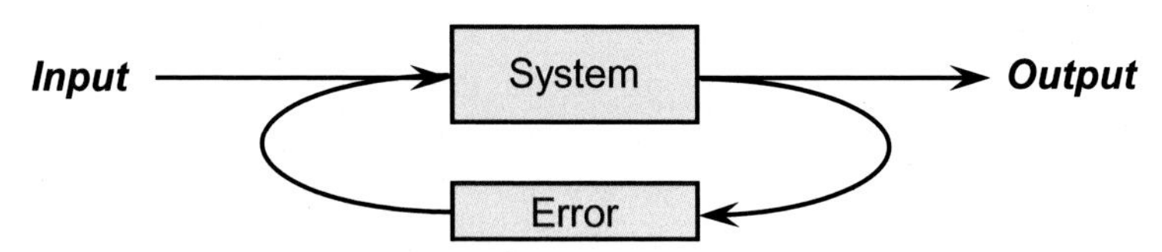

Figure 2.1 Cybernetics feedback control model.

[1] Weiner's comment was quoted in Klir (1991), *Facets of Systems Science*.

According to George Klir (Klir, 1991), in December 1954, Bertalanffy (a biologist), Kenneth Boulding (an economist), Ralph Gerard (a physiologist), and Anatol Rapoport (a mathematical biologist) formed the first organization devoted to systems thinking—the Society for General Systems Research (SGSR). The organization was founded with four objectives:

1. To investigate the isomorphy of concepts, laws, and models from various fields, and to help in useful transfers from one field to another
2. To encourage development of adequate theoretical models in fields which lack them
3. To minimize the duplication of theoretical effort in different fields
4. To promote the unity of science through improving communication among specialists

The name of the organization was later changed to the International Society for the Systems Sciences (ISSS).

Beginning in the late 1960s, the emphasis of the general systems effort was broadened from "hard" (mathematical) approaches to also include "soft" approaches, via the work of C. West Churchman, Russell L. Ackoff, Peter B. Checkland, and others. For example, the soft systems methodology (SSM) of Checkland (1981) is illustrated in Figure 2.2.

Checkland's work is comprehensive and has value and merit, but it also has limitations. SSM mostly provides a loosely structured template for thinking about systems and associated problem situations. The template, in fact, specifically addresses just one category of system investigations. Checkland certainly includes the reflective rationality that is required in complex system investigations, but he tends to avoid the technical rationality that is also required. Checkland's work is helpful, but not sufficient. The methodology has limited usefulness. (To be fair, most of the soft approaches have similar limitations.) In my view, regarding highly complex systems, the term "methodology" is not even appropriate. There is no distinct singular method for addressing such systems. Complex systems work procedures are very much dependent on the system being addressed and on its environment. My systems work spans the hard and soft domains. I do not rely on any general-purpose methodologies. My approach is to identify and define major tenets and tools of systems thinking and then apply them in an appropriate manner to the problem at hand. We will discuss that approach in Chapter 3.

There have been many well-intentioned pronouncements about the potential benefits of a general systems discipline. We are told that general systems efforts can provide cross-disciplinary tools for solving systems problems. Principles developed in one domain can be applied in other domains.

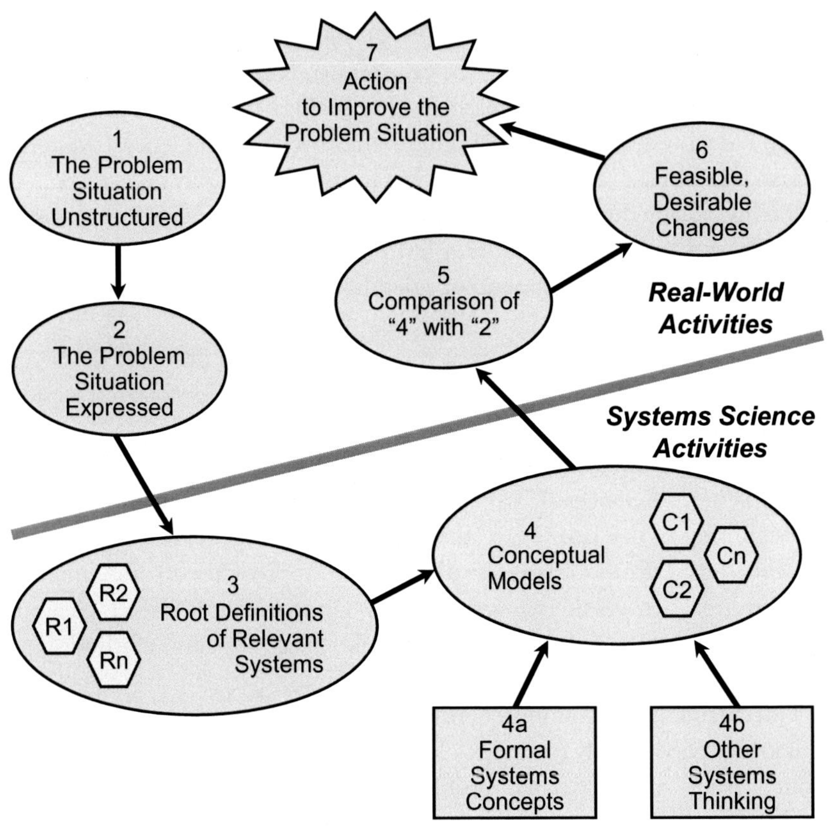

Figure 2.2 Soft systems methodology. *(Credit: Checkland (1981), publisher is John Wiley and Sons. Copyright © 1981 by John Wiley & Sons Ltd).*

"By elevating particular disciplinary insights to a general framework and language, we make some ideas of each discipline available for the use of all" (Weinberg, 1975). We can make "generally applicable insights available to the most general audience possible" (Weinberg and Weinberg, 1988). It seems that these potential benefits are mostly yet to be realized.

2.1.3 Related Efforts

2.1.3.1 Isomorphies

Quite consistent with the general systems mantra, Klir (1991) has written about isomorphic[2] systems. He explains that there are isomorphies or analogies across disciplines. When an isomorphic/analogous correspondence

[2] Merriam-Webster online defines isomorphic as "being of identical or similar form, shape, or structure." I would add "and having analogous relationships among parts."

exists, principles developed for phenomena in one discipline become readily available to corresponding phenomena in other disciplines. Isomorphies "made it possible, for example, to transfer methods from a methodologically well-developed area to areas methodologically less developed." Klir discusses the area of generalized circuits, where well-developed methods for analyzing electric circuits can be applied to less-developed areas of mechanical, acoustic, magnetic, and thermal systems. Isomorphic thinking has "created a growing awareness among some scholars that certain concepts, ideas, principles, and methods were applicable to systems in general, regardless of their disciplinary categorization."

2.1.3.2 Systems Repertoire

How do we make insights available across problem domains and disciplinary boundaries? Schön (1983) says we should develop and apply a systems repertoire:

- Bring your past experience to bear on the current, unique situation.
- Over time, build a repertoire of knowledge, approaches, and solutions to problems and problem subsets.
- View the current situation as both similar to and different from a situation in your repertoire.

We model one problem solution on another; we model the unfamiliar on the familiar. Thomas Kuhn has called this approach *thinking from exemplars*.

Maxwell (1873) long ago expressed a similar sentiment: "But though the professed aim of all scientific work is to unravel the secrets of nature, it has another effect, not less valuable, on the mind of the worker. It leaves him in possession of methods which nothing but scientific work could have led him to invent, and it places him in a position from which many regions of nature, besides that which he has been studying, appear under a new aspect."

2.1.3.3 Patterns

Christopher Alexander would not call himself a systems theorist. He is a building architect and architectural theorist. Many of his ideas, however, are consistent with systems thinking. The formal notion of *patterns* originated with the building architecture patterns of Alexander (1977, 1979). A pattern is a named problem/solution pair that can be applied to familiar systems or to new systems in new contexts, with advice on how to apply it. Each pattern is a generic solution to some system problem.

Generic patterns have been applied successfully to software systems, beginning in the 1980s with the work of Kent Beck and Ward Cunningham.

One set of software patterns that has been defined is the general responsibility assignment software patterns (GRASP[3]) set that is used to determine which responsibilities should be associated with which software components and the collaboration that is required among the components. (Such patterns might also prove useful for understanding process mechanisms of natural systems.)

The central idea here is that in architectural systems or in software systems—or in *any system domain*—patterns can be identified, learned, and applied within the domain or across domains and disciplines to build good human-made systems. This patterns approach could also help us better understand natural systems. Structure and process patterns are frequently observed in nature. The formal concept of patterns was developed in one discipline but can be applied in others. Patterns provide a cross-disciplinary vehicle for solving systems problems and, therefore, can contribute to general systems thinking methods.

2.1.4 Assessment

I have not been a direct or formal participant in the evolution and development of the systems thinking movements we've discussed, so I cannot offer anything like a comprehensive personal assessment. I can say that I agree with the spirit of much of the high-level guidance that is offered, although translating the guidance into action requires a lot of additional work in any given system context. I can also say that I have applied the regulation aspects of cybernetics to natural systems in my ecological work and I have used software patterns in my engineering work, both to good effect. Overall, however, the development of "systems science" seems to have been uneven, with periods of excitement and periods of disillusionment and criticism. This is still an immature field.

Other investigators have offered their views. Here's a sampling. Simon (1996) has recognized that, in science and engineering, the study of systems is increasing because of the need to deal with system complexity. Accordingly, we need to develop a body of knowledge and techniques in the systems area; we need to "provide substance to go with the name." Warfield (1976) has said, "The so-called systems methods have often been criticized as perpetually promising that something good is just around the corner, though not yet attained. . . . There are important believers who recognize that there is 'enough there' to make it eminently worth pursuing, while on the other

[3] The source is Larman (1998), *Applying UML and Patterns*.

hand, there are experts who feel that if there is something there, it is not well-organized or teachable ... and that further development and better practice are required ...." Flood and Carson (1993) say "there is universal agreement within the systems movement that systems thinking is at least a good idea worth exploring and developing in order to ascertain whether a coherent body of knowledge can be developed." If that could be accomplished, systems thinking may then be "ready to contribute more effectively to other disciplines and real world application."

Goguen (1998), however, laments that too many scientists and engineers are not even aware of the need for and the value of systems thinking. He says, "But it was (and still is) disappointing to me that so few people felt any need for concepts and theories of such generality; they seem happy to have (more or less) precise ideas about specific systems or small class of systems, with little concern for what concepts like system, behavior and interconnection might actually mean." I have often observed this condition, and I share the disappointment. Effective systems thinking is difficult. Too many people prefer to avoid the difficulty—and the benefits.

2.2 WEAVER'S RANGES OF SYSTEM COMPLEXITY

Many years ago, Warren Weaver (1948) gave us a very useful and helpful way of thinking about system complexity. Figure 2.3 illustrates Weaver's ranges of complexity.

I have summarized some of the characteristics of the three ranges (see the lower portions of Figure 2.3). The organized simplicity range includes simple systems with a small number of components that are typically addressed by well-established laws of physics. System behavior (that is addressed) usually can be quantified via deterministic mathematical methods. By looking at the top row of Figure 2.3, you might assume that the disorganized complexity range is the most complex. It is not. For systems in this range, the large numbers of system components typically act independently and randomly. Individual component dynamics statistically tend to cancel each other. Such systems may be microscopically probabilistic, but macroscopically deterministic. The law of large numbers states that, for these large-number systems, the system probability function variance tends to zero and the mean tends to the arithmetic average of behaviors. Deterministic mathematical methods (e.g., Boyle's law or the ideal gas law for gaseous systems) can often be applied (macroscopically). The organized complexity range is the most complex. These medium-number systems have rich relationships among

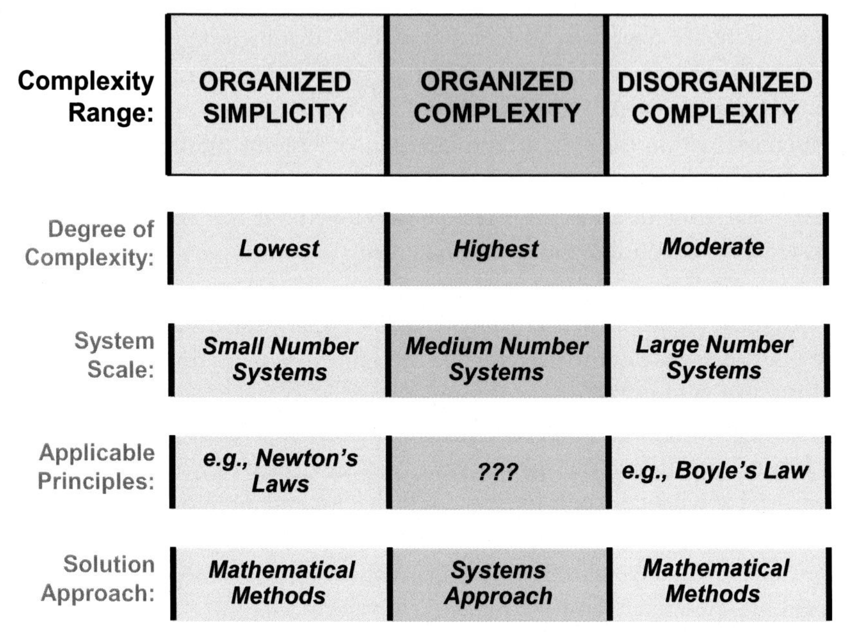

Figure 2.3 Weaver's ranges of system complexity.

components and a pluralistic mix of deterministic and stochastic properties. We are only in the very early stages of identifying principles that might apply to such complex entities. Identifying emerging principles for complex ecosystems, of course, is a major objective of this book. A systems approach (including mathematical methods) is necessary for understanding these systems. "The systems view is the emerging contemporary view of organized complexity" (Laszlo, 1972).

In his journal article titled *Science and Complexity*, Weaver (1948) offers more of his perspectives. He says that problems of organized simplicity and problems of disorganized complexity (for which solution methods exist) are only a tiny fraction of all systems problems. Most problems are between these two extremes—in the organized complexity range. "The really important characteristic of the problems of this middle region, which science has as yet little explored or conquered, lies in the fact that these problems . . . show the essential feature of organization." They "involve dealing simultaneously with a sizable number of factors which are interrelated into an organic whole. . . .These problems . . . are just too complicated to yield to the old nineteenth-century techniques which were so dramatically successful on two-, three-, or four-variable problems of simplicity. These new problems, moreover, cannot be handled with the statistical techniques so effective in describing average behavior in problems of disorganized complexity. ... These new problems, and the future of the world depends on many of them, requires science to make a third great advance, an advance that must be even greater than the nineteenth-century conquest of problems of simplicity or the twentieth-century victory over problems of disorganized complexity. Science must, over the next 50 years, learn to deal with these problems of organized complexity." Well, the fifty years is up, and sufficient progress has not yet been achieved.

George Klir also offers some perspectives. "Instances of systems with characteristics of organized complexity are abundant, particularly in the life, behavioral, social, and environmental sciences, as well as in applied fields such as modern technology or medicine" (Klir, 1985). "There is little doubt that, in the foreseeable future, systems science will continue to be the principal intellectual base for making advances into the territory of organized complexity, a territory that still remains, by and large, virtually unexplored" (Klir, 1991).

Using Weaver's work and some subsequent work by Flood and Carson (1993), I have constructed the diagram of Figure 2.4.

Consider the complexity spectrum of the sciences. Proceeding from least complex to most complex, we find Physics and then Chemistry at the low end, and Ecological Sciences and Social Sciences at the high end. Physics, chemistry, and some of traditional biology deal principally with problems/systems in the organized simplicity and disorganized complexity ranges. Systems biology and, to an even greater extent, the ecological sciences and social sciences must deal with the highly complex systems of the organized complexity range. As indicated in Figure 2.4, a systems approach is required in this range.

Here is my perspective on progress and success in the natural sciences: Sciences at the low end of the complexity spectrum (physics, chemistry) are usually associated with the most progress and success. This is because they are (relatively) simpler. At the high end of the spectrum, disciplines in the ecological and the social sciences are sometimes denigrated for a lack of progress. This is, of course, because these disciplines and the systems they deal with are dramatically more complex. Similarly, systems science often has been regarded as an admirable concept that never quite delivers. I think that these perceptions are at least partly due to inappropriate expectations. Our society expects neatly packaged, straightforward solutions to every problem. For our most complex systems (e.g., ecological systems), such expectations

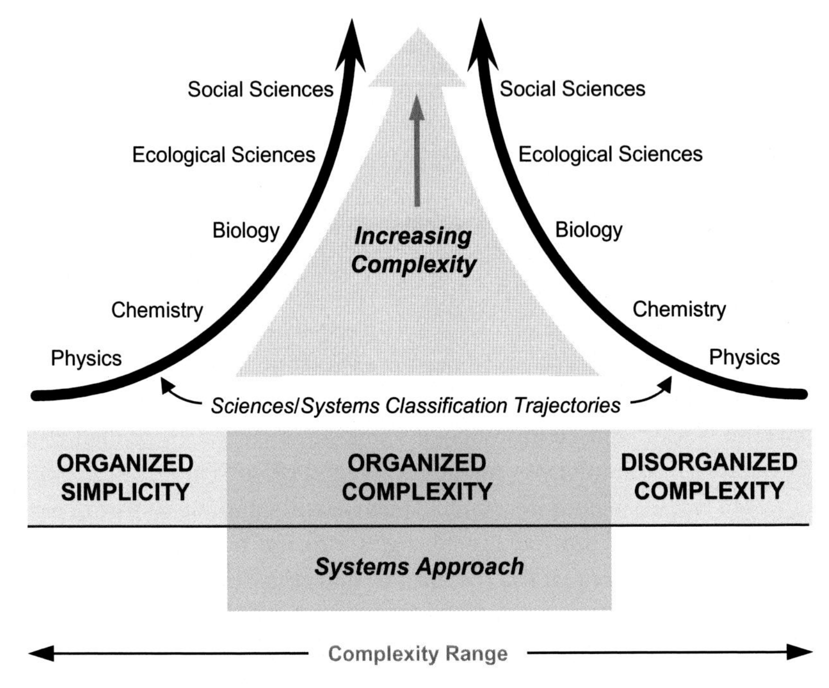

Figure 2.4 Complexity and the natural sciences. *(Credit: Flood and Carson (1993), p. 252, Springer. Copyright © Springer Science+Business Media, New York 1993).*

are not consistent with the reality of the natural world. Progress and success in the organized complexity range are much harder to come by. That said, a major objective of this book is, in fact, to make significant progress in the organized complexity arena of ecological system dynamics.

We'll revisit Weaver's ranges of complexity in Chapter 14—as we investigate the relationship of order and complexity in ecological systems.

2.3 THE MANY WAYS OF KNOWING

In Chapter 1 (Section 1.4), I introduced the terms *technical rationality* and *reflective rationality* and made the point that the systems approach is fundamentally multidisciplinary. The disciplines that I am talking about are not limited to the technical disciplines. The more human-oriented, reflective disciplines have a very important role as well. For successful complex systems work, we need to take advantage of the modes of knowing inherent in the sciences as well as in the humanities. We need to employ the overall spectrum of the ways of knowing.

There are a variety of views regarding the modes of knowing. Some of those views, both ancient (Aristotelian) and contemporary, are described in Section 2.3.1. I will then synthesize a composite view of the ways of knowing and discuss its properties and implications in Sections 2.3.2–2.3.5.

2.3.1 Ancient and Contemporary Views

2.3.1.1 Aristotelian View

Barnes (2000) discusses Aristotle's scientific and philosophical work and his classification of human knowledge that guided the work. The Aristotelian classification of knowledge is depicted in Figure 2.5.

The *theoretical* category was intended to include "all of what we now think of as science" as well as philosophy.[4] Aristotle maintained that this category contained "by far the greatest part of the sum of human knowledge." According to Aristotle, the primary philosophy is *theology*. Theology should

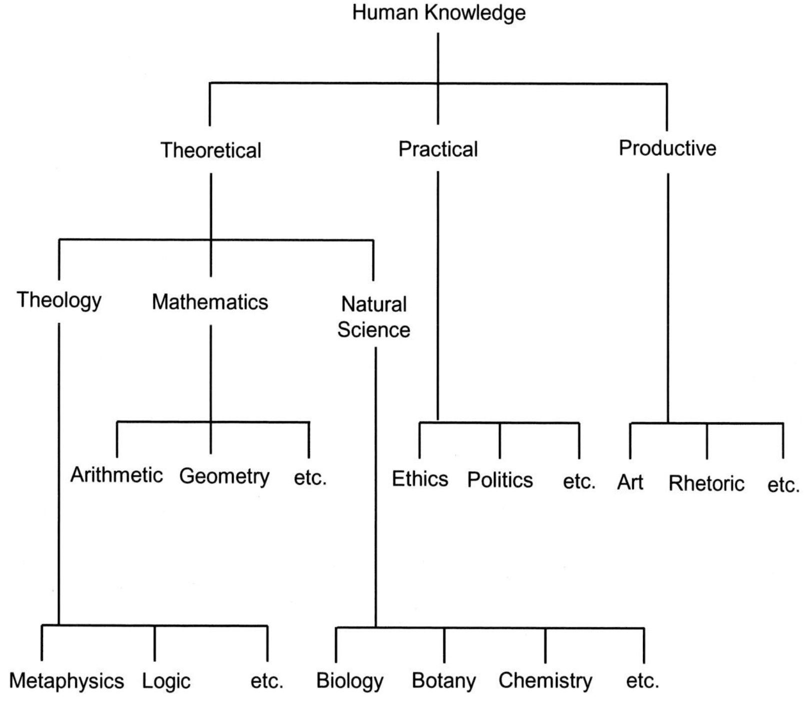

Figure 2.5 Aristotelian classification of human knowledge. *(Credit: Barnes (2000), diagram "The structure of human knowledge" p. 45. By permission of Oxford University Press, http://www.oup.com).*

[4] All of the quotes concerning the Aristotelian view are from Barnes (2000).

be considered "divine" but not necessarily religious; it identifies with the "heavens." Per Aristotle, *metaphysics*—meaning beyond physics (natural science)—is the study of *beings qua being*; that is, it is (roughly) the study of human existence.[5] *Logic* is "both a part and a tool of philosophy." The other two major categories of the Aristotelian classification of human knowledge can be summarized as follows: The *productive* category is "concerned with the making of things." The *practical* category is "concerned with action" and "how we ought to act."

2.3.1.2 Contemporary Views

Bohm and Peat (2000) offer their views on the ways of knowing. They say that there seems to be "three basic attitudes of mind to the whole of life: the scientific, the artistic, and the religious." The scientific is concerned with "rational perception through the mind and with testing and perceptions against actual fact, in the form of experiments and observations." The artistic "includes music, drama, literature, poetry, dancing, and the visual arts." It "is strongly concerned with beauty, harmony, and vitality." The artistic attitude emphasizes "the role of the imagination"—and reflection and creativity. "An artistic attitude is needed by all, in every phase of life." Regarding the religious, Bohm and Peat observe that "throughout history . . . it must be admitted that religions have tended to be caught up in all kinds of self-deceptions and in the exploitations of others." However, because the religious "is concerned primarily with wholeness [from the word *holy*]," there are potential opportunities for positive contributions to the ways of knowing.

Bohm and Peat emphasize that the three ways of knowing (their "three basic attitudes of mind") are not currently integrated, but rather fragmented and isolated. Each applies only to a particular subset of life. They say, "humanity has become conditioned to accept . . . a rigid separation" among these three attitudes. "What is clearly needed is a dialogue between these attitudes, in which sooner or later they can all come into the 'middle ground' between them, which will make available a new order of operation of the mind with rich possibilities for creativity." All three attitudes must be integrated to solve the problems of the world.

2.3.2 A Synthesis

From the ideas we've covered and from my own ideas, I have synthesized a composite view of the many ways of knowing. It is provided in Figure 2.6.

[5] *qua* (from the Latin): in the capacity or character of.

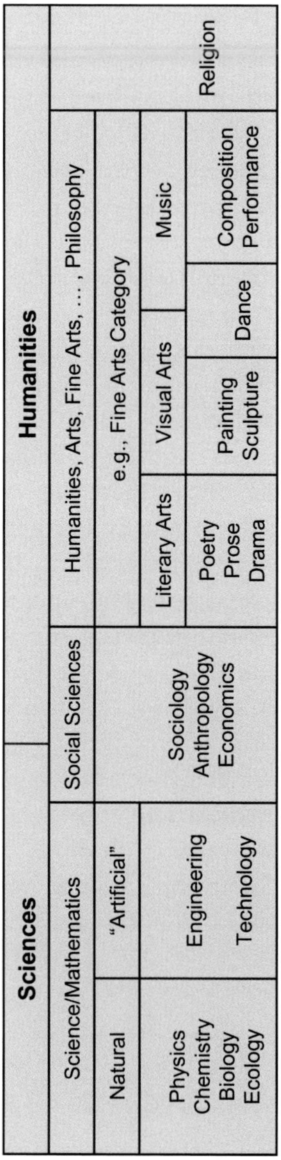

Sciences			Humanities				
Science/Mathematics		Social Sciences	Humanities, Arts, Fine Arts, ..., Philosophy				Religion
Natural	"Artificial"		e.g., Fine Arts Category				
Physics Chemistry Biology Ecology	Engineering Technology	Sociology Anthropology Economics	Literary Arts	Visual Arts	Music		
			Poetry Prose Drama	Painting Sculpture	Dance	Composition Performance	

Figure 2.6 The ways of knowing.

There are two broad domains: the Sciences and the Humanities. Within these domains, I identify categories and then disciplines within those categories. You will notice that this synthesized view is not exhaustive. Certainly, not all disciplines are mentioned. The intention, rather, is to show a representative set of domains, categories, and disciplines.

In my opinion, there is a very important additional, *transcendent* domain—the Systems domain—that transcends and can potentially utilize the entire spectrum of the ways of knowing. That is indicated in Figure 2.7. We will talk more about the systems domain and the systems approach later in the current chapter and then provide further detail in Chapter 3.

Refer to Figure 2.7 (or Figure 2.6) for the following discussion of the synthesized view of the ways of knowing. Consider first the Sciences domain and the science/mathematics category. The list of natural science disciplines proceeds from least complex (physics) to most complex (ecology). All of the most important ecological systems are highly complex. The "artificial" sciences involve human-made "artifacts." Engineering and related technology pursuits are the major subsets here. The dominant *knowing* approach in the Sciences domain is the scientific method. The approach is well-defined, rigorous, and evidence-based—and quite compelling for those reasons. (The scientific method is described in some detail in Chapter 7.) Note that the social sciences span the Sciences and the Humanities domains and, at times, the social sciences apply (portions of) the scientific method.

The Humanities domain includes a wide spectrum of disciplines that can be divided into disciplinary categories (i.e., humanities, arts, fine arts, . . ., philosophy, and religion) with differing ways of knowing. In Figure 2.7, as an example, we further expand the fine arts category into its subcategories and identify some of the disciplines within those subcategories. We will have more to say about the fine arts category shortly.

All of the many ways of knowing are important and all have validity. They can and should be used in a complementary fashion. *E pluribus unum*: from many, one.

2.3.3 The Ways of Knowing Are Complementary

Figure 2.8 illustrates some of the properties of the ways of knowing, and how and where these properties apply.

Let's start with a discussion of rationality modes (technical and reflective) and their complementary relationship. Technical rationality is a sciences-oriented competency that is based fundamentally on Western materialist

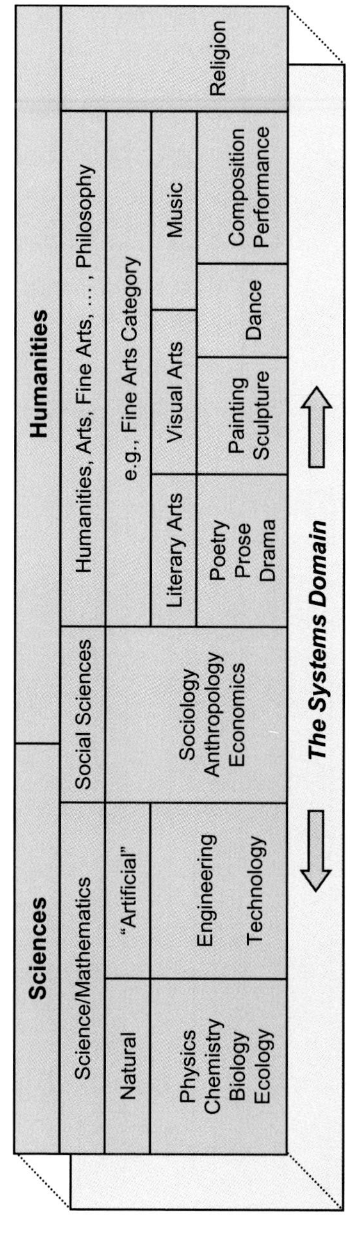

Figure 2.7 The ways of knowing and the systems domain.

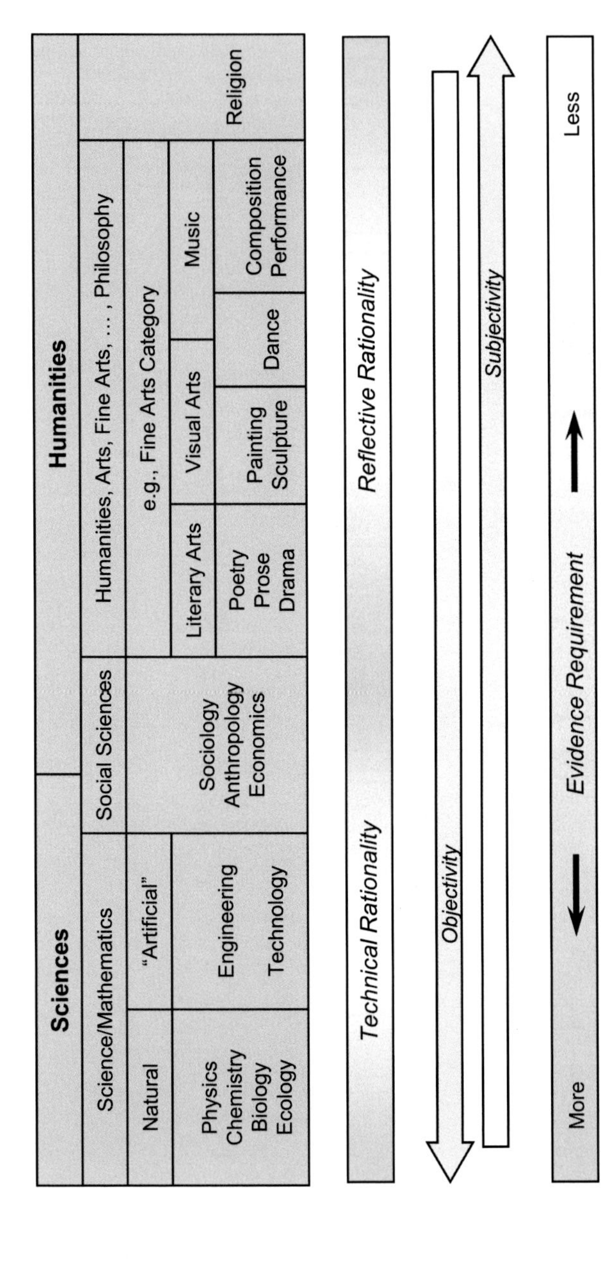

Figure 2.8 Properties of the ways of knowing.

philosophy, reductionism, and the scientific method. Reflective rationality can be viewed as a humanities-oriented competency. It includes the human ability to bring together sometimes disparate concepts and ideas to construct a unified whole, and the ability to view a system not just as parts, but rather as parts with rich interactions. (It is the interactions that mostly define a complex system.)

As indicated in Figure 2.8, the two rationality modes encompass the spectrum of disciplines that comprise the many ways of knowing. There are no well-defined boundaries between technical rationality and reflective rationality; however, we can say that the technical focus is the left side of the spectrum of the ways of knowing and the reflective focus is the right side. Technical rationality operates in the Sciences domain and extends into the Humanities domain. Reflective rationality operates in both the Humanities domain and the Sciences domain.

Einstein (1950) noted that we are very good at "means" (techniques and methods), but not so good at understanding how, when, or even whether to apply those means. In 1974, Alfred Kyle said, "We know how to teach people how to build ships but not how to figure out what ships to build" (see Schön, 1990). Generally speaking, technical rationality addresses the first issue and reflective rationality addresses the second. We clearly need both rationality modes. The systems approach, in fact, *requires* a complementary blend of technical rationality and reflective rationality (see Figure 2.7).

Referring again to Figure 2.8, let's discuss some specific examples of how insights from the fine arts disciplines can support the systems approach and the understanding of systems. We will see that fine-arts reflective rationality can make a very significant contribution to systems knowledge and to systems investigations.

The literary arts and music have provided very useful insights about systems and the systems approach. With respect to the literary arts, here are three examples. Henry David Thoreau clearly understood the importance of reflective rationality. He was concerned about too much emphasis on reductionist analysis. Thoreau said, "I fear that the character of my knowledge is from year to year becoming more detailed and scientific; that, in exchange for views wide as heaven's cope, I am being narrowed down to the field of the microscope. I see details, not wholes nor the shadow of the whole. I count some parts, and say 'I know.'"[6] Thoreau could see that the complementary application of reflective rationality and technical rationality

[6] Henry David Thoreau, *Journals*, August 1851.

was necessary. Ralph Waldo Emerson understood the limitations of human perception of reality and the potential impact on systems investigations (both reflective and technical aspects). Emerson said that we see reality through "many-colored lenses which paint the world their own hue, and each shows only what lies in its focus."[7] We see what we want to see, what we expect to see. (Chapter 13 of this book is devoted to a discussion of issues of human perception and their implications for systems work.) Oliver Wendell Holmes Jr. made a very insightful observation about the role of *function* in (living) systems. He said, "For to live is to function. That is all there is in living."[8] This observation supports the contention that *function* should indeed be the ultimate vertex of the systems triad. (It is; see Chapter 3.)

With respect to music, consider the insights provided by Daniel Barenboim, world-famous music director, conductor, and pianist. Barenboim presented the British Broadcasting Corporation (BBC) Reith Lectures in 2006. He delivered a series of five lectures titled *In the Beginning Was Sound*. In his fifth lecture, Barenboim made some points in the context of political conflict. These important points, however, have much broader applicability. He said, "when you play in an orchestra everybody is constantly aware of everybody else. . . . Music teaches us that everything is connected." These insights clearly suggest a major element of the systems approach: *network thinking*. (See Chapter 3.) Barenboim's insights apply to all kinds of systems, whether musical, political, ecological, or more generally, any highly complex system.

Leonardo da Vinci, the Renaissance-era genius in both art and in science, very clearly saw the intimate relationships between the two. Fritjof Capra (2007) discusses Leonardo's insights. Leonardo himself declared, "Painting embraces within itself all the forms of nature." Capra explains that Leonardo knew "that painting involves the study of natural forms [and that there is an] intimate connection between the artistic representation of those forms and the intellectual understanding of their intrinsic nature and underlying principles." Leonardo recognized the clear and strong relationship between his drawings and paintings of natural forms and the underlying processes that shaped and transformed the actual objects in nature.

Leonardo could also see that the patterns and processes of the microcosm (e.g., the human anatomy) are analogous to the patterns and processes of the macrocosm (e.g., the ecosphere). He recognized self-similarity across spatial

[7] Emerson (1983).

[8] Oliver Wendell Holmes Jr., quoted in *The Practical Cogitator*, Houghton Mifflin Company, 1962.

and temporal scales. That phenomenon much later became known as fractal behavior.

Leonardo knew that insights in one area of intellectual understanding often applied to other areas—and that was true within art, within science, and across art and science. He was able to see the universality. As Capra (2007) puts it, Leonardo had an "exceptional ability to interconnect observations and ideas from different disciplines." His "science cannot be understood without his art, nor his art without his science." Leonardo's work was a synthesis of art and science. Art facilitates systems thinking. Leonardo was clearly a systems thinker.

Benoit Mandelbrot (2012) has suggested that the visual arts and music can provide additional important insights regarding the fractal structure and behavior of complex natural systems. We can observe that many paintings, music compositions, and decorative arts are comprised of patterns—recurring patterns—that repeat again and again at different scales. Such works of art, according to Mandelbrot, exhibit fractal behavior. Mandelbrot provides the following examples and commentary. Leonardo da Vinci's well-known drawing, *A Deluge*, is unquestionably fractal. Claude Lorrain, a seventeenth-century European artist, painted landscapes that are "easily interpreted in fractal terms." Hokusai's famous print *The Great Wave*, which features Mount Fuji, is fractal. Additionally, fractal structures are common in Italian architecture and are present in Persian and Indian art. With respect to music, composer György Ligeti realized that fine music "must be fractal." According to Mandelbrot (2012), contemporary composer Charles Wuorinen stated that he uses "a fractal approach to composition. . . . He was well aware that much of Western music exhibits similar structures over different time scales." Wuorinen and Mandelbrot presented a *Music and Fractals* show at the Guggenheim Museum in New York in 1990.

Given the realization that fractals can be widely observed in great works of art, wouldn't it be reasonable to expect that actual great works of nature very often exhibit fractal behavior? Yes—Mandelbrot (2012) has found that fractals are "ubiquitous in nature and culture." (See also Mandelbrot, 1982, 2004, 2010a, 2010b.) Further, wouldn't it be reasonable to expect that complex ecosystems (and complex ecosystem dynamics) exhibit fractal behavior? Yes—and that is a principal finding of the work of this book.

Reflective rationality strongly supports the *synthesis* activities of the systems approach and systems understanding. Technical rationality has for centuries ably supported the *analysis* activities. Complementary application of both is necessary.

2.3.4 Additional Properties of the Ways of Knowing

2.3.4.1 Objectivity and Subjectivity

Objectivity is typically considered a property of the Sciences domain and subjectivity is typically considered a property of the Humanities domain. As indicated in Figure 2.8, however, the reality is not that straightforward. While objectivity is most prevalent in the Sciences domain and subjectivity is most prevalent in the Humanities domain, there are ranges of objectivity and subjectivity, and the ranges extend beyond the domain boundary, in both directions. Objectivity generally increases as we move to the left (in the diagram of Figure 2.8) across the spectrum of the ways of knowing. Subjectivity generally increases as we move to the right. As we move to the right across the Humanities disciplines, we transition from partially objective to primarily subjective.

In my opinion, Stephen J. Gould's views on this topic are pertinent and important. According to Gould (2003), we need to acknowledge that there are objective behaviors in the humanities and there are subjective behaviors in the sciences. There is a powerful myth about scientific procedure. It says that science is an objective activity, with no mental or social biases as in the humanities. This is not true. Science does have a subjective component. Gould says, "The peculiar notion that science utilizes pure and unbiased observations as the only and ultimate method for discovering nature's truth, operates as the foundational (and . . . pernicious) myth of my profession." In fact, we look for patterns consistent with (for or against) some hypothesis. Bias enters when we are unwilling to abandon preferences that observation refutes. "Universal cognitive biases affect the work of scientists as strongly as they impact any other human activity." Humanists know this, but scientists mostly can't see it. Note that a very good analysis of mental and social biases in scientific work was done by Francis Bacon, which is ironic because Bacon's name is most-often associated with objectivity. In his *Novum Organum*, Bacon discussed "mental and social impediments" that prevent pure objectivism. He called these impediments "idols" and famously warned "beware the idols of the mind." Gould says scientists must recognize that science is "a quintessentially human enterprise" that strives to reach "a more adequate and deeper understanding of material reality." Science has "an essential tie with humanistic studies." Perhaps an awareness and understanding of the commonalities of science and the humanities can help engender mutual respect for the differences, "in order to achieve an *integral* excellence."

Other investigators also have made pertinent observations. Bohm and Peat (2000) note that even quantum theory is susceptible to subjective influences. They say, "In essence, *all* the available interpretations of the quantum theory, and indeed of any other physical theory, depend fundamentally on implicit or explicit philosophical assumptions, as well as on assumptions that arise in countless other ways from beyond the field of physics." No interpretations "give the *final* 'truth' about the subject." Capra (2002) observes that the reluctance of scientists to acknowledge subjective components in science is a result of our Cartesian heritage.

2.3.4.2 Evidence Requirement

Figure 2.8 illustrates the evidence property of the ways of knowing. Evidence is a necessary and essential part of the Sciences domain, where the scientific method is the dominant knowing approach. The scientific method consists of hypothesis generation and factual evidence-based hypothesis testing that results in either refutation or corroboration of the hypothesis. As we proceed from left to right (in the diagram of Figure 2.8) across the Humanities disciplines, the evidence requirement decreases. Religion has the least need of evidence. For religion, beliefs are primary and evidence is secondary. Hypothesis testing is not required (and not possible). Faith is intended to substitute for evidence, testing, and verification/corroboration. This explains why religion could possibly be exploitive, at least in some cases.

2.3.5 Integrating the Ways of Knowing

Successful and comprehensive system investigation requires that we take advantage of the spectrum of the many ways of knowing. To deal effectively with highly complex systems, we utilize the sciences domain, the humanities domain, and of course the transcendent systems domain and the systems approach. The systems approach requires a complementary blend of technical rationality and reflective rationality. Reflective rationality strongly supports the *synthesis* activities of problem framing and systems understanding. Technical rationality has for centuries supported the *analysis* activities of the sciences.

The systems approach, at its best, depends on the integration of the multidisciplinary ways of knowing. I hope that my work in this book demonstrates that. The really tough problems in the world are highly complex systems problems that require this integrated multidisciplinary approach. As mentioned earlier, Bohm and Peat (2000) concur and say that all modes of knowing must be integrated to solve the problems of the world. Schön (1983) says we need "an epistemology of practice which places technical problem solving within a broader context of reflective inquiry." Gould (2003) says that science and the humanities together can "achieve an *integral* excellence."

A debate on multidisciplinary integration has been conducted by two eminent scholars/researchers/authors, Stephen J. Gould and Edward O. Wilson. They both agree that integration of the sciences and the humanities is desirable and necessary. They do not, however, agree on the method of integration. I will describe the debate in the next section.

After that, in Chapter 3, I will describe the systems approach and its major elements in detail.

2.4 A DEBATE ON THE INTEGRATION OF THE SCIENCES AND THE HUMANITIES

We have argued that the full set of competencies required for success in complex systems work encompasses the spectrum of the many ways of knowing. E. O. Wilson and S. J. Gould agreed that the sciences and the humanities should be integrated. Their debate concerned the *method* of integration. I find this debate and the issues it raises interesting and very relevant to the study of highly complex systems. Plus, we get to consider the views of two very well-respected scholars. Please refer to Figure 2.9.

The debate occurred near the early 2000s in two important books: first in E. O. Wilson's *Consilience* (Wilson, 1999) and then in a rebuttal in

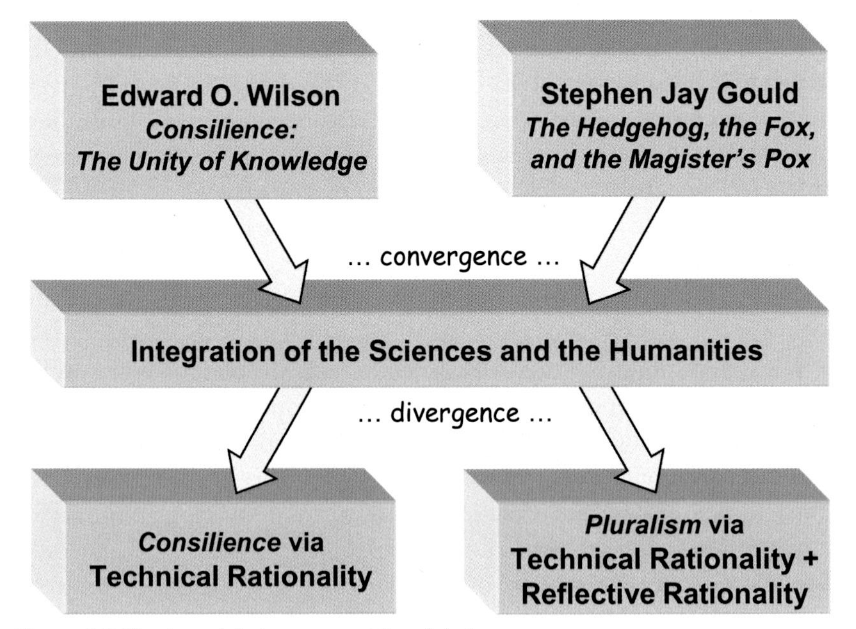

Figure 2.9 The two debaters . . . and the debate.

S. J. Gould's *The Hedgehog, the Fox, and the Magister's Pox* (Gould, 2003). Wilson believes that the method of integration should be based solely on the methods of science (technical rationality, in my terms). Gould disagrees and advocates a combination of the methods of science and the methods of the humanities (technical rationality plus reflective rationality, in my terms). We'll discuss Wilson's views and Gould's views and highlights of the debate in the following paragraphs. Keep in mind that this "debate" was not real-time interactive. It takes the form of claims by Wilson and then later rebuttal by Gould. Unfortunately, Stephen J. Gould died prior to the publication of his rebuttal in *The Hedgehog, the Fox, and the Magister's Pox*. There was no opportunity for further interaction.

The titles of the two books rather interestingly convey the authors' viewpoints. What is *consilience*? Wilson defines consilience as a "jumping together" of knowledge by the linking of facts and fact-based theory across disciplines to create a common groundwork of explanation. Wilson says, "a balanced perspective cannot be acquired by studying disciplines in pieces but through pursuit of the consilience among them." According to Wilson, consilience means the unification of knowledge across disciplines, and the single method of unification is technical rationality. (I should mention that Gould disagrees with this definition and use of the term consilience. He claims that William Whewell, who originated the concept of consilience in his 1840 treatise on *The Philosophy of the Inductive Sciences*, did not intend consilience to mean the unification of science and the humanities using a single mode of explanation. According to Gould, Whewell "forcefully rebuts the . . . argument for unification of all knowledge along a single chain of rising complexity, with all phenomena subject to one style of explanation.") Gould explains the title of his book as follows: The *hedgehog* and *fox* metaphors are taken from a proverb by Archilochus, a seventh-century BC Greek soldier-poet. Gould relates, "The fox devises many strategies; the hedgehog knows one great and effective strategy." When hunted, the fox can employ many strategies to escape. The hedgehog has just one strategy; it rolls itself into a ball with spines outward so that any attempt to bite or strike it would be painful for the attacker. Some would say "normal science" employs the style of the hedgehog, and the humanities employ the style of the fox. But surely, according to Gould, a "fruitful union" of both styles (i.e., technical rationality and reflective rationality) is required for the unification of knowledge across disciplines. Continuing with Gould's title explanation, the *magister* represents an authority figure and the *pox* represents a misuse of authority. Gould makes no further comment about the relevance of these last two metaphors to the debate (but one might imagine).

Happily, Wilson and Gould do agree on some very important points. As noted, they both agree that the sciences and the humanities should be integrated. Wilson explains that the relationship between science and the humanities is crucial for human welfare. "Most of the issues that vex humanity daily—ethnic conflict, arms escalation, overpopulation, abortion, environment, endemic poverty, to cite several most persistently before us—cannot be solved without integrating knowledge from the natural sciences with that of the social sciences and humanities. . . . The greatest enterprise of the mind has always been and always will be the attempted linkage of the sciences and humanities. . . . The ongoing fragmentation of knowledge and resulting chaos in philosophy are not reflections of the real world but artifacts of scholarship." Gould agrees with all of this.

Wilson and Gould also agree on the importance of technical rationality. Technical rationality is based, in large part, on reductionism and the scientific method. Wilson notes that "science . . . is the organized, systematic enterprise that gathers knowledge about the world and condenses the knowledge into testable laws and principles. . . . The cutting edge of science is reductionism, the breaking apart of nature into its natural constituents." Gould understands that technical rationality is good, but it is not enough. We cannot "deny the extraordinary power and achievements of reductionism since the beginning of the Scientific Revolution." The methods of science are "the standard experimental and quantitative procedures so well suited for simple, timeless, and repeatable events in conventional science." But we need more when dealing with highly complex systems. "Reductionism may not, despite its triumphs in a large domain of appropriate places, be universally extendable as an optimal path to complete scientific understanding. . . . Reductionism—a powerful method that should be used whenever appropriate, and that has been employed triumphantly throughout the history of modern science—must fail as a generality. . . . The pure reductionist program . . . represents the wrong pathway toward such a worthy goal of integration between the sciences and humanities."

As stated earlier, the fundamental disagreement of Wilson and Gould concerns the *method of integration* of the sciences and the humanities. We will first consider Wilson's approach and arguments, and then Gould's.

2.4.1 Discussion of Wilson's Position

Wilson says, "Consilience is the key to unification." Technical rationality is the means. With respect to the work of Sir Isaac Newton, Wilson talks about the "Ionian enchantment," an expression that comes from physicist and historian Gerald Horton. Its roots apparently go back to Thales of Miletus, sixth century BC. "It means a belief in the unity of the sciences—a conviction, far deeper than a mere working proposition, that the world is orderly and can be explained by a small number of natural laws." During the Enlightenment in Europe in the seventeenth century, scholars began suggesting that if the world could be explained by scientific laws, then "why not a Newtonian solution to the affairs of men? ... Why shouldn't their behavior and social institutions conform to certain still-undefined natural laws?" Perhaps *everything* could be understood in scientific terms. Wilson argues that the mind and conscious experience are physical phenomena that potentially have a scientific explanation. If mental processes have a scientific explanation, then the social sciences and the humanities must also have a scientific explanation. "Belief in the intrinsic unity of knowledge ... rides ultimately on the hypothesis that every mental process has a physical grounding and is consistent with the natural sciences." Regarding consilience across the natural sciences and the social sciences, Wilson says, "The social sciences are intrinsically compatible with the natural sciences. The two great branches of learning will benefit to the extent that their modes of causal explanation are made consistent." Regarding consilience across the natural sciences and the creative arts, Wilson says, "Gene-culture coevolution is, I believe, the underlying process by which the brain evolved and the arts originated." (See also Dutton, 2009.) If so, the arts potentially have a scientific explanation.

Edward O. Wilson summarizes his conclusions for us:

- "I have argued that there is intrinsically only one class of explanation [scientific explanation]. It ... unite[s] the disparate facts of the disciplines by consilience ...."
- "The central idea of the consilience world view is that all tangible phenomena, from the birth of stars to the workings of social institutions, are based on material processes that are ultimately reducible ... to the laws of physics."
- "The main thrust of the consilience world view ... is that culture and hence the unique qualities of the human species will make complete sense only when linked in causal explanation to the natural sciences."

2.4.2 Discussion of Gould's Position

Next, let's look further at Gould's position on the integration of the sciences and the humanities. Gould believes that both technical rationality and reflective rationality (my terms) are required for successful integration. Here are some overall comments from Gould: The sciences and the humanities "work in different ways and cannot be morphed into one simple coherence." Strengths from both disciplinary domains are needed to realize "the common goal of human wisdom." We cannot be "shearing off the legitimate differences. . . . In our increasingly complex and confusing world, we need all the help we can get from each distinct domain of our emotional and intellectual being." Science has been stunningly successful in its domain, but "each [science and the humanities] has so much to learn from the successes of the other." Proper integration "stresses respect for preciously different insights, inherent to the various crafts, and rejects the language (and practice) of hierarchical worth and subsumption." It is a combination of the two, not a dominance of one (science) over the other (humanities).

Gould argues that a pluralistic approach to the integration of the sciences and the humanities is necessary. We need to overcome the contentious attitudes and forget the superiority claims. Gould says, "I see no conceivable justification, other than human narrowness and the weight of 'traditional' practice, for continued contention between science and the humanities. . . . Yet science has shown a regrettable tendency . . . to claim superiority . . . as a 'better' way of knowing. . . . Scientists have tended to depict their own history as a steady march to truth, mediated by successful application of a universal and unchanging 'scientific method.'" But both the sciences and the humanities are important and valuable ways of knowing. We "must find a way to mediate and merge these two great and truthful ways." The solution is to employ Aristotle's "golden mean" and employ the best traits of both.

How about a "consilience of equal regard"? Gould wonders "why the dream of unification holds such power over the scholarly mind. . . . I find nothing . . . appealing in such neatly and symmetrically honed structures. . . . I want the sciences and humanities . . . to keep their ineluctable discovery aims and logics separate as they ply their joint projects and learn from each other." How about a different and better interpretation of the concept of consilience that recognizes "different, and equally valid, ways of knowing"? Gould advocates a consilience "that respects the rich, inevitable, and worthy differences, but that also seeks to define the broader properties shared by any creative intellectual activity. . . . We all want to enjoy the differences, yet find some meaningful order in the totality."

I have great respect for both E. O. Wilson and S. J. Gould, but in this case I find Gould's position much more compelling. (His views and my views are consistent.)

CHAPTER 3

The Major Elements of the Systems Approach

As stated in Chapter 1, I have found three concepts to be crucially important in my systems work—both my earlier work on human-made systems and my current work on natural systems. As illustrated in Figure 3.1, they are

- A Blend of Synthesis and Analysis
- Network Thinking
- The Systems Triad

I consider these concepts major elements of systems thinking and the systems approach.

The three concepts are discussed in detail in Sections 3.1–3.3, respectively, and these concepts are utilized throughout the book.

3.1 A BLEND OF SYNTHESIS AND ANALYSIS

3.1.1 An Incremental, Cyclical, and Iterative Process

A blend of synthesis and analysis is necessary. Not only do we have to know how to break things apart (analysis), we also have to know how to put things together (synthesis) in order to understand a system fully. Let's start with some simple definitions. The term *synthesis* comes from the Greek *syntithenai:* to put together. The term *analysis* comes from the Greek *analyein:* to break apart.

In systems investigations, reductionist analysis alone is not enough. I have observed, however, that much doctoral and other advanced research in the sciences focuses on analysis and follows an essentially *reductionist approach.* The researcher identifies an area of interest; drills down to some narrow, specialized subset; and then investigates that subset in depth—all with the objective of discovery. The resulting discovery, however, is likely to be isolated in the sense that it is not adequately "connected" to a larger context. (I'll give a further explanation of the potential problems with the reductionist approach in Section 3.1.2.).

My research takes a *systems approach* to scientific discovery. It does not focus solely on analysis, but rather is a blend of synthesis and analysis.

Understanding Complex Ecosystem Dynamics
http://dx.doi.org/10.1016/B978-0-12-802031-9.00003-6

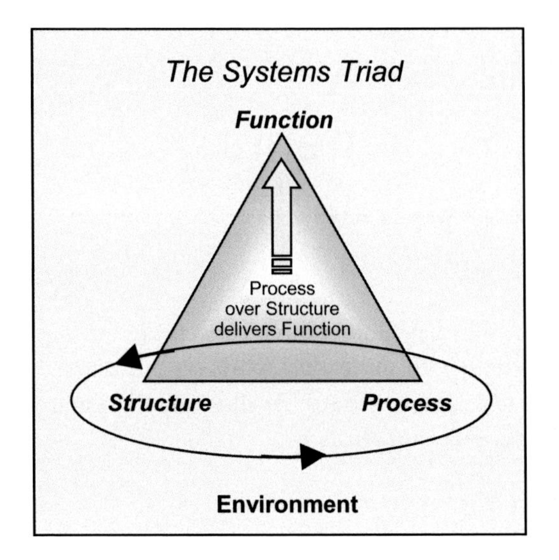

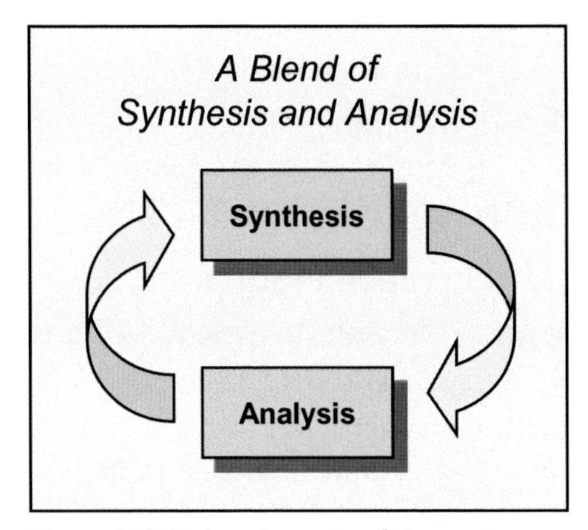

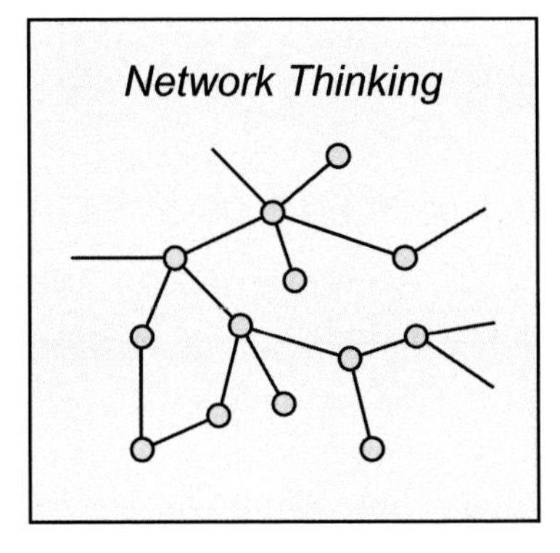

Figure 3.1 Major elements of the systems approach.

The synthesis work provides the required higher level system connections and context. Research results are not isolated, but rather are appropriately integrated into a total system picture.

All effective systems efforts include both synthesis and analysis to some degree. The process is incremental, cyclical, and iterative, as illustrated in Figure 3.2.

In the work described in this book, synthesis is most prominent in

- Creating a functionality-driven view of the function-structure-process landscape of ecological systems (Part II)
- Creating a network-centric view of the characteristics of ecological system dynamics (Part IV)

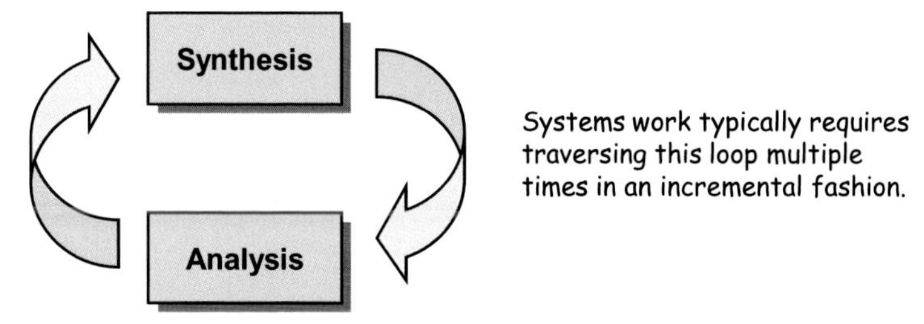

Figure 3.2 Combined synthesis and analysis.

Analysis is most prominent in Part V of the book:

- Developing an ecological network model to test my hypothesized view of ecosystem network dynamics
- Analyzing the model simulation output and generating results

Notice the similarity between Figure 3.2 and Figure 1.2 of Chapter 1 (which illustrates the engineering design model). In Figure 1.2, "design candidate generation" is essentially a synthesis process, while "design candidate evaluation" is an analysis process. Engineering design is an iterative synthesis and analysis procedure that converges on an appropriate design solution. In Chapter 7, we will devise a high-level evolution model. It has the same form as Figures 3.2 and 1.2. Perhaps evolution is the ultimate synthesis/analysis process.

3.1.2 Potential Problems with the Reductionist Approach

Reductionism is analysis-based: it breaks systems apart. The reductionist approach is predicated on "the belief that in every complex system the behavior of the whole can be understood entirely from the properties of its parts" (Capra, 1996). "The past three centuries of science have been predominantly reductionist, attempting to break complex systems into simple parts, and those parts, in turn, into simpler parts" (Kauffman, 1995). "In recent years the practice of science has become increasingly reductionist in seeking to understand phenomena by detailed study of smaller and smaller components" (Odum and Barrett, 2005).

The structure of ecological systems (and other complex systems) is a network, and a network can be represented as a hierarchy. Figure 3.3 depicts the reductionist approach in terms of a network hierarchy.

Part (a) of Figure 3.3 shows the typical analysis-based reductionist approach of drilling down the system network hierarchy in order to conduct a detailed investigation of an increasingly specialized network node. At each higher level of the hierarchy, a subsystem (an aggregate network node) is

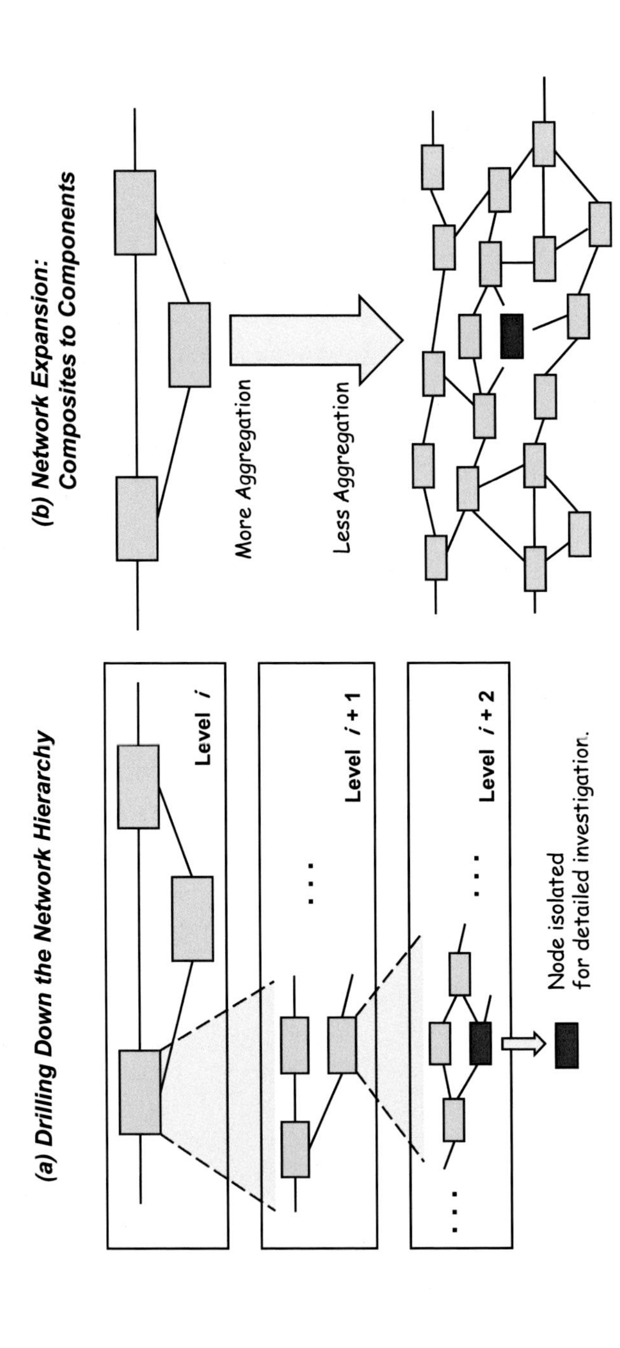

Figure 3.3 Reductionist approach and network hierarchy. (a) Drilling down the network hierarchy and (b) network expansion: composites to components.

selected for further study; network connections (links) are broken with associated information loss; and then the drilling down continues. "The reductionist likes to move from the top down, gaining precision of information about fragments as he descends, but losing information content about the larger orders he leaves behind."[1] Part (b) of Figure 3.3 depicts the network expansion as the top-level set of composite network nodes and links are decomposed into their component network nodes and links. The top-level network is highly aggregated. Each time we move one level down the hierarchy, de-aggregation occurs. As we proceed down, level by level, the network view rapidly expands and the number of network nodes and links rapidly increases. The bottom-level view of the network shown in part (b) of Figure 3.3 suggests that, when a node is isolated for detailed investigation, the total number of broken network connections can be very substantial. Not only are there several broken *direct* connections at this level (three are shown in the figure), but there is also a very large number of broken *indirect* connections to many other nodes in the network. The resulting information loss can be substantial. This information loss can be quantified. That topic is covered in Chapter 4: *Reductionism and Information Loss*.

Here are two additional comments about the reductionist problems of isolation and information loss from noted authors. Odum and Barrett (2005) say that reductionism is "specialization in isolation." Bar-Yam (2004) states, "Indeed, one of the main difficulties in answering questions or solving problems ... is that we think the problem is in the parts, when it is really in the relationships between them." The relationships/interactions among the parts—the connections—make all the difference.

3.1.3 More Views on Synthesis and Analysis

Over the years, many respected thinkers and investigators (in both humanistic and scientific areas) have expressed their views on synthesis and analysis.

Plato apparently understood the need for a blend of synthesis and analysis when he discussed problem solving in *Phaedrus*. He seemed to advocate synthesis followed by analysis: "First, the taking in of scattered particulars under one Idea, so that everyone understands what is being talked about .... Second, the separation of the Idea into parts, by dividing it at the joints, as nature directs, not breaking any limb in half as a bad carver might."

H. D. Thoreau seemed concerned about too much emphasis on reductionist analysis: "I fear that the character of my knowledge is from year to year

[1] Paul Weiss, 1969, quoted in Klir (1991), *Facets of Systems Science*.

becoming more detailed and scientific; that, in exchange for views wide as heaven's cope, I am being narrowed down to the field of the microscope. I see details, not wholes nor the shadow of the whole. I count some parts, and say 'I know.'"[2] Thoreau saw the advantages of abstraction and synthesis. "There is some advantage, intellectually and spiritually, in taking wide views with the bodily eye and not pursuing an occupation which holds the body prone. There is some advantage, perhaps, in attending to the general features of the landscape over studying the particular plants and animals which inhabit it. A man may walk abroad and no more see the sky than if he walked under a shed."[3] ... "Many a man, when I tell him that I have been on to a mountain, asks if I took a glass with me. No doubt, I could have seen farther with a glass, and particular objects more distinctly—could have counted more meeting-houses; but that has nothing to do with the peculiar beauty and grandeur of the view which an elevated position affords. It was not to see a few particular objects, as if they were near at hand, as I had been accustomed to see them, that I ascended the mountain, but to see an infinite variety far and near in their relation to each other, thus reduced to a single picture."[4]

Somewhat more recently, Whitehead (1959) said that the scientist must have a "union of passionate interest in the detailed facts with equal devotion to abstract generalisation." On the subject of systems architecting, Rechtin and Maier (1997) say that understanding a system requires "a structuring of [the] unstructured ... an inspired synthesizing ...." While discussing software system development skills, Mowbray and Zahavi (1995) say, "Only about one out of five software developers has the capability to create good abstractions. ... In order to create good software systems, the people who have the abstraction ability should be recognized and promoted into positions of responsibility and authority as software architects." (Grudin (1990) defines *abstraction* as "the refinement of raw data into coherent ideas and the organization of individual ideas into more inclusive categories.")

Here in Section 3.1, we have highlighted potential problems with the exclusive use of the reductionist approach. Reductionism, of course, has been a valuable tool for hundreds of years and continues to be a valuable tool. The key, in my opinion, is to use it prudently in the context of a *synthesis plus analysis* systems approach.

[2] Henry David Thoreau, *Journals*, August 1851.
[3] Ibid.
[4] Henry David Thoreau, *Journals*, October 20, 1852.

There is yet another related problem with reductionism. Look again at the bottom of Figure 3.3, part (b). The reductionist approach can leave us with a "node-centric" view of the system under investigation. System behavior, however, is primarily determined by network characteristics, not node characteristics. Networks trump nodes. We'll discuss that in the next section.

3.2 NETWORK THINKING

"Network thinking means focusing on relationships between entities rather than the entities themselves" (Mitchell, 2009). Complex systems most often take the form of networks. What most determines the behavior of these complex systems? Although they have importance, it is not the individual system nodes. It is the even more important *interactions* among the nodes—the dynamic interplay of the network as a whole. Reductionist scientific research is conducted by drilling down and isolating a system component (node) for detailed analysis. It is assumed that understanding a system is all about understanding the nodes. That is incorrect. It is not all about the nodes; it is about the network. To understand a complex system, network thinking is required (see Figure 3.4).

3.2.1 West and East

First, let's consider the mindset we bring to our scientific pursuits, and why that might cause us to reach an incorrect conclusion about the importance of nodes vs. networks.

Westerners mostly see *things*. Easterners mostly see *relationships*. Those are the general observations of Nisbett (2003) and Ji, Peng and Nisbett (2000). In the Ji et al. (2000) article, the authors note that "scholars in many

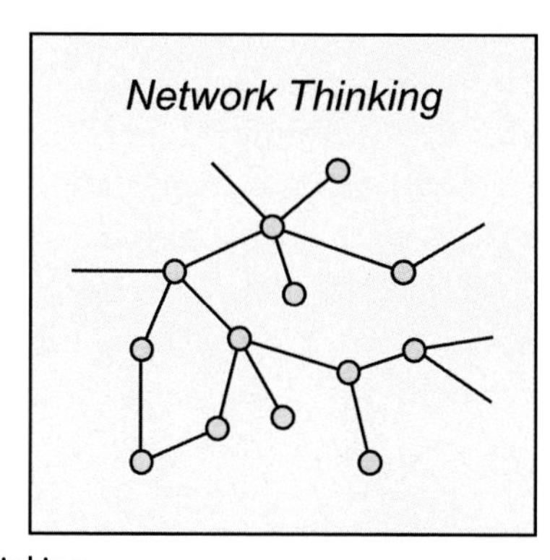

Figure 3.4 Network thinking.

disciplines have maintained that people in Asian cultures, especially the East Asian cultures of China, Korea and Japan, have a relatively holistic cognitive orientation, emphasizing relationships and connectedness." The Asian "stance contrasts with the analytic Western world view," which can be traced back "to the ancient Greeks, who saw the world as composed of objects which are understood as individuals." While the ancient Chinese "focused on relationships among objects in the field, the [ancient] Greeks were prone to focus more exclusively on the object."

Ji, Peng and Nisbett refer to a recent study "requiring participants to watch underwater scenes in which a focal fish moved across a field of less salient fish and animals and static seascape features such as rocks and coral. When asked to describe what they had seen, Japanese participants reported more observations about the background and more relationships between the focal fish and the background than did American participants." Ji, Peng and Nisbett say "the results support our contention that East Asians orient themselves to the environment as a whole and to relationships among objects within it."

Apparently, Western cognition tends to be reductionist and Eastern cognition tends to be holistic. Here's the problem: our scientific pursuits (and the scientific method) follow the Western perspective. While the reductionist perspective is valuable, we need to add the holistic perspective to our scientific investigations. If we do not, we risk reaching incorrect conclusions.

3.2.2 Network Thinking Trumps Node Thinking

As already noted, reductionist scientific research typically assumes that understanding a system is all about understanding the parts (the nodes). But that thinking is flawed. Networks trump nodes.

System behavior depends primarily on network dynamics, not on node details. Many respected investigators and authors agree. Simon (1996) declares that system behavior is determined more by the organization and relationships of the system components than by the detailed properties of each of the components. Solé and Goodwin (2000) say, "Interactions, not individuals, are the key ingredients of behavioral complexity." The behavior of an ecological system or any complex system has much more to do with the collective behavior of the network and less to do with the details of individual nodes. Vicsek (2002) suggests, "The laws that describe the behavior of a complex system are qualitatively different from those that govern its units." Krugman (1996) relates that Philip Anderson—a Bell Labs scientist, a Nobel laureate in physics, and sometimes called the "father" of complexity theory—believes that system collective behavior cannot be determined from individual unit behavior.

3.2.3 The Importance of Network Dynamics

Two examples of the importance of network (rather than node) dynamics follow.

3.2.3.1 The Human Genome

Human genome research has produced wonderful accomplishments, but it is still in a nascent stage. Consider a simplified, yet useful, view of the genome as a network of genes (the nodes). There has been impressive progress in describing and detailing the nodes, but there is still little understanding of the network relationships—the relationships among genes and between genes and the environment. Our existing knowledge, therefore, does not represent the culmination of our understanding of genetic behavior, rather only the beginning. We still do not know much about how the nodes (genes) interact via their network connections to achieve genetic functions. Ultimately, what we seek to understand is the behavior of the very complex genetic networks and not just the behavior of the nodes. Capra (2002) notes that biologists "know very little of the ways in which genes communicate and cooperate in the development of an organism. In other words, they know the alphabet of the genetic code but have almost no idea of its syntax." Much additional work is necessary and is ongoing. "The current cascade of complete genome sequences ... now compels a major shift in bioscience research toward integration and system behaviour."[5] In 2015, the National Human Genome Research Institute[6] is implementing that shift and is making good progress.

Networks trump nodes. The failure to appreciate and observe that reality can result in incorrect conclusions and misguided initiatives. The hopes of "genetic determinism" for one-to-one relationships between genes and human characteristics are *not* generally true (Gould, 2003). "Scientists have often claimed to have found *the gene* for a given process, as if whole processes could be represented by elementary genetic interactions" (Solé and Goodwin, 2000). The relationships between, say, genes and disease or genes and other human traits are network-based and complex.

[5] James Bailey, geneticist at the Institute for Biotechnology in Zurich, quoted in Capra (2002).
[6] See the National Human Genome Research Institute website at http://www.genome.gov/.

Genetic determinism seems to be the conceptual basis for so-called genetic engineering. In my view, "genetic engineering" is mostly a for-profit commercial endeavor and the term is surely a misnomer. It is not sound engineering and it does not properly account for gene network effects. Back in 1999, there was a relevant story in Time magazine.[7] The story said that work by scientists at Princeton, MIT, and Washington University—reported in the journal *Nature* in September 1999—suggested that intelligence may be related to a gene named *NR2B*. The press began to report that the IQ gene had been found, that intelligence was a property of this gene, and that in the future we might be able to make a person more intelligent by "engineering" his or her genes. This, of course, has not happened. Such claims reflect a serious lack of understanding with regard to network relationships, that is, the relationships among genes and between genes and the environment.

3.2.3.2 Node Replacement

Because system behavior depends primarily on network dynamics and not on node details, we might ask whether, under some circumstances, nodes can be replaced without substantially changing network behavior. The answer seems to be yes.

In fall 2005, the ECOL 8000 class at the University of Georgia's Odum School of Ecology discussed this topic. The discussion participants recognized that the class was a network and the nodes were the students and instructors. Despite "node replacements" (and additions and subtractions) from semester to semester or even week to week, the expected and observed class behavior (network behavior) was maintained. This was true even under conditions of stress. For a class field trip, a "keystone" node (a lead instructor) unexpectedly was unable to attend. At the last minute, another competent instructor stepped in as a "node replacement." The field investigations proceeded as usual and were successful. Network behavior was not altered.

Laszlo (1972) has observed that a range of complex "people" systems (e.g., consumers, investors, teams, societies, cultures) change individual "parts" all the time, yet seem to maintain essentially the same properties.

Holland (1996) has said (in the context of system diversity), "If we remove one kind of agent [node] from the system, creating a 'hole,' the system typically responds with a cascade of adaptations resulting in a new agent that 'fills the hole.' ... The ecosystem interactions are largely re-created, although the agents are quite different." The network maintains coherence. In a complex adaptive system, "a pattern of interactions disturbed by the

[7] TIME Magazine, September 1999.

extinction of component agents often reasserts itself, though the new agents may differ in detail from the old."

Here's something interesting to think about. Node replacement may increase ecosystem diversity in an additive fashion (a new and different node is added to the system). Beyond that, node replacement might actually *accelerate* ecosystem diversity in a multiplicative fashion. Here's how: new agents (nodes) fill "holes" left by departing agents. Each new agent brings with it possible opportunities for further (new and different) interactions and new niches. This might encourage still other new agents to become part of the network. The network dynamics at work here might accelerate ecosystem diversity. Human interference with these dynamics may be a mistake. Trying to maintain the status quo of an ecosystem may actually impede diversity. In some cases, it may be exactly the wrong thing to do. Natural changes are inevitable and very often positive. Beware the human impacts—especially the rampantly destructive ones, but sometimes also the well-intentioned ones.

Network thinking is crucial. "One of the key insights of the systems approach has been the realization that the network is a pattern that is common to all life. Whenever we see life, we see networks" (Capra, 2002). The structure of ecological systems is a network. Networks and network thinking occupy a central role in my work, and in this book.

3.3 THE SYSTEMS TRIAD

Function, *structure*, and *process* are three extremely important concepts in any system investigation. Unfortunately, these concepts are generally not well understood. The terms are very frequently used, but very rarely defined. Clarity is sorely needed. In this section, we clearly define, describe, and discuss these concepts and their relationships. The *systems triad* is a vehicle that I have developed and use to increase understanding of the roles and relationships of function, structure, and process (the triad vertices) in complex system dynamics; see Figure 3.5.

3.3.1 Definition of Terms

Function describes what a system does. The term comes from the Latin *fungi*, to perform. Function is the fundamental property of any system. For human-made systems, function is the *raison d'etre*—the reason or purpose for existence—of the system. In natural (nonteleological) systems, there is function but not purpose. Examples of ecological system functions include production, consumption, respiration, decomposition, and storage.

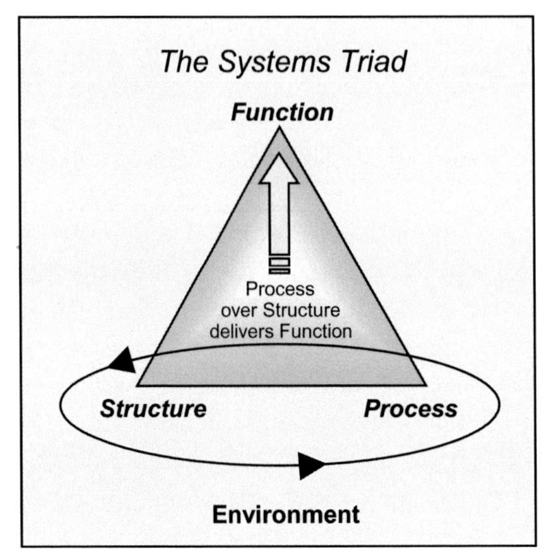

Figure 3.5 The systems triad.

Structure is derived from function and specifies a system's physical form. Merriam-Webster online defines structure as "the arrangement of particles or parts" in a system. Complex systems (including ecological systems) consistently take the form of networks. Network structure is comprised of nodes (structural elements) and links (node structural connections).

Process is similarly derived from function. Processes execute on the platform of structure. The resulting process-over-structure active flows deliver function. The word process comes from the Latin *processus* and *procedere,* meaning procedure. Merriam-Webster's definition of process includes "something going on" and "a series of actions" that "lead toward a particular result."

It is useful to define another term that is closely related to function, structure, and process. An *event* is an outcome of a process-over-structure flow. A set of events comprise a function. A function, therefore, is a composite event—a composite outcome of process-over-structure flow. (We will use the term *event* in later portions of the book.)

3.3.2 Function and Form

Function is the fundamental[8] property of any system. With respect to the human-being system, Oliver Wendell Holmes Jr. has said, "For to live is

[8] Merriam-Webster online defines *fundamental* as "serving as a basis supporting existence."

to function. That is all there is in living."[9] A group of prominent ecologists have said that system "functioning means showing activity. . . . Ecosystem functioning reflects the collective life activities of plants, animals, and microbes."[10] Function is the system driver; everything about the system follows from it. Architect Sullivan (1896) coined the phrase "form ever follows function" more than a century ago. This very well-known system principle has been applied across system design disciplines (building architecture, engineering, and many others) for years. Function and form, and their relationships, are depicted in the systems triad diagram of Figure 3.6.

System *form* represents a particular implementation of the system. It includes both the system physical structure and the associated system processes that operate on that structure. The two are a matched set: structure + process = form. Form is derived from function, and function is delivered by form. Structure is the derived platform upon which derived processes execute, in order to produce function. Bar-Yam (2004) notes that "the function of the system dictates its structure." (In my view, function dictates structure *and* process.) There is another very important point illustrated in Figure 3.6. The cardinality of the relationship between function and form is one-to-many. There are many physical implementations that can deliver

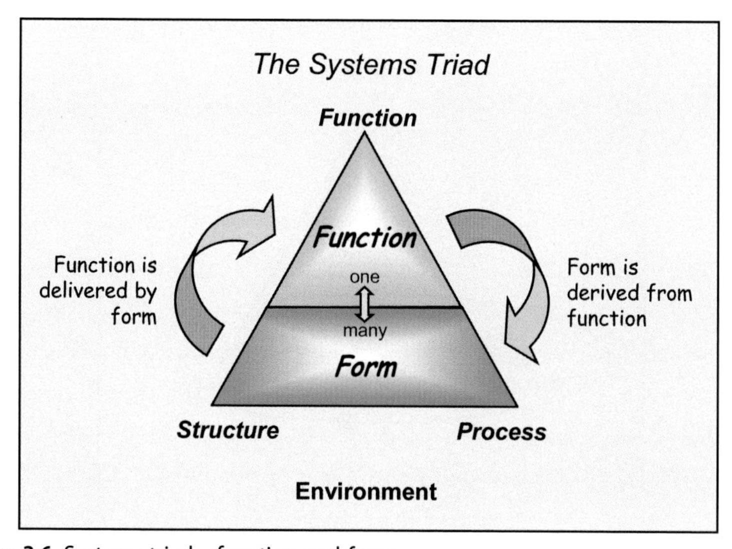

Figure 3.6 Systems triad—function and form.

[9] Oliver Wendell Holmes Jr., quoted in *The Practical Cogitator*, Houghton Mifflin Company, 1962.
[10] These two quotes are from Naeem et al. (1999) (http://www.epa.gov/watertrain/pdf/issue4.pdf).

a given set of functions. The particular implemented form often depends on the system environment.

3.3.3 Process-Over-Structure Flow

Refer next to the systems triad diagram provided in Figure 3.7. The emphasis in this figure is on process-over-structure flow. As indicated, process over structure delivers function in the context of environment. (I recently discovered, in the book *Philosophy and Design,*[11] a very similar statement: "process and structure co-produce function in the context of environment.") Process-over-structure *flow currency* (that which is flowing) varies by system type. For financial systems, the flow currency could be money. For software systems, the currency is usually information. For ecological systems, the flow currency is matter and/or energy.

When developing a human-made system, we can essentially control all aspects of the systems triad, within the constraints of the system environment. The engineers, in concert with the system stakeholders, specify the system functions, architect the structure, and design the processes. In my experience, a key focus during system development is

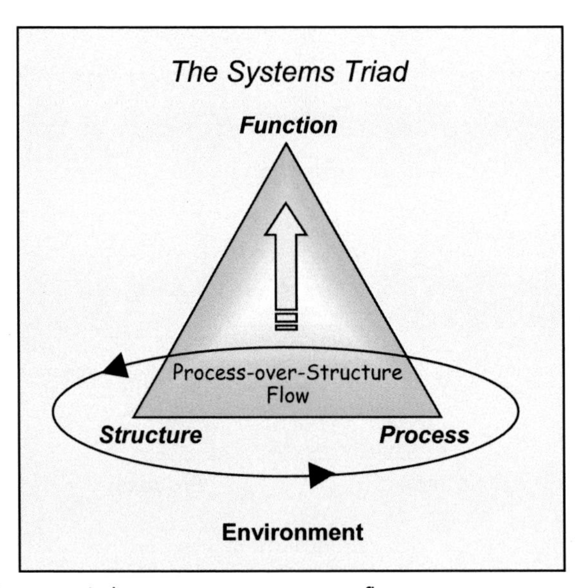

Figure 3.7 Systems triad—process-over-structure flow.

[11] *Philosophy and Design—From Engineering to Architecture*, edited by Vermaas et al., Part 2, chapter by Kristo Miettinen, Springer Netherlands, 2008.

process-over-structure flow. We must make certain that the flow outcomes yield the desired system functions.

When attempting to understand natural systems, we do not control the systems triad. When we observe an existing natural system, all we can see directly are some aspects of the implemented form. We can see physical structure and we can observe some process-over-structure flow outcomes. Process mechanisms can only be inferred. System functions can only be inferred. Again, as in the human-made system development case, a key focus for understanding natural systems is (or should be) process-over-structure flow.

Bohm (1983) has proposed that the primary characteristic of reality is flow or flux. Flux—from the Latin *fluere*, to flow—can be defined as a continuous moving on or passing by; a continued flow. Bohm says that reality is "a set of forms in an underlying universal movement." He provides some historical background: "The notion that reality is to be understood as [flux] is an ancient one, going back at least to Heraclitus, who said that everything flows. In more modern times, Whitehead was the first to give this notion a systematic and extensive development." (Alfred North Whitehead, in the twentieth century, developed a *philosophy of organism,* which included the conviction that reality is composed principally of flows rather than objects.) Throughout Bohm's book, "the central underlying theme [is] the unbroken wholeness of the totality of existence as an undivided flowing movement without borders. ... Not only is everything changing, but all *is* flux." As the John Travolta character in the movie *Phenomenon* says, "Everything is on its way to somewhere."

The process-over-structure view is quite relevant to the long-standing mind/brain discussion. René Descartes, in the seventeenth century, hypothesized that mind and matter were separate phenomena. How then is the mind related to the brain? This question perplexed science and philosophy for many years. Fritjof Capra[12] writes that, according to the Santiago theory of cognition, "Mind is not a thing but a process—the process of cognition, which is identified with the process of life. The brain is a specific structure through which this process operates. The relationship between mind and brain, therefore, is one between process and structure." Cognitive activities can be described as process-over-structure flow. Capra further notes that "consciousness is a special kind of cognitive process that emerges when cognition reaches a certain level of complexity." ... "Mind and matter no longer appear to belong to two separate categories, but can be seen

[12] All of the quotes in this paragraph are from Capra (2002).

as representing two complementary aspects of the phenomenon of life—process and structure. At all levels of life, beginning with the simplest cell . . . process and structure are inseparably connected."

Here's an example of process-over-structure flow that will be quite familiar to ecologists. Consider an ecosystem *compartment model*. Such models traditionally have been used in ecological system network analysis. (We will look at a compartment model in detail in Chapter 16.) The compartments, which represent ecosystem components (usually a grouping of similar species or materials), are the network nodes and each simple flow between two compartments is a network link. Even though compartment models are steady-state models with static structure and constant simple flows, we can apply the concept of process-over-structure flow. A compartment model process-over-structure flow is

- A system (or subsystem) input/output relationship that generates an event that contributes to the realization of a function
- A "thread" through system compartments and links
- A "compound flow" that encompasses some set of system compartments, system "simple flows," and system inputs and outputs

In set-theoretic notation, the definition can be expressed as

$$\text{Process-over-structure flow} \equiv \{z_i, x_i, f_{ij}, y_i \text{ where } i, j \in 1, 2, \cdots, n\}$$
$$\textit{such that} \text{ a system event and ultimately a system function is achieved}$$

In the expression, the z variables are system inputs, the x variables are state variables (representing compartments), the f variables are simple flows (over links), the y variables are system outputs, i and j are indices in the range 1 to n, and n is the number of system compartments. A compartment model process-over-structure flow, therefore, is a sequentially ordered k-tuple of system inputs, compartments, simple flows, and outputs such that a system event and ultimately a system function is achieved.

We see that process-over-structure flow is an end-to-end flow path across system (model) structure often driven by input(s) from the environment and producing output(s) to the environment. A function is the net result of the occurrences of a set of process-over-structure flows.

3.3.4 New Perspective on Statics and Dynamics

Traditional system models are often steady-state models with static structure and constant flows. Yes, these models can be very useful under some circumstances, but it must be acknowledged that they are simplifications and not reality.[13] All three elements of the systems triad—function, structure, and process—are actually dynamic. Bohm (1983) explains that even structure is dynamic: all systems "are in a continual movement of growth and evolution of structure ...." Certainly in longer term "developmental" time frames, system structure changes with time. But even in shorter term "operational" time frames, system structure—represented as a network—is not static. Network node state changes continually occur, and network links continually become active or inactive as required by the system processes/functions and the environment. The network "flickers." (We'll have more to say about that later in the book.) Everything is changing all the time. Everything is dynamic. As David Bohm has said, "All *is* flux."

We will revisit the systems triad in Part II (and again in Parts IV and VI). In Part II, we will see that the systems triad is the basis for an ecological system dynamics framework.

[13] All system models incorporate simplifying assumptions. This is necessary to reduce a complex real-world situation to something that may be tractable and amenable to, for example, mathematical or numerical analysis. Statistician George E. P. Box famously said, "All theoretical models are wrong, but some are useful."

CHAPTER 4

Reductionism and Information Loss

Traditional scientific research is most often conducted using a reductionist approach rather than a systems approach. This chapter addresses a significant problem with the indiscriminate use of reductionism; that is, reductionism can break network connections, isolate the target of investigation from the larger system in which it resides, and thereby cause information loss. Complex system networks, network connections, and network connectivity information can be described mathematically. The connectivity information loss that results from broken connections in reductionist investigation can be quantified. That is the objective of this chapter. To accomplish the objective, the mathematics of ecological network connections are developed, pertinent principles and formulations of information theory are described, and the information theory formulations are applied to the ecological network connection mathematical representation in order to calculate the information loss from broken connections associated with reductionist ecological investigation.

4.1 INTRODUCTION

As discussed in Chapter 3, scientific research (including ecological research) is most often conducted using a reductionist approach. The researcher identifies an area of interest; drills down to some narrow, specialized subset; and then performs a detailed investigation of that subset. A major problem with the indiscriminate use of reductionism is that it results in *isolation* and *information loss*. It isolates the target of investigation from the larger system in which it resides. This isolation breaks connections with the system and causes information loss. Reductionism is "specialization in isolation" (Odum and Barrett, 2005).

Because complex systems take the form of networks, reductionism can be viewed from a network perspective. The reductionist researcher essentially drills down the system network hierarchy to reach the increasingly specialized target of investigation. (See Chapter 3, Figure 3.3.) At each level

Understanding Complex Ecosystem Dynamics
http://dx.doi.org/10.1016/B978-0-12-802031-9.00004-8

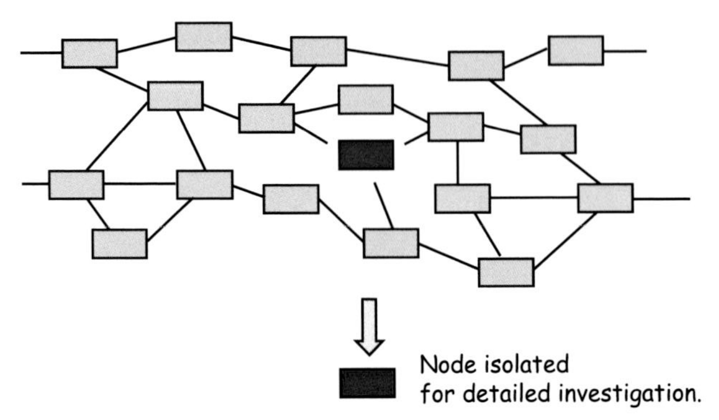

Figure 4.1 Isolating a node and breaking its connections.

of the hierarchy, a subsystem is selected for further study, network connections (links) are broken, and the associated information typically is lost. This procedure continues, level by level, until the target of investigation (an isolated node) is reached. "The reductionist likes to move from the top down, gaining precision of information about fragments as he descends, but losing information content about the larger orders he leaves behind."[1] After traversing multiple levels of the network hierarchy, the result of this reductionist process can be depicted as shown in Figure 4.1.

The number of broken network connections is substantial. When a node's *direct* connections are broken (three of these are shown in Figure 4.1), all manner of *indirect* connections to the other network nodes are also broken. The resulting information loss can be very significant.

Complex system networks, network connections, and network connectivity information can be described mathematically. The connectivity information loss that results from broken connections in reductionist investigation can be quantified. That is the objective of this chapter. We will account for both modes of information loss: broken *direct* connections and all manner of broken in*direct* connections. Further, we will show how the information loss results vary with network complexity and size.

Here's a map of the chapter. Section 4.2 addresses some of the mathematics of network connections. (The focus is on ecological networks.) Ecological network connections are characterized as random variables and their probability distributions are described. To determine realistic connection probability distribution functions, all available connections must be considered. In a complex ecological network, there can be multiple paths of different lengths available to connect any given pair of nodes. I reason that

[1] Paul Weiss 1969, quoted in Klir (1991), *Facets of Systems Science*.

the shorter available paths are more likely to be used. That behavior is consistent with a "long-tailed" power-law path length probability distribution. Section 4.3 provides evidence of such behavior, with examples. Principles and formulations of information theory are covered in Section 4.4. Finally, in Section 4.5, the information theory formulations from Section 4.4 are applied to the ecological network connection representation developed in Section 4.2, in order to calculate the information loss from broken connections associated with reductionist ecological investigation.

4.2 NETWORKS AND CONNECTIONS

Ecological systems take the form of networks. Network connections (links) between network nodes can be characterized as random variables. Any two nodes in the network (any given pair) are either *connected* or *not connected*. The connections can be either *direct* or *indirect*. The (potential) connection can be represented by a discrete random variable Y whose probability distribution function $P\{y\}$ takes on two values:

$$P\{Y = y_i\} \quad y_i \in \{y_c, y_{nc}\}$$

where

$$y_c = \text{node pair is connected}$$
$$y_{nc} = \text{node pair is not connected}$$

and

$$P(Y = y_c) = p_c$$
$$P(Y = y_{nc}) = 1 - p_c$$

where

$$p_c = \text{probability of connection}$$

Typically, for real-world ecological networks that have achieved critical connectivity, $p_c > 0.8$. When such a complex network reaches a critical threshold (1–2 direct connections per node) and *percolates*, it rather abruptly transitions from unconnected to connected—from separate node clusters to a *giant* node cluster. The clusters grow, at first by linking to individual nodes and later by coalescing with other clusters. The so-called giant cluster contains on the order of N nodes, where N is the total number of nodes in the network. All nodes or nearly all nodes are joined together. The node connections are via all manner of direct and indirect paths. See Chapter 9 for a more complete description of this network behavior. Investigators Strogatz (2001), Albert and Barabási (2002), and Newman (2003) are the sources for the description.

If a node pair is connected, the connection path length is 1 for a direct connection and greater than 1 for an indirect connection. Let's represent the connection path length by a discrete random variable X with probability distribution function $P\{x\}$:

$$P\{X = x_i\} \quad x_i \in \{1, 2, 3, \ldots\}$$

In Chapter 12, a calculation procedure (based on the work of Song et al., 2005, 2007) that determines a probability distribution function for node-to-node connection path lengths will be described. The results there indicate that, in many real-world networks, the distribution is a power-law/fractal distribution. I claim (and provide further evidence in Section 4.3 of the current chapter) that a power-law distribution is an appropriate and suitable choice for our path length probability distribution function[2] $P\{x\}$:

$$P\{X = x_i\} = Cx_i^{-\gamma} \quad x_i \in \{1, 2, 3, \ldots, N-1\}$$

where

C is a constant

$\gamma =$ the network path length fractal dimension (scaling exponent)

and

$N =$ the number of nodes in the network

Using data from complex biological networks, Song et al. have calculated values of γ that are approximately equal to 2.

4.3 SHORT PATHS AND POWER-LAW DISTRIBUTIONS

Multiple paths (with differing path lengths) can exist between any given pair of nodes in a complex ecological network. There is considerable evidence to suggest that the paths utilized to carry out system functions (involving propagation of energy, matter, and information) are more likely to be the shorter paths. This behavior is consistent with power-law path length distributions. Two specific real-world examples of short paths and power laws are provided in Sections 4.3.1 and 4.3.2, respectively.

4.3.1 Lévy Flights

Consider an ecological network of species. A crucial function of that system, of course, is food consumption, and the *foraging* component of consumption. In foraging, what is the best statistical strategy to adopt in order to search

[2] Why did I choose a maximum x_i value of $N-1$ for the path length probability distribution function? Without cycling, the largest possible value of path length is $x_i = N-1$. With cycling (which is often observed in real-world ecological networks), it is theoretically possible to have a path length value greater than $N-1$. In actual data, from empirical investigations (mine and others), I have not seen any values that large. Regardless, such an occurrence would have a very low probability and would have negligible impact on this analysis.

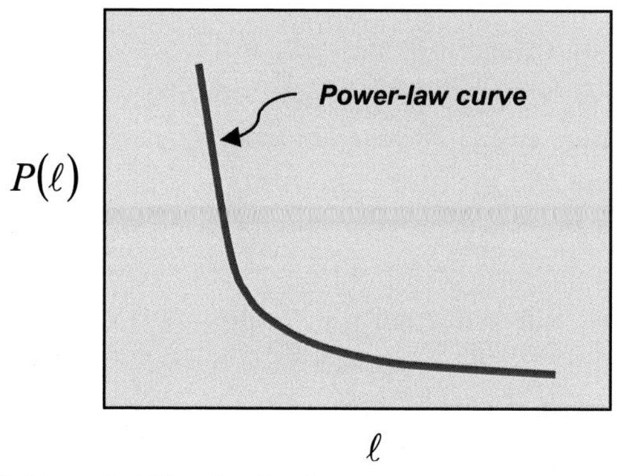

Figure 4.2 Lévy flight probability distribution.

efficiently for food "targets" (other network nodes)? Search activities are reflected in the probability distribution of flight lengths (connection path lengths) taken by a forager over time. Analysis results (Vishwanathan et al., 1999) indicate that the optimal foraging strategy corresponds to Lévy flight motion, depicted in Figure 4.2.

The Lévy flight probability distribution function is given by

$$P(\ell) \propto \ell^{-\mu}$$

where $\ell =$ the flight length and $\mu = 2$

This, of course, is a power-law distribution. The shortest flight lengths have the highest probability. Optimal foraging flight lengths tend to be shortest path lengths.[3]

Energy efficiency is the driver here. Odum and Barrett (2005) state, "Through natural selection, organisms achieve as favorable a benefit-cost ratio of energy input minus energy costs of maintenance as possible. ... For animals, the critical factor is the ratio of usable energy in food minus the energy cost of searching for and feeding on food items. ... Optimal foraging is defined as the maximum possible energy return under a given set of foraging and habitat conditions."

4.3.2 Foraging Ant Colonies

Over time, ant colonies find and use the shortest paths in their foraging. Here we have further evidence that when multiple paths (with differing path lengths) exist between node pairs in an ecological network, the shortest paths are more likely to be used.

[3] The Lévy flight probability distribution is (yet) another power-law distribution. As we proceed in this book, we will see that power-law distributions are pervasive in nature.

Ant colonies solve the "shortest route" problem (also known as the Operations Research traveling salesman problem) when traveling to/from a food source. (See Schiff, 2008.) When foraging for food, an ant secretes a chemical substance called *pheromone*[4] to mark its trail. Shorter routes are traveled in less time and get more pheromone deposited more quickly than in longer routes. Other ants prefer to travel a trail richer in pheromones, so the shorter routes get reinforced. This process eventually finds the shortest route, which most ants will then use. Following the same process, an ant colony will find the shortest route from the nest to each food source as well as the shortest route between food sources; that is, the ant colony solves the shortest route/traveling salesman problem.

This is no small feat. To visit each of the destination sites (food sites), and then return to the origin site (nest), there are effectively $(n-1)!/2$ possible routes. With, say, $n=15$ sites, there are more than 43 billion possible routes.

A cellular automata model has been developed by Dorigo and Gambardella (1997) that mimics the behavior of ants and achieves near-optimal to optimal solutions to the traveling salesman problem. This cellular automata model is as good as or better than competing Operations Research solution approaches.

4.4 INFORMATION THEORY

Information theory is a branch of applied mathematics and electrical engineering involving the quantification of information. It was originated at Bell Labs by Shannon (1948) in his seminal work, "A Mathematical Theory of Communication." Shannon's initial application of the theory was communication over transmission channels in telecommunications systems. Information theory has since been applied to quantify information content in a wide variety of complex systems, including ecological systems. Odum and Barrett (2005) note that information theory is "widely used in assessing the complexity and information content of all kinds of systems."

Consider a system characterized by a discrete random variable Z with a probability distribution function of the form

$$P\{Z=z\} \text{ vs. } z$$
$$\text{where } z \in \{z_1, z_2, \cdots, z_n\}$$
$$\text{and } n = \text{the finite number of possible values (outcomes) of } z$$

The Shannon formulation of the information content of such a system is given by

[4] Per Merriam-Webster online (http://www.merriam-webster.com/dictionary): *phero mone* is a chemical substance that is usually produced by an animal and serves especially as a stimulus to other individuals of the same species for one or more behavioral responses.

$$H(Z)=\sum_{i=1}^{n}P_i(z)\log_2\left[\frac{1}{P_i(z)}\right]=-\sum_{i=1}^{n}P_i(z)\log_2 P_i(z)$$

$$\text{where } P_i(z)=P(Z=z_i) \text{ and } \sum_{i=1}^{n}P_i(z)=1$$

Information content $H(Z)$ is nonnegative and resides in the range

$$0\leq H(Z)\leq \log_2 n$$

The minimum $H(Z)=0$ is reached if and only if $P_i\ (z)=1$ for a single value of z_i and $P_i\ (z)=0$ for all other values of z_i. The maximum $H(Z)=\log_2 n$ is reached if and only if all z_i are equally likely, that is, $P_i\ (z)=1/n$ for all values of z_i. $H(Z)$ has units of *bits*.

Information content $H(Z)$ is very often referred to as information entropy $H(Z)$, because it deals with probabilistic outcomes and is similar to thermodynamic entropy in other ways as well. There is a close resemblance between Shannon's formula and entropy formulae from thermodynamics. In statistical thermodynamics, the most general formula for the entropy S of a thermodynamic system is the Gibbs entropy:

$$S=-k_{\mathrm{B}}\sum_{i=1}^{n}P_i\log_e P_i$$

where $k_{\mathrm{B}}=$ the *Boltzmann* constant

If the thermodynamic system under consideration is a system of gas molecules, then P_i would be the probability of gas molecules being in one of n possible states. Except for the constant k_{B} and the change of log base, this formula matches Shannon's formula.

There is also a physical relationship between thermodynamic entropy[5] and information entropy. As the thermodynamic entropy of a system increases, the Shannon information entropy that describes the informational state of the system increases in a corresponding manner.

A special case of information entropy for a discrete random variable with two outcomes is the binary entropy function H_{b}:

$$H_b=-p\log_2 p-(1-p)\log_2(1-p)$$

Recall that the connection random variable Y defined in Section 4.2 has two outcomes; that is, a network node pair is either connected or not connected. The previous binary entropy expression applies (with $p=p_{\mathrm{c}}=$the probability of connection). This expression will be used in our calculations of Section 4.5. The binary entropy function, with its "humped" shape, is shown graphically in Figure 4.3.

[5] We will address thermodynamic entropy in ecological systems in Chapter 5.

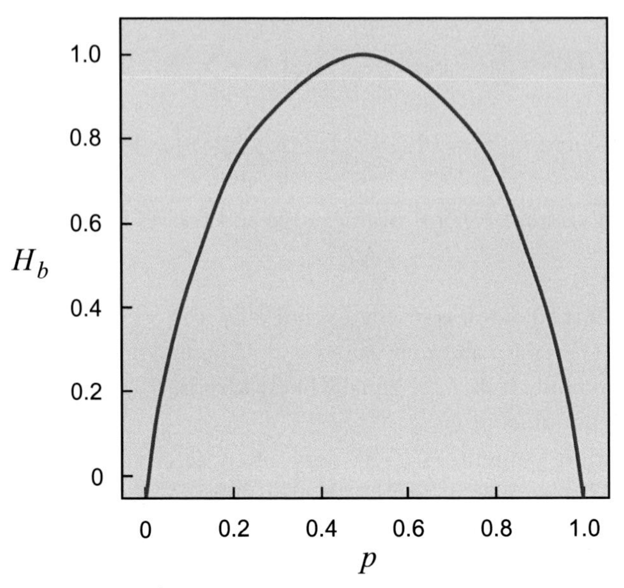

Figure 4.3 Binary entropy function.

Next, we will provide definitions of conditional entropy, joint entropy, and the entropy "chain rule" in terms of arbitrary random variables X and Y. In Section 4.5, we'll apply these concepts to the specific network connectivity discrete random variables X and Y that we defined in Section 4.2.

4.4.1 Conditional Entropy

The conditional entropy of discrete random variable X given discrete random variable Y is defined as

$$H(X/Y) = \sum_{y \in A_Y} P(y) \left[\sum_{x \in A_X} P(x/y) \log_2 \frac{1}{P(x/y)} \right]$$

$$= -\sum_{y \in A_Y} P(y) \left[\sum_{x \in A_X} P(x/y) \log_2 P(x/y) \right]$$

where $A_Y = \{y_1, y_2, \ldots, y_m\}$ and $A_X = \{x_1, x_2, \ldots, x_n\}$

Conditional entropy values are in the range

$$0 \le H(X/Y) \le H(X)$$

The lower limit is reached if X is determined by Y, that is, $P(x/y) = 0$ or 1 only. The upper limit is reached if X is independent of Y, that is, $P(x/y) = P(x)$ for all y.

4.4.2 Joint Entropy

The joint entropy of discrete random variables X and Y is defined as

$$H(X, Y) = \sum_{x, y \in A_X, A_Y} P(x, y) \log_2 \frac{1}{P(x, y)} = - \sum_{x, y \in A_X, A_Y} P(x, y) \log_2 P(x, y)$$

4.4.3 Entropy "Chain Rule"

The chain rule for entropy relates entropy, conditional entropy, and joint entropy:

$$H(X, Y) = H(X) + H(Y/X) = H(Y) + H(X/Y)$$

Finally, note that the total information entropy of a system can be calculated from the information entropies of its subsystems. If the information entropies are independent across subsystems, then the total system entropy is the sum of the individual subsystem entropies. If dependencies exist, then this sum is an upper bound on total system information entropy.

4.5 INFORMATION LOSS FROM BROKEN CONNECTIONS

In this section, the information theory formulations from Section 4.4 are applied to the random variable representations of ecological network connectivity developed in Section 4.2, in order to calculate the information loss from broken connections associated with reductionist ecological investigation.

Here's the setup and procedure that will be used. Consider a complex ecological network with N nodes. One of the nodes is targeted for detailed reductionist investigation. (Please refer back to Figure 4.1 for a pictorial view.) If E is an event that occurs with probability $P\{E\}$, then $H(E)$ bits is the quantity of information associated with the occurrence of E. If E represents the existence and path length value of a connection path from the target node to some other network node, then $H(E)$ provides the amount of network connectivity information in bits associated with the occurrence of that connection. If we could sum the information contributions over all target-node-to-other-node pairs, we could quantify the total amount of network connectivity information for the target node and, therefore, the amount of connectivity information (in bits) that would be lost if all connections to the target node were broken. Note that this application of information theory does not address the actual message information conveyed over the connections (e.g., the information inherent in mass and energy flows in an ecological network). Rather, it addresses just the network connectivity information. In general, information theory is concerned with occurrence possibilities and not with contextual meaning of messages.

In Section 4.2, the discrete random variables that are necessary to carry out this procedure have been defined. The connection random variable Y models whether or not a connection exists between a pair of nodes. If there is a connection, the path length random variable X accounts for all manner of direct and indirect connections that could exist between the two nodes. To get the connectivity information content of the node pair, we need to determine the joint information statistics of X and Y. As described in Section 4.4, that joint behavior is given by

$$H(X, Y) = H(Y) + H(X/Y)$$

where

$$H(Y) = -\sum_{y \in A_Y} P(y)\log_2 P(y)$$

and

$$H(X/Y) = -\sum_{y \in A_Y} P(y)\left[\sum_{x \in A_X} P(x/y)\log_2 P(x/y)\right]$$

and

$$A_Y = \{y_c, y_{nc}\}$$
$$A_X = \{1, 2, \ldots, N-1\}$$

Using our definitions of the probability distribution functions of random variables Y and X, we have

$$H(X, Y) = H(Y) + H(X/Y)$$

$$H(Y) = -[p_c \log_2 p_c + (1 - p_c)\log_2(1 - p_c)]$$

where $p_c =$ the probability of connection of a node pair

$$H(X/Y) = -\left[P(y_c)\left[\sum_{x \in A_X} P(x/y_c)\log_2 P(x/y_c)\right] + P(y_{nc})\left[\sum_{x \in A_X} P(x/y_{nc})\log_2 P(x/y_{nc})\right]\right]$$

$$H(X/Y) = -\left[P(y_c)\left[\sum_{x \in A_X} P(x/y_c)\log_2 P(x/y_c)\right] - P(y_{nc})\left[\sum_{x \in A_X} P(x/y_{nc})\log_2 \frac{1}{P(x/y_{nc})}\right]\right]$$

but $P(x/y_{nc}) = 0$ and $0 \times \log_2 \frac{1}{0} \equiv 0$ because $\lim_{\theta \to 0^+} \theta \log_2 \frac{1}{\theta} = 0$

Therefore

$$H(X/Y) = -p_c \sum_{x \in A_X} P(x/y_c) \log_2 P(x/y_c)$$

$$H(X/Y) = -p_c \sum_{i=1}^{N-1} Cx_i^{-\gamma} \log_2 Cx_i^{-\gamma}$$

where

C is a constant

N = the number of nodes in the ecological network

and

γ = the network path length fractal dimension (scaling exponent)

We also know from the Shannon formulation of information entropy, and because we are dealing with a probability distribution function, that

$$\sum_{i=1}^{N-1} P_i(x) = \sum_{i=1}^{N-1} Cx_i^{-\gamma} = 1$$

$$C \sum_{i=1}^{N-1} x_i^{-\gamma} = 1$$

This relationship will be used later to calculate the value of the constant C.

Putting all of these pieces together, the expression that is needed to calculate the information content of a target-node-to-other-node pair is obtained:

$$\text{Information content of node pair} = H(X, Y) = H(Y) + H(X/Y)$$

$$H(X, Y) = -p_c \log_2 p_c - (1 - p_c) \log_2 (1 - p_c) - p_c \sum_{i=1}^{N-1} Cx_i^{-\gamma} \log_2 Cx_i^{-\gamma}$$

So far, we have derived the connectivity information content/entropy of a pair of network nodes—the target node and any other node in the network. Let's refer to such a node pair as a target-to-other pair. In an ecological network of N nodes, there are $N-1$ "other" nodes and, therefore, $N-1$ such pairs. If the specific information content of each pair is independent of the information content of every other pair, then the total information content of all of the target-to-other node pairs is $N-1$ times the information content of an individual pair. (Information entropy is additive for independent information events.) This "independence" total information content value will be calculated. Information dependencies, however, may exist among some target-to-other node pairs in a given ecological network. In that case, the "independence" calculated value will serve as an upper bound on total information content.[6]

Calculating the upper bound, therefore, is straightforward. Calculating a lower bound on total information content is potentially much more difficult. We would have to consider a specific system network, determine the information content associated with each target-to-other node pair, determine the physical dependencies among the network paths that can connect all the target-to-other node pairs, estimate the resulting information dependencies among all those target-to-other node pairs, and then attempt to compute the total information content associated with target node connections in light of these information content values and dependencies. This would be a very difficult procedure, and the result would apply just to a specific system network. Such a detailed approach would not be very helpful. We want a more general result that could apply across a spectrum of real systems.

So, I'll use some engineering judgment and make some engineering approximations. (That is, I will employ *reflective rationality*; see Chapters 1 and 2.) I'll generate a range of possible information content values—upper bound to "reasonable" lower bound—as a function of system size. As described previously, the upper bound on information content assumes full independence of all target-to-other pairs (no redundant information). As just discussed, the lower bound is not so easy. In the set of network paths that connect a target node to other nodes, we know that any given path can have links in common with some other path, which could yield information dependencies among paths (redundant information). The problem is that, in general, we do not know the "dependency distribution." We do, however, know that, for real systems of interest, the path length distribution follows a power law. Shorter paths are more likely than longer paths. Shorter paths have fewer links, fewer opportunities for links in common with other paths, and therefore perhaps less information dependency with other paths. We might expect, then, that the "dependency distribution" is skewed more toward independence and away from dependence. Given that, I will make the following (apparently conservative) numerical assumption. To obtain an approximate lower bound on information content, I assume that half of the target-to-other node paths are fully independent (no redundant

[6] If events A and B are mutually exclusive, then $P(A \cup B) = P(A) + P(B)$. If events A and B are not mutually exclusive, then $P(A \cup B) = P(A) + P(B) - P(A \cap B)$. Information entropy behaves in an analogous manner.

information) and half are fully dependent (all redundant information). I strongly suspect that for real systems of interest, this is a reasonable lower bound. If you think my numerical assumption is somewhat arbitrary, you are right. I am not concerned with the exact accuracy of numbers here; I am concerned with general behaviors.

The development of the engineering results follows. Information content is denoted IC. The upper bound on total information content of all connections to the target node is given by

$$IC_{max} = (N-1)\left[-p_c \log_2 p_c - (1-p_c)\log_2(1-p_c) - p_c \sum_{i=1}^{N-1} Cx_i^{-\gamma} \log_2 Cx_i^{-\gamma}\right]$$

The lower bound on total information content of all connections to the target node is given by

$$IC_{min} = \frac{1}{2}(N-1)\left[-p_c \log_2 p_c - (1-p_c)\log_2(1-p_c) - p_c \sum_{i=1}^{N-1} Cx_i^{-\gamma} \log_2 Cx_i^{-\gamma}\right]$$

In reductionist ecological investigation, the connections to the target node are broken. The resulting information loss IL is equal to the information content IC of the connections. The upper and lower bounds on total information loss are given by

$$IL_{max} = IC_{max}$$

$$IL_{max} = (N-1)\left[-p_c \log_2 p_c - (1-p_c)\log_2(1-p_c) - p_c \sum_{i=1}^{N-1} Cx_i^{-\gamma} \log_2 Cx_i^{-\gamma}\right]$$

$$IL_{min} = IC_{min}$$

$$IL_{min} = \frac{1}{2}(N-1)\left[-p_c \log_2 p_c - (1-p_c)\log_2(1-p_c) - p_c \sum_{i=1}^{N-1} Cx_i^{-\gamma} \log_2 Cx_i^{-\gamma}\right]$$

I suggested earlier in this chapter that, for realistic complex ecological networks, parameter values of $p_c = 0.9$ and $\gamma = 2$ should be representative. I use these values to calculate information loss IL for the following network size range: $N = 10$–1000. The results are plotted in Figure 4.4. The plot abscissa uses a log (base 2) scale.

The larger and more complex an ecological network, the greater the number of connections (direct and indirect) to a target node, the greater the total information content of the target node interactions, and the greater

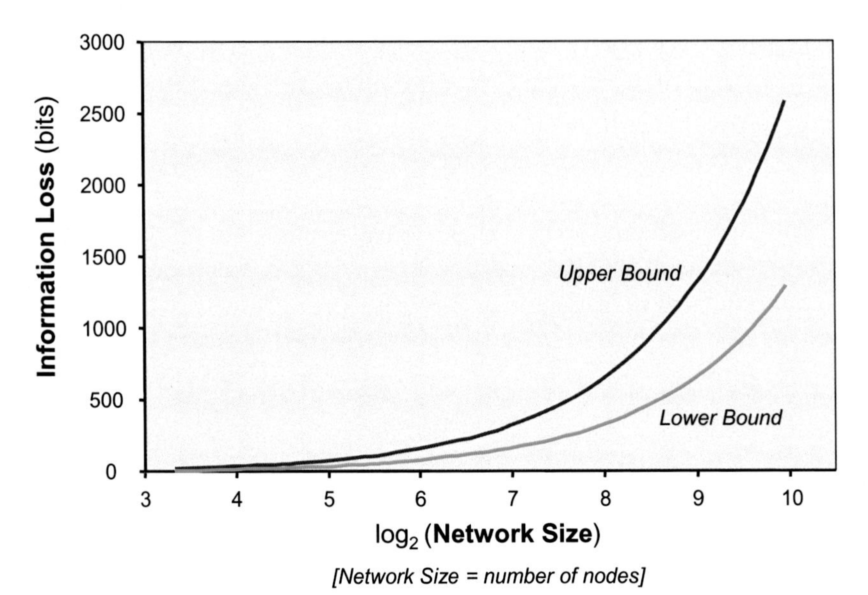

Figure 4.4 Information loss versus ecological network size/complexity/maturity.

the information loss when the target node is isolated for reductionist investigation and its connections are broken.

Odum and Barrett (2005) suggest that more mature ecosystems—in later seral stages—are likely to comprise a more complex, fully developed network with a larger number of nodes and node interactions and, therefore, higher connectivity information content.[7] Reductionist analysis may be especially problematic for such ecosystems.

The chart of Figure 4.4 is not about the precision of the numbers. The point is that there is *significant* information loss for many real systems, and reductionist analysis that ignores connections with the rest of the system can yield incorrect conclusions.

Back in Chapter 3, in the section on network thinking, we discussed an example involving "genetic engineering," which, at least at this time, is essentially a reductionist practice. Gene networks and information propagation in gene networks are not understood, and so they are mostly ignored. Important connectivity information is no doubt being ignored. Marketing claims are made about the benefits of "engineering" individual genes. These claims are likely incorrect. System behavior is not determined by genes

[7] See Odum and Barrett (2005, chapter 8 and table 8-1).

(nodes) in isolation. In particular, claims about the low risk and safety of genetically modified (GMO) foods and other products are suspect. Without knowledge of the effects of one node (one gene) on other nodes (other genes) and overall system behavior, those claims could be quite incorrect.

Did you notice that, in this chapter, I used reductionist analysis tools to explain shortcomings of reductionism? Reductionism, therefore, must be valuable—and of course it is. My point is that reductionist analysis and technical rationality must be used prudently, together with synthesis and reflective rationality.

PART II

A Function-Structure-Process Framework for Ecological System Dynamics

In this part, we synthesize and describe a functionality-based dynamics framework to set the context and direction for our ecosystem work. In my view, such a framework is essential for understanding natural systems. The framework consists of operational, developmental, and core functional tiers. The framework elements are discussed here (and throughout the book).

CHAPTER 5

Overview of an Ecological System Dynamics Framework

In Chapter 3, the function-structure-process systems triad was described. The triad is the basis for the functionality-driven ecological system dynamics framework that is synthesized and described here in this chapter. The framework sets the context and direction for our ecosystem work, and can be the vehicle that guides us toward a comprehensive and disciplined understanding of ecological system functions and the implemented form (structure and process) that delivers the functions. The ecosystem dynamics framework consists of operational, developmental, and core functional tiers. The extremely important core functional tier spans the operational and developmental tiers. The small interacting set of *core functions*—comprised of self-organization, regulation/adaptation, and propagation—is involved in or contributes to essentially all developmental and operational functions. Self-organization (and thermodynamic entropy in ecological systems) is discussed in some detail in this chapter. Other important aspects of the core functions and the dynamics framework are discussed in the subsequent two chapters, and throughout the book.

At this point in time, I have achieved a top-level view of the framework. This view, which provides the crucial total system picture, is sufficient for the work at hand. More detailed development of the framework represents an important follow-on effort—by me and/or interested others.

5.1 TWO ESSENTIAL QUESTIONS

There are two essential questions that must be addressed in order to *understand* an ecological (natural) system or to understand and *build* a human-made (artificial) system. *What* does the system do? *How* does the system do it? See Figure 5.1. Yaneer Bar-Yam (2004) agrees and echoes the sentiment when he says, "Scientists look at something and want to understand what it does, and how it does it." As depicted in the systems triad diagram of Figure 5.2, the "what" is described by the system's *functions*. The "how" is described by

Understanding Complex Ecosystem Dynamics
http://dx.doi.org/10.1016/B978-0-12-802031-9.00005-X

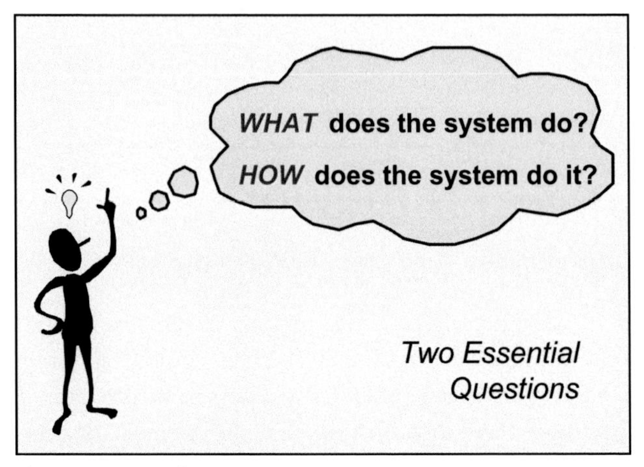

Figure 5.1 The two essential questions.

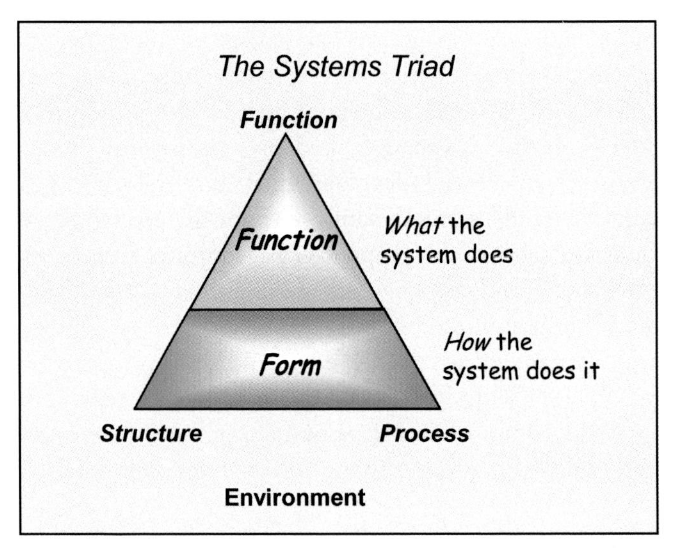

Figure 5.2 Systems triad—what and how.

the system's implementation, i.e., the implemented *form* (structure and process) that realizes the system functions.

There is also a third question: Why? This question can be viewed as teleological—having to do with the purpose of the system. When building an artificial (human-made) system, we can answer this question. If we didn't know why we were building a system, we wouldn't build it. In a commercial/industrial setting, for example, it is the system stakeholders that provide the answer. For a natural ecological system, on the other hand, we are not in a position to answer the "why" question, or even to know (scientifically) if there is an answer. To understand an ecological system, at best we can begin to address and answer the "what" (function) question and the "how" (structure and process) question.

5.2 DEVELOPING THE FRAMEWORK

So let's address the *what* and *how* questions in a systematic way. Let's develop a framework that addresses the questions; embodies the concepts of function, structure, and process; and therefore can serve as a basis for a comprehensive understanding of ecological system dynamics. The top-level view of the framework is provided in Figure 5.3.

As shown, the two major partitions of the framework address *what* and *how*, respectively. The left-hand partition covers the what, i.e., the system functions. The right-hand partition covers the how—the system implementation, i.e., the implemented structure and processes that realize the system functions.

First consider the left-hand ecological system function partition. In his writing on complex adaptive systems, John Holland (1996) recommends a two-tier model of system dynamics: an upper tier that represents the "slow dynamics" of long-term system development and a lower tier that represents the "fast dynamics" of short-term system operation. David Bohm (1983) has said that to understand a system, we must understand the way "in which it forms, maintains itself, and ultimately dissolves." As shown in Figure 5.3, I take Holland's and Bohm's advice and partition ecological system functions into *developmental functions* (long-term dynamics) and *operational functions* (short-term dynamics). The developmental functions involve "forming and dissolving." The operational functions mostly involve "maintaining," and they apply in time frames for which the long-term dynamics are usually assumed to be negligible. Ecological system developmental functions focus primarily on evolution-related functions—evolution of ecosystem components (e.g., species) as well as evolution of ecosystem structure and processes. An example of the latter is ecosystem succession. Top-level ecological system operational functions include production, consumption, respiration, and decomposition.

There is a third function partition shown in Figure 5.3, which is extremely important. This partition consists of a small interacting set of *core functions* that span the slow and fast dynamics regimes. I believe that an

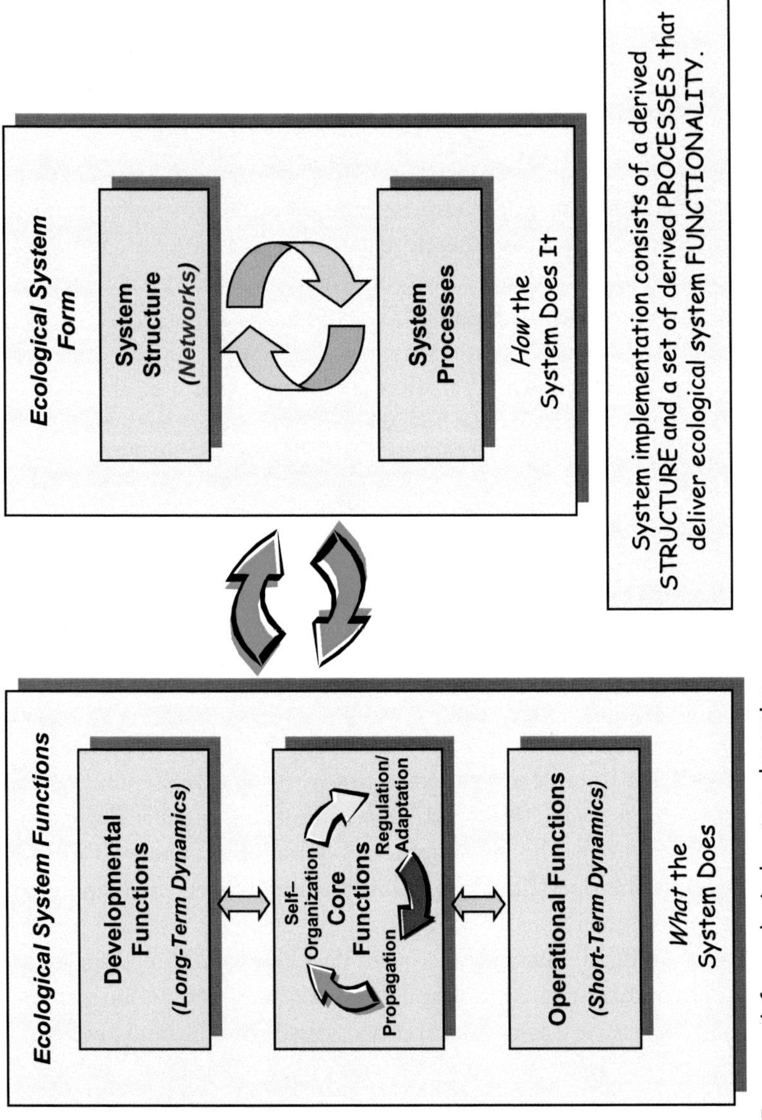

Figure 5.3 Framework for ecological system dynamics.

understanding of these core functions of self-organization, regulation/adaptation, and propagation is crucial for a comprehensive understanding of ecological systems. *Self-organization* is perhaps the key underlying enabler of all ecosystem functionality. Further, it is my view that an ecological system's *regulation/adaptation* with respect to its environment and also its *propagation* of energy/matter are involved in or contribute to essentially all developmental and operational functions. Ecological systems self-organize, adapt, and propagate in support of operational and developmental functions.

Next, consider the right-hand side of Figure 5.3—ecological system implemented form, i.e., structure and process. The structure of complex ecological systems is best represented by networks. System processes execute on network structure to yield the active end-to-end flows that deliver ecological system functions. There are exceptions, but end-to-end process flows tend to get neglected in traditional complex systems investigation. In my survey of the classic and recent network literature, for example, I have found that most network investigation to date in the scientific community has focused on the structure of networks. Investigation of end-to-end process flows on networks lags way behind. This focus on structure is typical of many Western scientific pursuits. As discussed in Part I, reductionist research isolates a system component (node) for detailed analysis, and in so doing breaks network connections and severs end-to-end process flows. Process concepts are also underutilized in traditional ecological modeling. The term "process" is employed primarily in a fragmented pairwise sense. Process and process equations are used mostly to explain and calculate compartment-to-compartment pairwise simple flows of the model currency. System end-to-end process-over-structure flow must become an important focus of our work if we are to achieve a comprehensive understanding of ecological system dynamics.

The framework discussed here has and will continue to provide guidance to me in my work and in my efforts to better understand ecological systems. Other investigators are certainly welcome to use the framework, and to offer suggestions for its elaboration and improvement. It is still a "work in progress."

5.3 THE THREE CORE FUNCTIONS OF THE FRAMEWORK

5.3.1 Self-organization

Self-organization is the fundamental core function of complex ecological systems. It is, in a sense, the underlying enabling function that makes all other functions possible. Stuart Kauffman (1995) has explored this function extensively and seeks its general principles. He says that order "arises naturally and spontaneously because of these principles of self-organization—laws of complexity that we are just beginning to understand." "Self-organization . . . may be the ultimate wellspring of [the] dynamical order" that underlies the origin, development, and operation of living systems. Several other prominent authors and researchers have explored the subject as well, in the context of a variety of both living systems and nonliving systems. A sampling is provided in the following paragraphs.

Fritjof Capra (1996) notes that philosopher Immanuel Kant (in 1790) was "the first to use the term 'self-organization' to define the nature of living organisms." Capra says the hallmark of self-organization is "the striking emergence of new structures and new forms of behavior." The work of Ilya Prigogine (1977 Nobel laureate in chemistry) demonstrates that. In Capra's (1996) words, "The first, and perhaps most influential, detailed description of self-organizing systems was the theory of 'dissipative structures' by the Russian-born chemist and physicist Ilya Prigogine." Prigogine's 1970s investigations included Bénard cells. These are hexagonal convection cells that appear spontaneously when a thin layer of liquid is heated from below. This is a nonequilibrium (temperature nonuniformity) threshold phenomenon, which exhibits spontaneous self-organization. Prigogine also studied the Belousov-Zhabotinskii chemical reaction, in which a combination of appropriate chemicals plus energy yields a reaction solution whose color oscillates from red to blue to red at regular time intervals. It's a self-organized chemical clock. Note that both of these phenomena require an input of energy that is dissipated by the system. Prigogine introduced the term *dissipative structures* to describe the principle that "in open systems dissipation becomes a source of order."

Eugene Odum and Gary Barrett (2005) weigh in on the subject of self-organization. They cite the work of Prigogine as well as the work of Bob Ulanowicz (1997). "A major key to ecosystem development is the concept of *self-organization*, based on Prigogine's theory of non-equilibrium thermodynamics. Self-organization can be defined as the process whereby complex systems consisting of many parts tend to organize to achieve some sort of stable, pulsing state in the absence of external interference. . . . Self-organized ecosystems can only be maintained by a constant flow of energy through them; therefore, they are not in thermodynamic equilibrium. . . . Ulanowicz used the term ascendency for the tendency for self-organizing, dissipative systems to develop complexity of biomass and network flows over time."

Paul Krugman (2008 Nobel laureate in economics and a professor at Princeton University) discusses the importance of the concept of self-organization—across disciplines. In his book *The Self-organizing Economy*, Krugman (1996) says, "The most provocative claim of the prophets of complexity is that complex systems often exhibit spontaneous properties of self-organization." "I believe that the ideas of self-organization theory can add substantially to our understanding of the economy." "In the last few years the concept of *self-organizing systems* . . . has become an increasingly influential idea that links together researchers in many fields, from artificial intelligence to chemistry, from evolution to geology."

5.3.2 Regulation/Adaptation and Propagation

Regulation/adaptation is the system control function. The need for regulation/adaptation in ecological systems is profound; it is required for system survival. In ecosystems, regulation/adaptation is *homeorhetic*; i.e., it maintains stable, but flexible, fluctuating system dynamics. We will devote Chapter 6 to a discussion of regulation/adaptation and homeorhetic control.

Propagation is the fundamental and essential means of interaction among nodes in an ecological network and between the network and its environment. The interplay of material cycles and energy flows inherent in ecosystem regulation/adaptation is in fact provided by the propagation function. (That will also be discussed in Chapter 6.) In my more detailed work on complex ecosystem dynamics, the propagation function becomes the focus. In Parts IV and V of this book, propagation is the centerpiece of much of the synthesis, modeling, and analysis activities.

5.4 SELF-ORGANIZATION AND THERMODYNAMIC ENTROPY IN ECOLOGICAL SYSTEMS

Self-organization has a major impact on thermodynamic entropy in ecological systems. *Thermodynamic entropy* can be considered a measure of disorder. The second law of thermodynamics says that, in a *closed* system, entropy increases with time. Ecological systems, however, are *open* systems, with inputs from and outputs to the environment. In ecological systems, order (negentropy) increases with time as the system spontaneously self-organizes.

The "revised" second law of thermodynamics explains how order can increase in open ecological systems.

5.4.1 The "Revised" Second Law of Thermodynamics

Eric Schneider and James Kay (1995) and Felix Müller (1996) were among the investigators who "revised" the second law to apply to open ecological systems. In ecological systems, entropy decreases and order increases.

A depiction of the "revised" second law of thermodynamics is provided in Figure 5.4.

Note that *exergy* is defined as energy that is available to be used by the system. Although the Figure 5.4 illustration is mine, Odum and Barrett (2005) describe the figure and the concept very well. "Organisms, ecosystems, and the entire ecosphere possess the following essential thermodynamic characteristic: They can create and maintain a high state of internal order, or a condition of low entropy. Low entropy is achieved by continually and efficiently dissipating energy of high utility [e.g., sunlight or food] into energy of low utility [e.g., heat]. In the ecosystem, order in a complex biomass structure is maintained by the total community respiration, which continually 'pumps out disorder.' Accordingly, ecosystems and organisms are open, non-equilibrium thermodynamic systems that continually exchange energy and matter with the environment to decrease internal entropy but increase external entropy (thus conforming to the laws of thermodynamics)." Stephen Wolfram (2002) adds that ecological systems (and biological systems) "reduce their internal entropy only at the cost of increases in the entropy of their environment." (More detailed analysis of ecosystem thermodynamics can be found, for example, in S.E. Jørgensen and Y.M. Svirezhev, *Towards a Thermodynamic Theory for Ecological Systems*, Pergamon Press, July 2004.)

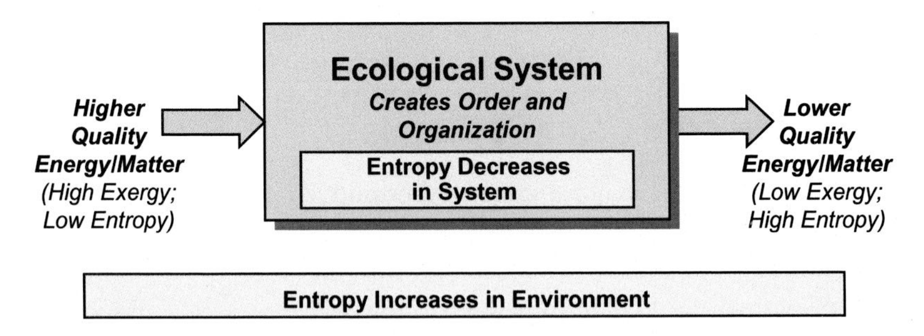

Figure 5.4 Ecological systems and the revised second law.

The ecosystem entropy decrease can be represented in very simple mathematical terms. If S = entropy, then the second law of thermodynamics requires that:

$$\frac{dS}{dt} \geq 0$$

For open ecological systems:

$$\frac{dS_T}{dt} = \frac{dS_s}{dt} + \frac{dS_e}{dt}$$

where

$$\frac{dS_T}{dt} = \textit{total change in entropy} \geq 0$$

$$\frac{dS_s}{dt} = \textit{change in system entropy}$$

$$\frac{dS_e}{dt} = \textit{change in environment entropy}$$

If the ecological system "pumps out disorder" such that dS_e/dt is sufficiently positive, then it can be true that:

$$\frac{dS_s}{dt} < 0$$

Ecological system entropy decreases and order increases. The ecosystem *self-organizes*. "By the 1930s physicists often considered local entropy decrease a defining feature of life" (Wolfram, 2002).

In the remaining two chapters of Part II, we address important elements of the dynamics framework in detail. Chapter 6 covers the core function regulation/adaptation. Chapter 7 discusses the framework developmental tier. (The framework operational tier is the focus of much of the synthesis, modeling, and analysis work in Parts IV and V of the book.)

CHAPTER 6

Regulation/Adaptation: Control Aspects of Ecological Systems

This chapter addresses regulation/adaptation, one of the core ecological system functions, in detail. Regulation/adaptation provides system control. The chapter begins with a discussion of the general principles of system control, and then focuses on the control of biological and ecological systems. Biological systems exhibit *homeostatic* control (i.e., maintaining the same standing or remaining the same). Ecological systems, however, are controlled in a *homeorhetic* sense (i.e., maintaining similar, but pulsing, dynamical behavior). As noted in Chapter 5, the three core functions of the ecological system dynamics framework are an interacting set. The interactions among the regulation/adaptation function and the other core functions are explored. The interaction outcomes strongly suggest and support the prevalence of fluctuating dynamics in ecosystems.

6.1 INTRODUCTION

Regulation/adaptation provides system control. The need for the regulation/adaptation function in ecological systems is profound: it is required for system survival. Ecosystem control differs somewhat from the control processes found in artificial systems or even in some biological systems. Those differences will be described. Ecosystem control processes allow an ecosystem to respond and adapt very effectively to perturbations from the environment.

Classical *control theory* was developed in the context of human-made mechanical and electrical systems. In the late 1700s, James Watt controlled the speed of his steam engine with a governor—a centrifugal feedback mechanism. In 1868, James Clerk Maxwell published a formal analysis of centrifugal governor dynamics. Harold S. Black, a Bell Labs scientist, invented negative feedback amplifiers in 1927 and originated the concept of electronic control in electrical/electronic systems.

In the spirit of general systems thinking, *cybernetics* expanded the scope of classical control theory to include biological systems. Cybernetics was defined

Understanding Complex Ecosystem Dynamics
http://dx.doi.org/10.1016/B978-0-12-802031-9.00006-1

by Norbert Wiener (1948) as the study of control and communication in the animal and the machine. The term *cybernetics* comes from the Greek *kybernētēs*, meaning steersman, governor, pilot, or rudder. The field of cybernetics has broadened in the last several decades to encompass additional system types, including social, economic, and business management systems.

In this chapter, I endeavor to answer the following questions: What are the general principles of system control? How do these principles apply to ecological systems? Is ecosystem control local or global or both? What are the relationships among the regulation/adaptation function and the other core functions of the ecological system dynamics framework? What is the nature of the resulting ecosystem dynamics?

6.2 GENERAL PRINCIPLES AND CONTROL OF HUMAN-MADE SYSTEMS

There are several types of control or regulation mechanisms that apply to human-made systems, depending on whether the control is applied internally on the system or externally on the environment, and whether it is feedback control or anticipatory control (see, e.g., Weinberg and Weinberg, 1988). A regulation classification tree is shown in Figure 6.1 and corresponding system diagrams are shown in Figure 6.2.

As you can see from the diagrams of Figure 6.2, for both internal and external regulation:

- When controller inputs are received from the system (or subsystem) state variables, the regulation mechanism is called feedback control.
- When controller inputs are received from the environment, the regulation mechanism is called anticipatory control.

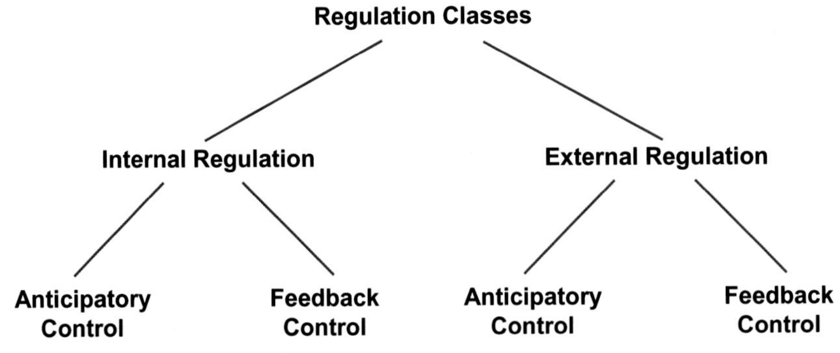

Figure 6.1 Regulation classification tree.

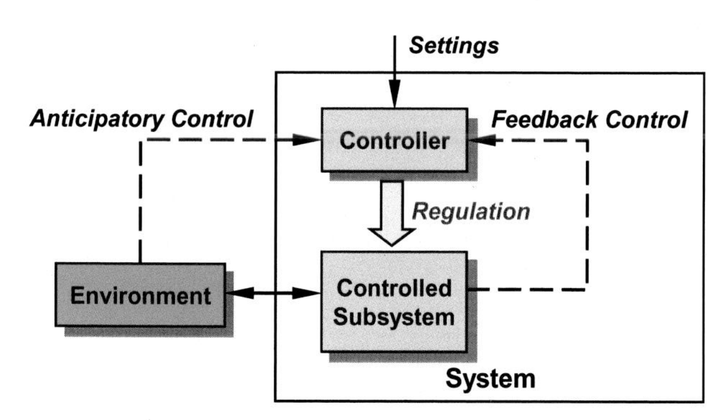

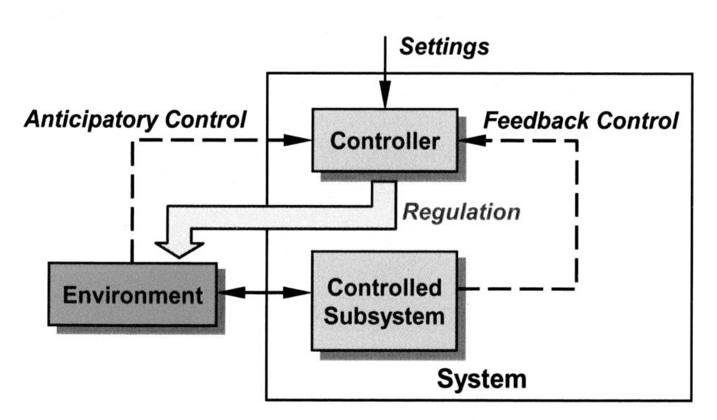

Figure 6.2 System regulation diagrams.

The "settings" inputs shown on the diagrams specify the teleological set points for the controllers, i.e., the desired values of system state variables to be achieved by the regulation process.

6.3 CONTROL OF BIOLOGICAL/ECOLOGICAL SYSTEMS

Generally, all of the aforementioned types of regulation/adaptation also apply to biological and ecological systems, with some additional considerations.

Odum and Barrett (2005) have developed a biological/ecological hierarchy; my diagram of the hierarchy is provided in Figure 6.3. The lower hierarchy levels (organism level and below) correspond to biological systems. The upper hierarchy levels (population level and above) correspond to ecological systems. System control in these two regions can differ considerably. Odum and Barrett (2005) say, "Compared with the strong set-point controls at the organism level and below . . . the population level and above are much less tightly regulated, with more pulsing . . . behavior, but they are controlled nevertheless."

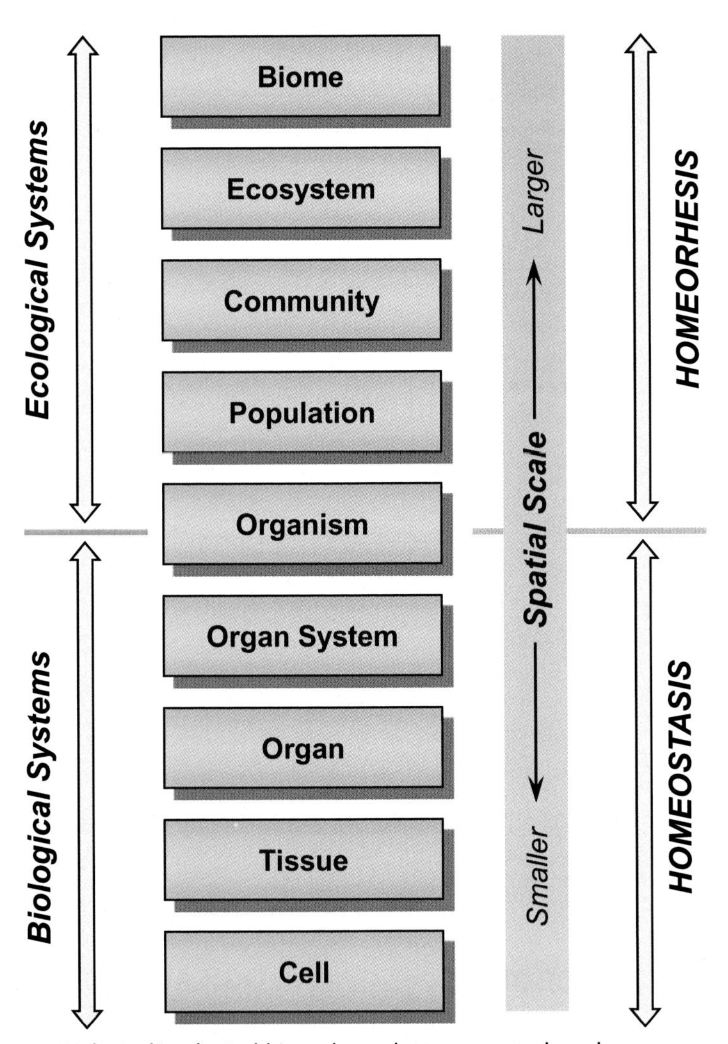

Figure 6.3 Biological/ecological hierarchy and system control modes.

Biological system control and ecological system control are covered in Sections 6.3.1 and 6.3.2, respectively.

6.3.1 Biological Systems

According to Odum and Barrett (2005), biological systems are "homeostatic, goal-seeking organismic systems." Figure 6.4 illustrates biological system control. The focus is on internal regulation and the feedback control case.

Distinct *homeostatic* set points are used to regulate biological systems. Homeostasis, from *homeo* (same) and *stasis* (standing), means maintaining the same standing or remaining the same, in the face of system fluxes in and out. Odum and Barrett say homeostasis maintains "steady states within limits."

6.3.2 Ecological Systems

Per Odum and Barrett (2005), "ecosystems can be considered cybernetic in nature" but in a different way than biological systems (or human-made systems). Ecological systems are controlled, not in a homeostatic sense, but in a *homeorhetic* sense. Homeorhesis is a term from the Greek that means maintaining similar flow or dynamical behavior. Homeorhesis maintains "pulsing states within limits."

A diagram of ecological system control is provided in Figure 6.5.

Ecological systems are nonteleological and, as depicted in the figure, they have no specific set points. Odum and Barrett say, "The lack of set-point

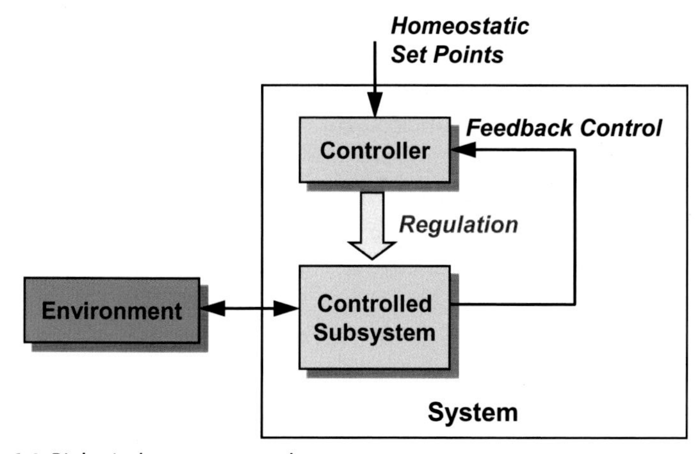

Figure 6.4 Biological system control.

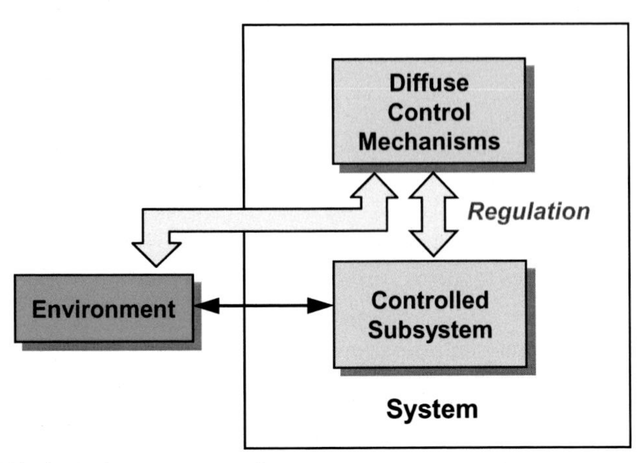

Figure 6.5 Ecological system control.

controls results in a *pulsing state* rather than a *steady state*." "There are no equilibriums [in ecological systems], but there are *pulsing balances*."

Ecological system control mechanisms are diffuse. Patten and Odum (1981) made that case decades ago. They explained, in their article *The Cybernetic Nature of Ecosystems*, that ecosystem control mechanisms "are all the factors, processes and interactions . . . which serve to mediate the movement or transformation of energy-matter." These mechanisms control the ecosystem internally and interact externally with the environment. The interplay of material cycles and energy flows generates the self-controlling homeorhesis that allows an ecological system to adapt and respond very effectively to system perturbations.

John Holland's work on complex adaptive systems (Holland, 1996) seems consistent with the aforementioned characterization of complex ecological system control. Holland says, for example, that complex adaptive systems (including ecological systems) "exhibit coherence under change, via conditional action and anticipation, and they do so without central direction."

Comment: I must note that I have some disagreement with the Odum and Barrett (2005) terminology. They use the term "steady states" to describe the homeostasis phenomenon in biological systems and the term "pulsing balances" to describe the homeorhesis phenomenon in ecological systems. In my view, these phenomena are neither "steady" nor "balanced." It is a matter of degree only: homeostasis typically involves less fluctuation and homeorhesis typically involves more fluctuation. Both phenomena are quite dynamic. In Chapter 15, I refer to a heart rate dynamics study to illustrate process fractal dynamical behaviors (frequency-spectrum behaviors and time-scale behaviors). The human circulatory system is certainly a biological system. The results of the heart rate study show that healthy heart rate is not at all steady; it shows significant spontaneous variation. As part of the body's biological network, a healthy circulatory subsystem is very responsive to what is going on in the body's other subsystems and in the body's environment.

6.4 CORE FUNCTION RELATIONSHIPS AND THE NATURE OF THE RESULTING DYNAMICS

Self-organization, regulation/adaptation, and propagation are the core functions of the ecological system dynamics framework. These three core functions are an interacting set. In Chapter 5, our discussion of the "revised" second law of thermodynamics explained that ecological system self-organization involves moving and transforming energy and matter. In the previous subsection of the current chapter, we established that ecosystem regulation/adaptation mechanisms are diffuse and also involve the movement and transformation of energy-matter. The movement of energy and matter, of course, requires the propagation function. Core function interactions, therefore, include a self-organization-to-propagation relationship and a regulation/adaptation-to-propagation relationship.

The regulation/adaptation-to-propagation relationship is addressed in Section 6.4.1, and the nature of the resulting system dynamics is discussed in Section 6.4.2.

6.4.1 The Regulation/Adaptation-to-Propagation Relationship

Two related views of the relationship are provided here.

6.4.1.1 Life as a Relaxation Phenomenon

This view was suggested by Csermely (2006). Peter Csermely agreed that ecological system regulation/adaptation requires effective propagation to dissipate perturbations that impinge upon an ecological network. To survive, every living system must respond (adapt) effectively to stimuli from its environment. The stimulus can be considered a source of tension, and the response as a relaxation. When an ecological network receives a stimulus (e.g., energy, biomass, information) from its environment, the input is often propagated and dissipated locally, yielding local relaxation in the system. When some stimulus cannot be dissipated locally, tension persists. As the stimulus inputs continue, tension gradually increases. Local tensions can accumulate and may develop to a point where global propagation suddenly occurs. The set of local and global relaxation events (propagation events) accomplishes the regulation/adaptation function. Any event can have small or large spatial extent and long or short temporal duration, as well as anything in between. Relaxation events are indicative of fluctuating dynamics.

Csermely (2006) says that living system functions "cannot be performed in the absence of widespread network communication." They cannot be performed without local-to-global dynamics.

6.4.1.2 Local-to-Global Dynamics

Many natural complex network systems, including ecological systems, self-organize into *small-world* networks—i.e., networks with *high clustering* and *short path lengths*. (We will comprehensively describe small-world networks and their characteristics in Chapter 8). The small-world characteristics facilitate local-to-global dynamics. High clustering is extremely effective for local propagation and processing. Short path lengths provide efficient propagation channels between distant parts of the system, thereby supporting dynamical processes taking place on the network that require global propagation and processing. As previously noted, in ecological systems, such propagation capabilities facilitate effective dissipation of stimuli from the environment; that is, they facilitate ecological system regulation/adaptation. In response to some stimulus, an ecological network propagates and dissipates locally and/or globally as appropriate until the "processing" of the stimulus is completed. Solé and Bascompte (2006) say it this way: In ecological systems, "flows of energy [or biomass, etc.] enter into the system and are dissipated at different, interconnected scales." These local-to-global processing events are indicative of fluctuating dynamics.

Further, network global propagation and processing potentially changes the state of each involved node, which can affect subsequent local processing. Global behavior can affect local behavior. We thereby have local-to-global-to-local dynamics.

6.4.2 Fluctuating Dynamics

The core function relationships discussed suggest and support the prevalence of fluctuating dynamics in ecosystems. Fritjof Capra (1996) says, "All the variables we can observe in an ecosystem . . . always fluctuate."

In my work, I have seen lots of evidence of ecosystem fluctuating dynamics. An illustration of event behavioral dynamics that can result from the types of core function interactions we have been discussing is provided in Figure 6.6. It is a preview of just a small portion of the results I present later in the book. It is based on my ecological network dynamics model.

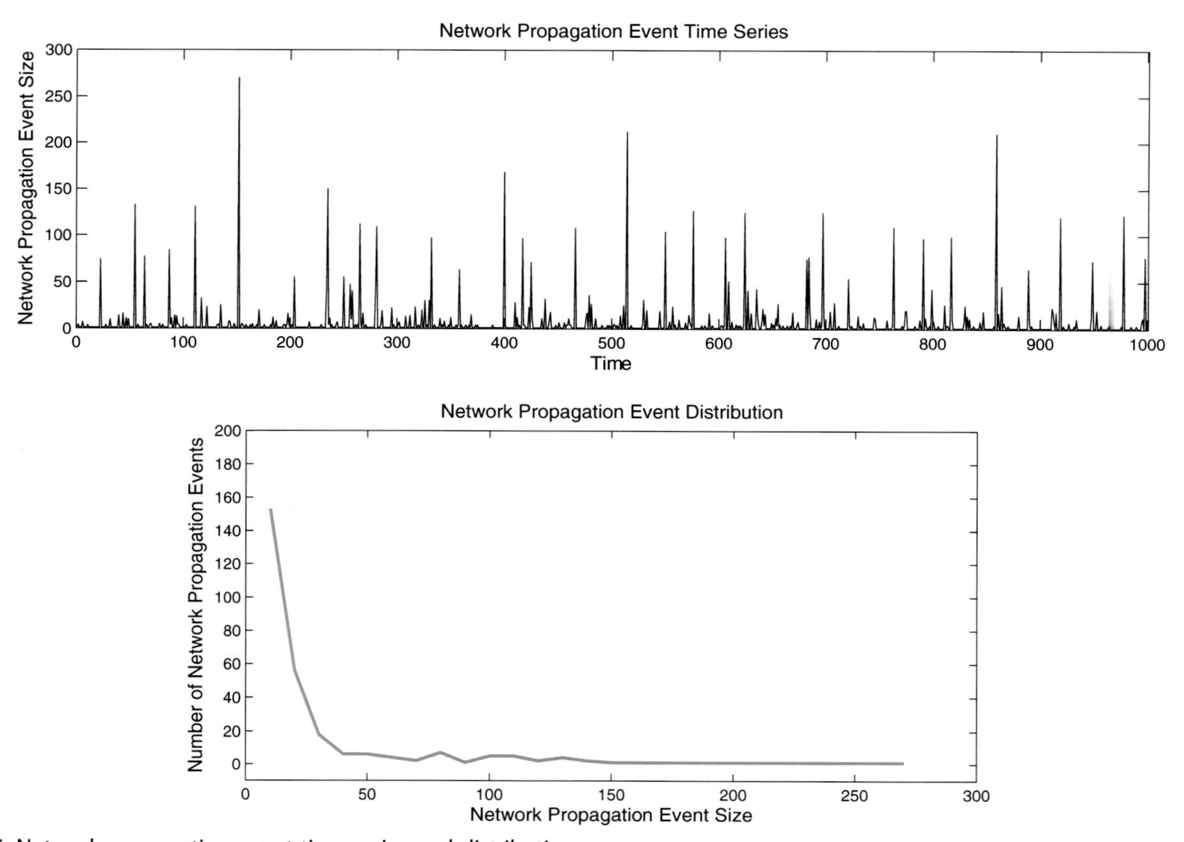

Figure 6.6 Network propagation event time series and distribution.

Figure 6.6 displays an ecological network propagation event time series and distribution.

The time series plots network event size vs. time. It clearly indicates *punctuated* fluctuating behavior. It also indicates *local-to-global* dynamics, with many small events but also medium, large, and even a few very large events that involve almost the entire network. Smaller events are mostly local and larger events are global. The associated network event distribution (number of events vs. size of events) shown in Figure 6.6 is derived from the time series. It is a long-tailed power-law distribution. The time series and distribution shown here are indicative of *fractal* behavior. In my view, such dynamics and behaviors are typical of highly complex ecological systems and fully consistent with the view of ecological network dynamics developed in Part IV and the associated results presented in Part V of this book.

CHAPTER 7

Evolution and Universal Development Concepts

This chapter addresses the developmental tier of the ecological system dynamics framework. The focus here is on the premier development function: evolution. Evolution theory is a centerpiece of the biological and ecological sciences, and likely ranks at or near the top of the list of influential theories in all of science. Not only does evolution theory describe the development of species, but it also seems to have much broader applicability. The chapter begins with a discussion of how evolution principles apply across spatial scales (from subsets of individual species to small networks of species to very large networks of species) and across time scales (from very long term to short term). Next, a high-level evolution process model for natural systems is formulated. The two major elements of this development process model are *choice generation* and *selection*. The evolution model is then applied to a broad array of development arenas—mostly human-made system arenas. In every case, the model fits. The evolution model appears to be a universal development model.

7.1 EVOLUTION: SPATIAL AND TEMPORAL PERSPECTIVES

Evolution results from a development process. I observe that the general evolution development process and its more specific instantiations take place across a wide range of spatial and temporal scales.

Darwin's origin of species context was small spatial scale (individual species) and large temporal scale (geological time). The term *microevolution* usually refers to developmental changes within a single population over a few generations. Microevolution takes place at small spatial scales over relatively short to moderate time frames. *Macroevolution*, on the other hand, describes the development of large networks of species over geological time. Macroevolution occurs at large spatial and large temporal scales. Ecological *succession* is an evolutionary development process that involves moderate

Understanding Complex Ecosystem Dynamics
http://dx.doi.org/10.1016/B978-0-12-802031-9.00007-3

spatial scales and time frames. A succession sere[1] might involve a spatial region of tens or hundreds of square miles and might occur over a period of decades to centuries. Odum and Barrett (2005) note that, at the Indiana Dunes National Lakeshore at the south end of Lake Michigan, succession has transformed dry, sterile sand dunes into a moist, productive closed canopy forest. The transformation from a sand dune system to a forest system took approximately 1000 years.

I suggest that evolutionary development processes occur even at small temporal scales and small to moderate spatial scales. Consider the real-time response of an ecological system network to an energy/matter stimulus received from the environment. The stimulus triggers what I will call *operational time-frame evolution* of the system network. Here is my description of the process: The ecological system nodes are initially unconnected, but are available for interaction. In response to the stimulus, connections between local neighboring nodes become active and begin formation of a network in order to "process" (transform, dissipate, cycle) the input. If local processing is not sufficient, progressively more network nodes and links become active to achieve global processing. The nascent unconnected network *evolves* to become a connected ecological network that operates locally and globally as appropriate to process the stimulus. If the system input is removed, the network connections eventually become inactive once more—until the next input pulse is received. Network connections are not persistent. They are dynamic: active when needed and inactive when not needed. This scenario is repeated over and over again in response to the system environment.

Here's my bottom-line view: Evolution principles have very broad applicability. Evolutionary development processes in natural biological/ecological systems occur at all spatial and temporal scales. And, as we will see later, these development processes apply to human-generated "artificial" systems as well.

7.2 FORMULATING AN EVOLUTION PROCESS MODEL

Evolutionary development processes—across space and time—appear to have common fundamental components. From a general perspective, they all have a *choice generation* mechanism and a *selection* mechanism. A high-level biological/ecological evolution model is illustrated in Figure 7.1.[2]

[1] Definitions: *succession*—a group, type, or series that succeeds or displaces another; *sere*—a series of ecological communities formed in ecological succession (from http://www.merriam-webster.com/).

[2] Howard T. Odum discussed such a model in his 1971 book (Odum, 1971).

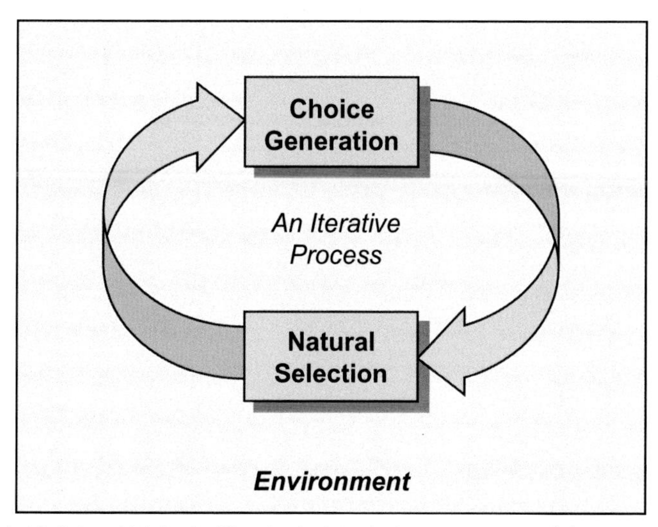

Figure 7.1 High-level biological/ecological evolution process model.

As indicated, the process of evolutionary development is an iterative process that occurs in the context of environment. Holland (1996) says that "evolution continually generates and selects building blocks at all levels." Odum and Barrett (2005) explain that, at the organism level, choice generation involves recurrent gene mutations and genetic drift (stochastic or chance changes in gene structure). Selection pressure comes from the environment and interacting species. At higher levels (e.g., communities), "coevolutionary and group selection processes" apply. Coevolution involves reciprocal selection without direct genetic exchanges. Group or community selection yields traits favorable to the group even when they may be disadvantageous to group members (in isolation). Whatever the biological/ecological level, natural selection selects traits that enhance survival and reproductive success, in the context of environment.

Several investigators believe that natural selection does not operate alone. Complex system self-organization also plays an important role. Kauffman (1995) claims that evolution is an order-seeking process. He argues that Darwin's natural selection is not the only source of order; it is not even the primary source. "Another source—self-organization—is the root source of order. . . . Laws of complexity spontaneously generate much of the order of the natural world. It is only then that selection comes into play, further molding and refining." The "underlying order [is] further honed by selection . . . ." Kauffman is suggesting "an evolutionary process that commingles both self-organization and selection. . . . Neither alone suffices. Life and its evolution have always depended on the mutual embrace of spontaneous order and selection's crafting of that order."

D'Arcy Thompson[3] seems to have reached a similar conclusion. In Thompson's view of evolution (which was considered heretical when he stated it in 1917), natural selection operates only to eliminate the unfit. It is not the (sole) driving force forward; there are also physical driving forces (consistent with Aristotle's "efficient cause") that should be mathematically expressible. Although Thompson did not use the term self-organization, the physical forces he discussed surely contribute to those dynamics.

Odum and Barrett (2005) say, "The organized complexity that has developed in the natural world is difficult to explain solely by selection at the individual and species level; hence, higher-level selection and the process of self-organization have to play major roles." Odum and Barrett also reference Wesson (1991): "Wesson argued that what we call *self-ordering* must be added to natural selection to explain the evolution of complex systems."

We can make an addition to the biological/ecological evolution process model of Figure 7.1 to reflect the importance of self-organization in evolution event dynamics. The result is shown in Figure 7.2. (I will have much more to say about ecological event dynamics later in the book.)

Simon (1962) has some interesting thoughts on the iterative nature of the evolution development process from a hierarchical development point

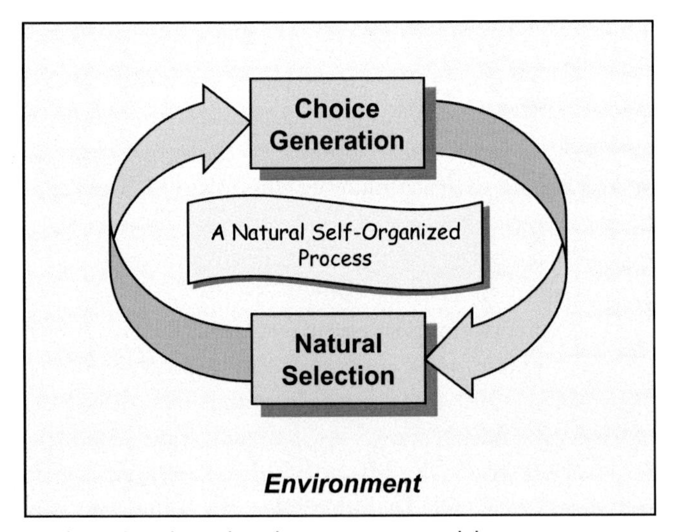

Figure 7.2 Biological/ecological evolution process model.

[3] Thompson (1992), *On Growth and Form*; first published in 1917.

of view. Simon says that, as the development process proceeds, "various complexes come into being, at least evanescently, and those that are stable provide new building blocks for further construction." The iterative hierarchical process can be illustrated as shown in Figure 7.3.

At a given hierarchy level, the evolution development process cycles iteratively until it converges on a *stable intermediate form* (SIF). That SIF becomes the basis for choice generation at the next higher hierarchy level, where a new higher level SIF is produced. The process then proceeds to the next hierarchy level, and so on. If the evolution development process generates more than one SIF at a given hierarchy level, then multiple SIFs move to the next level. There, each SIF may converge to another SIF or may not. If it does not, that SIF is abandoned. If it does, the hierarchy can expand horizontally, and then continue to move vertically. Simon's hierarchical development perspectives are not limited to natural systems; they have broader complex system applicability.

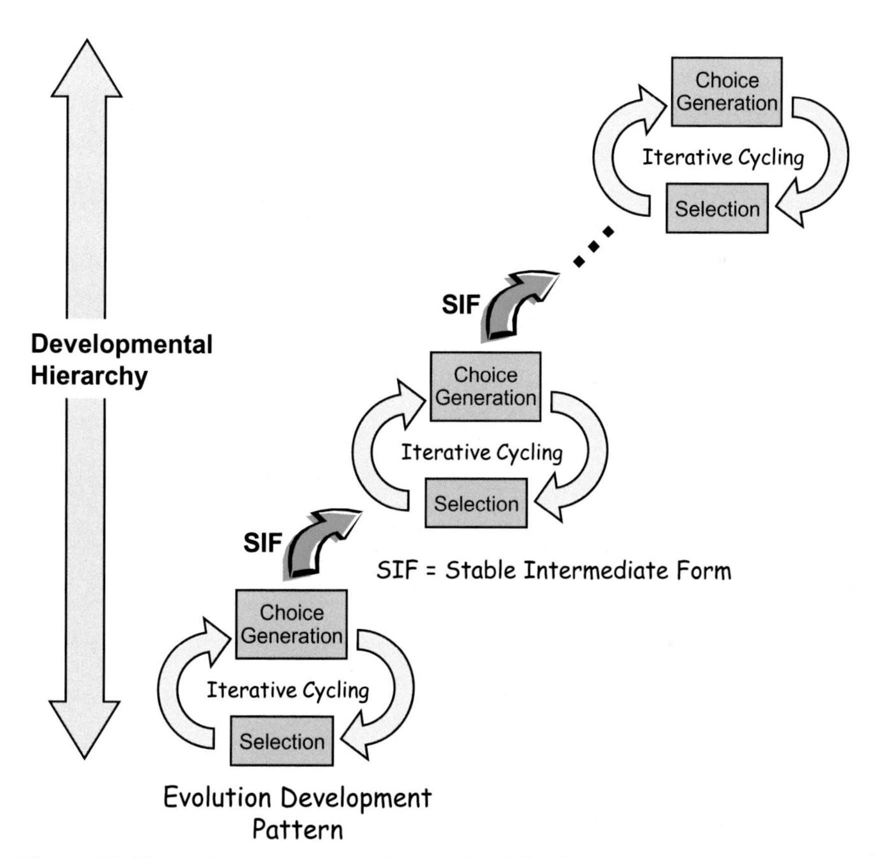

Figure 7.3 The evolution process and hierarchical development.

7.3 THE EVOLUTION MODEL AS A UNIVERSAL DEVELOPMENT MODEL

Does our natural (biological/ecological) system evolution model also apply to "artificial" (human-generated) systems? I say the answer is yes. There is universality here. Kauffman (1995) provides support; some pertinent quotes from Kauffman follow:

- "Organisms arise from the crafting of natural order and natural selection; artifacts [arise] from the crafting of *Homo sapiens*. Organisms and artifacts so different in scale, complexity, and grandeur, so different in the time scales over which they evolved, yet it is difficult not to see parallels."
- "It would not be astonishing if the same laws governed both biological and technological evolution. Tissue and terra-cotta may evolve by deeply similar laws."
- "Evolution explores its landscapes without the benefit of intention. We explore the landscapes of technological opportunity with intention, under the selective pressure of market forces."
- "General laws may govern the evolution of complex entities, whether they are works of nature or works of man."

The natural system evolution model does indeed apply to "artificial" systems. As shown in Figure 7.4, natural self-organization is replaced by human-generated organization. Artificial systems are teleological and the result of human intention. Human-generated organization drives their evolutionary behavior.

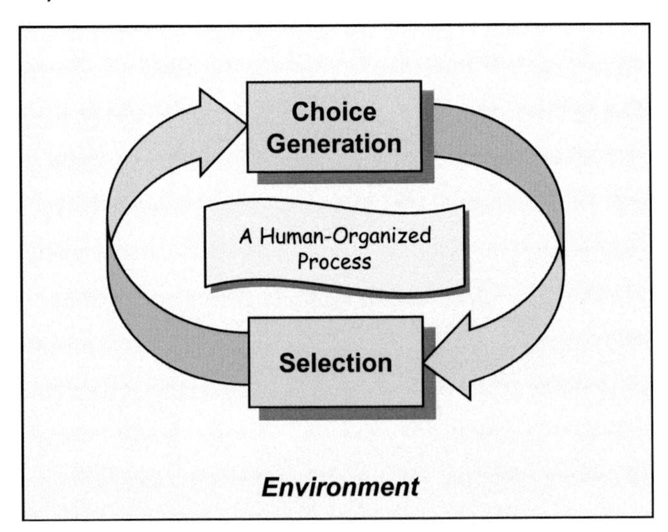

Figure 7.4 Evolution process model for human-made systems.

We will discuss how this model applies to human-made system design, social system development, evolutionary engineering, the scientific method, and the development of a scientific field in sections 7.3.1–7.3.5, respectively. We will demonstrate that *the evolution model is a universal model.*

7.3.1 Human-Made System Design

Artificial system development is often said to consist of requirements, design, and implementation. Design is a key component. I contend that the evolution model is a universal design model.

White (1998) says, "design is a creative, iterative, decision-making process." Weinberg (1975) puts it this way: "Learning to design is learning to generate and evaluate" candidate system configurations in an iterative fashion. Simon (1996) describes the design process as a generator-test cycle: "Think of the design process as involving, first, the generation of alternatives and then, the testing of those alternatives against a whole array of requirements and constraints." The common elements across these descriptions are the notions of choice generation, selection, and iterative cycling—all with underlying human-generated organization. These, of course, are the elements of our evolution model for artificial systems shown in Figure 7.4. (As you would expect, Figure 7.4 matches the conceptual design model of Figure 1.2 that I provided as part of our discussion of engineering design in Chapter 1.)

Let's take a look at Casaday's (1996) system design work, which focuses on the *choice generation* mechanism. Figure 7.5 depicts Casaday's construction-evaluation model of design.

The similarity of this model with the evolution model of Figure 7.4 is obvious. With Casaday's model, one iteratively *constructs* and *evaluates*

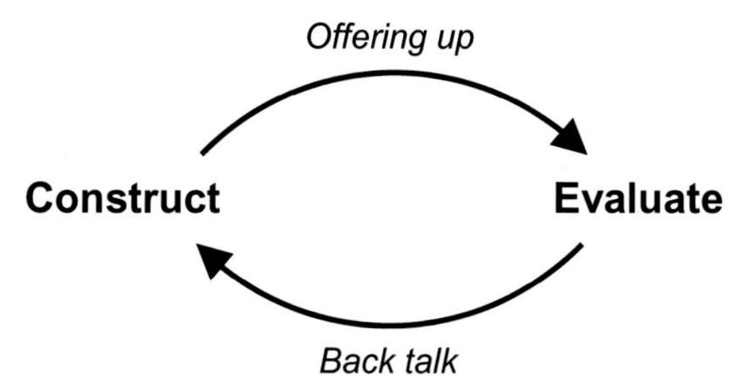

Figure 7.5 Construction—Evaluation model of design. *(Credit: Casaday (1996, p.354); publisher is CRC Press, Taylor and Francis Group LLC Books. Republished with permission of Taylor and Francis Group LLC Books; permission conveyed through Copyright Clearance Center, Inc).*

candidate designs until convergence on an appropriate design for the given problem situation is achieved. Choice generation (construction) is guided by templates. Selection (evaluation) is based on human judgment, and is a human decision-making process.

Casaday's choice generation mechanism utilizes templates (very similar to *patterns*, which we discussed in Chapter 2). The templates provide principles and abstractions from successful practice; they embody successful existing designs. Here is what Casaday has to say: "Design is viewed as a process of constructing and evaluating a succession of more and more complete design models until a final design is produced. In this view, an important part of support for design is a powerful method for constructing design models. . . . Generic forms, called templates . . . can be instantiated to produce specific design models, greatly facilitating construction. . . . Templates can be extracted from designs, from artifacts, and from design process. They can then be evolved, fine-tuned, and adapted." The work "is aimed at creating useful techniques, grounded in successful practice and sound theory, and transferable to other designers. . . . Templates . . . appear promising as a way of capturing, applying, and transferring knowledge of design."

Next, here are some thoughts on the *selection* mechanism in artificial system design from two other investigators. Engineer Buede (2000) discusses two pertinent methods (from the field of decision analysis) for selecting among candidate designs. *Value theory* determines the value of each alternative candidate to the system stakeholders. *Utility theory* determines the value of each candidate to the stakeholders when risk preference (uncertainty) is included. Selection is based on the value results. Simon (1996) points out that an optimization approach (finding the "best" design) is often not realistically feasible. Sometimes *satisficing* (finding the "good enough") is necessary. Simon suggests an approach:

- Define an *aspiration level* for each dimension of system satisfaction and use these levels to evaluate system design candidates.
- If a design candidate meets or exceeds the aspiration level along each dimension, then it *satisfices*.
- If not, then it doesn't and the candidate must be modified or rejected. (Or the aspiration levels may need to be lowered.)

Simon notes that this is a human decision-making approach that "fits our empirical observations of human decision making far better than the utility maximization theory. . . . In the face of real-world complexity, the business firm turns to procedures that find good enough answers to questions whose best answers are unknowable."

The main point of this subsection is that human-generated system design follows natural system evolution principles. Of course, so does "design" in natural systems. We'll conclude the subsection with a couple of examples:[4] (1) The human immune system uses the evolution model—choice generation and selection—to create antibodies "designed" to find, attack, and eradicate specific pathogens/antigens that have invaded the body and are causing sickness. The time frame for this process is a few days to weeks. In simple terms, lymphocytes have receptors that can recognize (bind with) a particular pathogen (then called an antigen), and then produce antibodies to destroy that type of pathogen/antigen. Because it is not known a priori which type of pathogen will invade the body, the immune system randomly produces many types of lymphocytes with differing "recognition" capabilities. (This is choice generation.) The lymphocytes that are successful survive (selection) and produce more and more of that type until the invading pathogen/antigen is eradicated. (2) Ant colonies[5] use the evolution model to "design" food foraging trails. When foraging, ants move randomly in different directions looking for food (trail choice generation). Successful ants return to the nest, while depositing a trail of pheromones (i.e., signaling chemicals). Other ants follow the pheromone trails and reinforce them with more pheromones (trail selection). The resulting trail network is near optimum. *Note that, in each example, the process rules are simple and local, and repeated over and over.*

7.3.2 Social System Development

This subsection on social system development and the next subsection on evolutionary engineering are based on the work of Yaneer Bar-Yam. All of the quotes in this subsection are from Bar-Yam (2004).

Social system development follows the evolution process model. "Human organizations (governments, corporations and other social organizations) may be thought of as undergoing a kind of evolutionary change through survival of the most effective organizations." To illustrate that social system development processes correspond to biological evolution processes, Bar-Yam uses a competitive sports analogy. He makes the following points: There is a large reservoir of candidate athletes (choice generation). There are competitions to select the best among them (selection). Also, in biology, there is parent-to-child heredity. In sports, there is coach-to-athlete

[4] The subject matter comes from Mitchell (2009).

[5] The food foraging procedures of ant colonies are also discussed in Chapter 4.

heredity. This "is a different kind of heredity, however, through transmission of knowledge—knowledge of how to prepare and train, physically and mentally, for competition, as well as how to compete effectively during an event." How about reproduction? "Just as in biology, where selection involves increased reproduction as a measure of success, the process of evolution by selection in sports involves learning by copying or emulating [i.e., reproducing] the most successful competitors." Diversity analogies exist as well. There is a diverse array of athletes that occupies niches, where a niche is defined as "a particular environment and set of resources." In biology and in sports, there are many different ways to contribute.

Bar-Yam explains that competition has always been a central force in evolution theory. Neo-Darwinists say that gene competition is the primary factor in biological evolution. Social Darwinists take the more extreme view that having losers (the poor, the outcasts, etc.) in societal competition is expected and acceptable. Bar-Yam states that competition, however, is just one factor in evolution. Cooperation is another. "Competition and cooperation always coexist" in evolution processes.

In the sports analogy, the development process involves both competition and cooperation. There is a constructive relationship between cooperation and competition when the two operate at different levels of organization (Figure 7.6).

Effective cooperation at one level enables effective competition at the next higher level. Bar-Yam says, "This interaction between competition and cooperation at different levels has been surprisingly absent from much of the scientific dialogue about evolution." In evolution, generally there are "elaborate schemes of cooperation and competition at all levels of organization."

Bar-Yam believes that the evolution development process applies to all sorts of complex systems in society. He also believes that teams are an

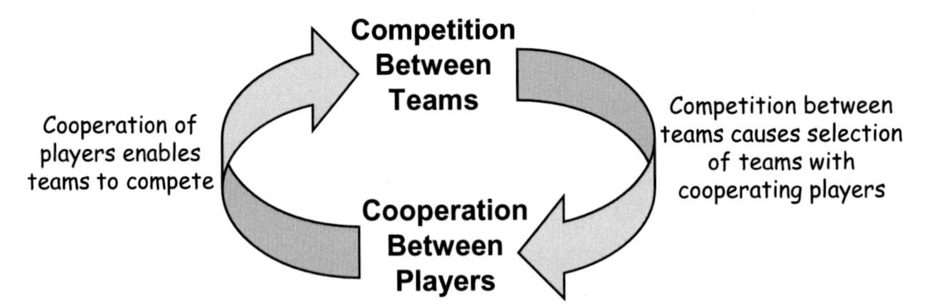

Figure 7.6 Competition and cooperation in sports. *(Credit: Bar-Yam (2004, p. 81); publisher is NECSI Knowledge Press. http://necsi.edu/publications/mtw/).*

effective approach for solving complex societal problems. Bar-Yam says that the multilevel competition/cooperation perspective "can teach us how to improve the effectiveness of teams in any context."

7.3.3 Evolutionary Engineering

In this subsection, based on the work of Yaneer Bar-Yam, we apply the evolution process model to engineering. All of the quotes here are from Bar-Yam (2004).

In the past, engineers have often attempted to comprehensively describe complex human-made systems before they build them. If the degree of system complexity is directly related to the length of the system description (one acknowledged measure of complexity), then it may be futile to attempt to write such an a priori description. Perhaps a new process for engineering complex systems is needed. Bar-Yam suggests an adaptive process that evolves over time—an evolutionary development process. Evolution is "the only process that we know of that creates highly complex systems. . . . Operationally, the key to the creation of an evolutionary process is an agreement to compete and cooperate at different levels of organization." Let us develop "a new strategy for complex systems engineering based upon an understanding of how complex systems arise in nature."

Bar-Yam describes how an evolutionary engineering process would work. He considers an existing complex human-made system that is being upgraded. A given system component needs improvement, so "a new variant of the equipment is introduced in parallel with the old version" (choice generation). If the new version works better, it replaces the old version (selection)—perhaps gradually over time, taking more and more of the load. The old version may persist (e.g., for system robustness and/or safety reasons) until we're sure the new one works better in all aspects of the environment. If successful, the new version may spread and be similarly adopted in other parts of the system. The evolutionary engineering process is iterative and ongoing. The process can apply to equipment, software, humans (system users) and their training, and to the array of interactions among all these. There can be "multiple variants of equipment, software, training or human roles . . . in parallel." Our "large engineering projects should be managed as evolutionary processes undergoing continuous rapid improvement through adaptive innovation."

I would represent Bar-Yam's evolutionary engineering model as shown in Figure 7.7.

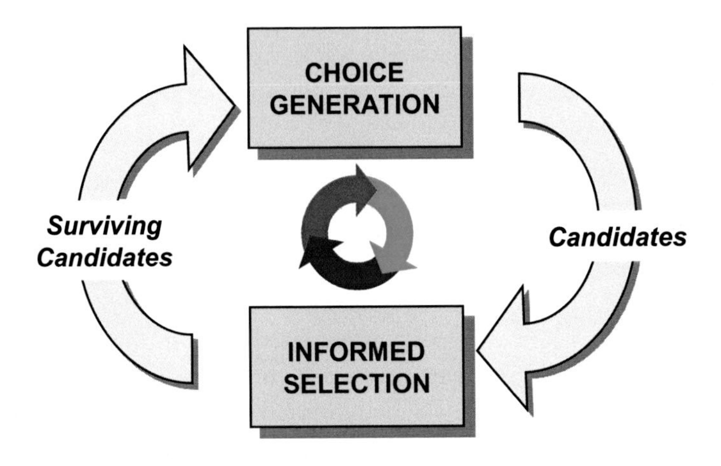

Figure 7.7 Evolutionary engineering model.

Engineering teams compete at the component level and cooperate at the system level. The process applies to initial system development as well as to system upgrade.

"The wide applicability of evolutionary change is a fundamental expression of the unique status it has as the only mechanism we know by which systems that are both effective and highly complex can arise." Bar-Yam says that the evolution model can be applied to all human-engineered systems, including health care systems, education systems, and ecologically sustainable systems. The evolution model is a *universal development model.*

7.3.4 The Evolution Model, the Scientific Method, and General Problem Solving

The scientific method has served as the primary method for developing scientific knowledge since at least the mid-1600s. Figure 7.8 displays a *kinematic graph* of the method. In the graph, the circles represent operands and the arrows represent operators. Box 7.1 provides an explanation of this kinematic-graph depiction of the scientific method.

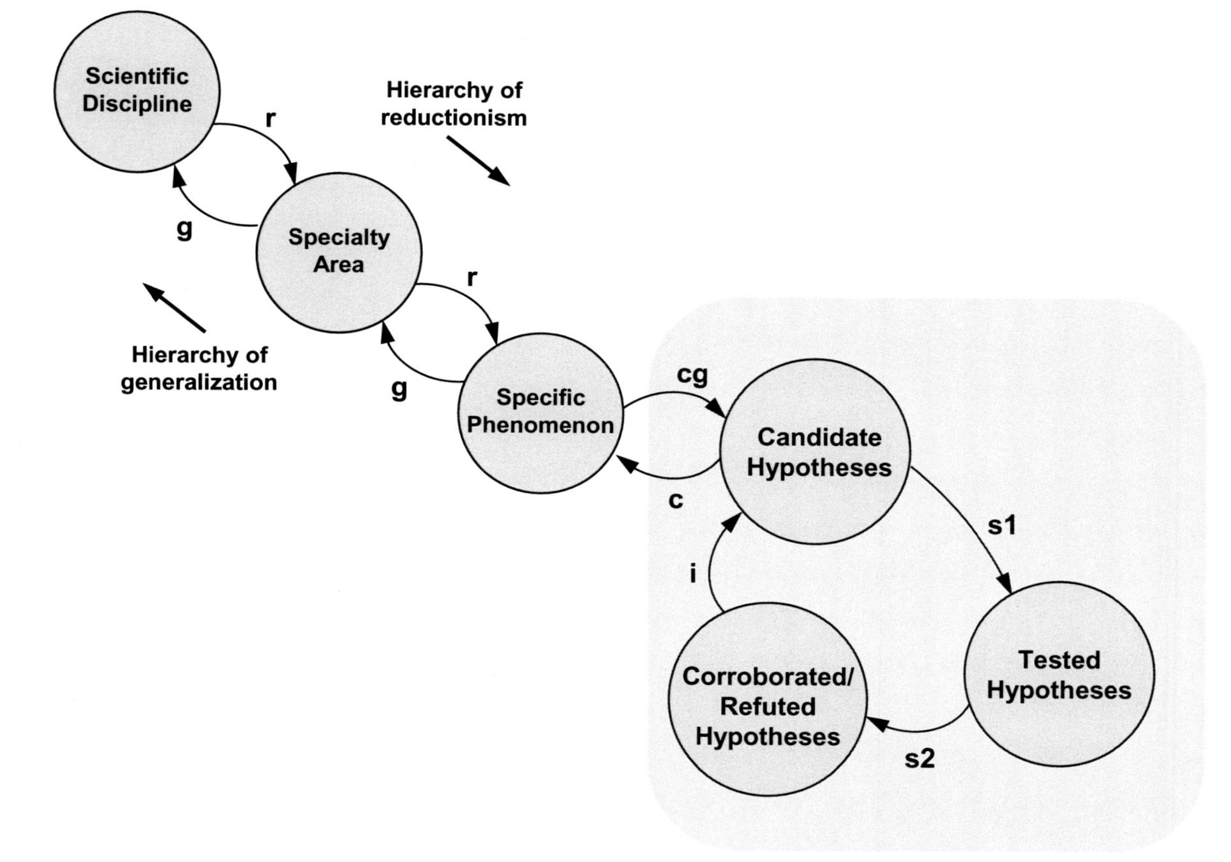

Figure 7.8 Kinematic graph of the scientific method.

BOX 7.1 Explanation of the Scientific Method

Explanation of the Kinematic Graph:

- The reduction operators (r) establish a hierarchy of reductionism. An investigator moves downward from a discipline-wide view to a specialty area within the discipline and then to a specific phenomenon within that specialty area. Although only three levels of hierarchy are shown here, there are often more levels.
- Study of the phenomenon yields candidate hypotheses (operator cg—for choice generation).
- Experimentation, modeling, computer simulation, and other methods (operator s1—s for selection) produce tested hypotheses.
- Analysis (operator s2—s for selection) of the empirical data and modeling results yields either corroboration or refutation of each candidate hypothesis.
- Iteration (operator i) occurs at this point. Examination of results may cause the investigator to go back to modify the hypotheses and then pass through the lower loop again. In some cases, the investigator may cycle back (operator c) to revisit the phenomenon, develop new and different candidate hypotheses, and then proceed through the process again from that point.
- If the investigator (and others in the relevant scientific community) becomes satisfied that a hypothesis is corroborated and that it accurately describes some aspect of the phenomenon, the investigator may move up through the hierarchy of generalization (operator g).
- Unfortunately, the generalization route is rarely followed. Too much information has been lost while traveling the earlier reductionism route. Hypotheses very often remain as unconnected piece-parts in a larger puzzle.

From Figure 7.8 and Box 7.1, we can determine that the operator labeled "cg" and the operand "candidate hypotheses" comprise *choice generation*. Operator "s1" and operand "tested hypotheses" together with operator "s2" and operand "corroborated or refuted hypotheses" comprise *selection*. The process iterates over choice generation and selection. Scientific discovery via the scientific method is an evolutionary development process that follows the evolution model of Figure 7.4.

General problem solving is also an evolutionary development process that similarly follows the evolution model. Simon (1962) provides the following example. To discover the proof for a difficult theorem, conduct a search through a solution space. Start with axioms and previously proved theorems. Try various transformations to generate candidate new expressions. (This is choice generation.) Evaluate the candidates to select an expression for further consideration. (This is selection.) Via more transformation, generate further candidate new expressions (iteration) until a path is discovered that leads to the solution. Some candidates are abandoned along the way if they don't show progress. (The "unfit" are eliminated.) The ones that show progress continue. Simon (1996) provides additional insights. The selection portion of the problem solving process "derives from various rules of thumb, or heuristics, that suggest which paths should be tried first and which leads are promising." The overall process "can always be equated with some kind of feedback of information from the environment." Human problem solving is "discovering a process description of the path that leads to a desired goal."

One of John Holland's[6] specialties is the *genetic algorithm* approach to solving systems problems. This area of investigation has a close relationship with the evolution model. A *choice generator* generates candidate algorithms for "solving a problem." The algorithms compete and take their places in a fitness landscape. The most fit algorithms are *selected* for further processing. Those algorithms are combined and "mutated" to yield a new set of candidates which then compete, and so on. The iterative procedure continues for several "generations," until a problem solution is obtained.

7.3.5 Development of a Scientific Field

Thomas Kuhn's very influential book, *The Structure of Scientific Revolutions* (Kuhn, 1996), addresses the evolutionary development of a scientific field.[7] Kuhn describes the development pattern this way: "Successive transition from one paradigm to another via [scientific] revolution is the usual developmental pattern of mature science." My overview depiction of Kuhn's thesis is provided in Figure 7.9.

Normal science is "research firmly based upon one or more past scientific achievements, achievements that some particular scientific community acknowledges for a time as supplying the foundation for its further practice." These achievements yield the paradigms of the normal science field. A paradigm is an accepted "framework of a scientific school or discipline within which theories, laws, and generalizations and the experiments performed in support of them are formulated."[8] Evolution of the scientific field occurs via the evolution of its paradigms, which occurs via scientific revolutions.

The overall scientific field evolutionary development process follows our evolution model. This is illustrated in Figure 7.10.

[6] John H. Holland, of the University of Michigan and the Santa Fe Institute.
[7] The quotes in this subsection are from Kuhn (1996), unless noted otherwise.
[8] This definition is from Merriam-Webster online (http://www.merriam-webster.com/).

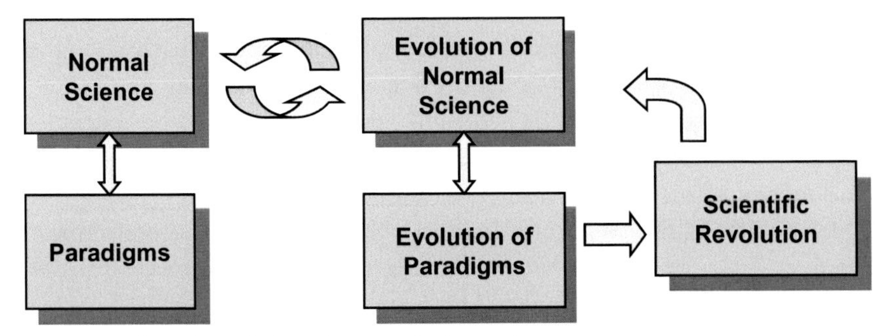

Figure 7.9 Overview of the science evolution pattern.

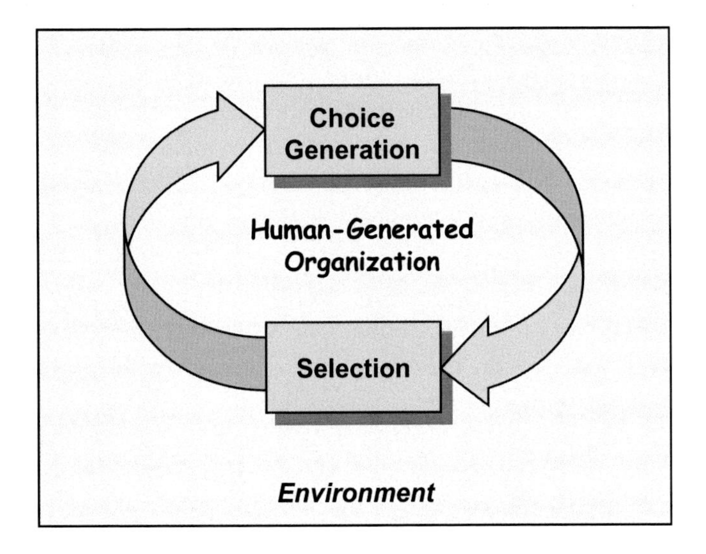

Chronology of the evolutionary development of a scientific field:

- Old paradigm
- Old paradigm exhibits anomalies
- New (incremental) choices are generated but are not effective
- Crisis state is reached
- New (revolutionary) choices are generated and one is selected
- ***New paradigm***

Figure 7.10 Evolutionary development of a scientific field.

The chronology provided in Figure 7.10 indicates that anomalies (failures of the scientific framework) show up in the old paradigm. There are repeated attempts to modify the old paradigm incrementally to resolve the anomalies. This is because deep-seated resistance to any substantial change typically exists within the normal science field. The incremental attempts do not work. The situation worsens and a crisis state is reached. A crisis seems necessary to accomplish substantial change.[9] "Retooling is an extravagance to be reserved for the occasion that demands it." But scientists typically "do not renounce the paradigm that has led them into crisis. . . . Once it has achieved the status of paradigm, a scientific [framework] is declared invalid only if an alternate candidate is available to take its place." Eventually, a suitable candidate becomes available, and a new paradigm replaces the old paradigm. A scientific revolution has occurred. The scientific-field development process is consistent with our evolution model.

The deep-seated resistance to substantial change that typically exists within "normal science" has also been addressed by Bohm and Peat (2000). They say that scientists (and others) try to maintain "the mind's habitual state of comfortable equilibrium. As a result, there is . . . a strong disposition to *impose* familiar ideas, even when there is evidence that they may be false. This, of course, creates the illusion that no fundamental change is required, when in fact the need for such a change may be crucial. If several people are involved, collusion will follow, as they mutually support one another in their false responses." (You may want to consider applying this, for example, to the climate change crisis.)

Kuhn clearly argues that the development of a scientific field is not a gradual, incremental acquisition of knowledge; rather, it is "a series of peaceful interludes punctuated by intellectually violent revolutions." These revolutions are "the tradition-shattering complements to the tradition-bound activity of normal science." A time series of knowledge acquisition events, therefore, would be *punctuated*. I would not be surprised if the event probability distribution function has a power-law form. Those types of time and distribution statistics characterize the behavioral dynamics of many natural system events, including evolution events.[10] They are *fractal* dynamics. (As I mentioned in previous chapters, we will be discussing and describing

[9] I think that is true more generally. It seems that a crisis is a prerequisite for substantial change of any institution or institutionalized belief. As documented by Klein (2007) in her book *The Shock Doctrine*, there are those who create crises in order to further their own agendas.

[10] See, for example, Gould and Eldredge (1977), *Punctuated Equilibria: The Tempo and Mode of Evolution Reconsidered*.

these dynamics extensively later in the book.) It may be that, in addition to following the natural system evolution process model, the development of a scientific field also follows natural system behavioral statistics.

Kuhn provides additional insights regarding the parallels between species evolution and science evolution. "The developmental process described in this essay [his book] has been a process of evolution *from* primitive beginnings—a process whose successive stages are characterized by an increasingly detailed and refined understanding of nature. But nothing that has been or will be said makes it a process of evolution *toward* anything. . . . Does it really help to imagine that there is some one full, objective, true account of nature and that the proper measure of scientific achievement is the extent to which it brings us closer to that ultimate goal?" Darwin "recognized no goal set either by God or nature." Species evolution "moved steadily *from* primitive beginnings but *toward* no goal."

There is an analogy relating "the evolution of organisms to the evolution of scientific ideas," and it is "very nearly perfect." The evolutionary process of science likely occurs "without benefit of a set goal, a permanent fixed scientific truth." New, alternative paradigms are generated, followed with "selection by conflict within the scientific community of the fittest way to practice future science." Science evolution follows species evolution. Again, *the evolution model is a universal development model.*

7.4 SUMMARY OBSERVATIONS AND THE PATH FORWARD

The evolution development process model discussed in this chapter is very similar to the synthesis/analysis process model that we discussed in Chapter 3. That model was illustrated in Figure 3.2, and is reproduced here as Figure 7.11. We noted in Chapter 3 that all comprehensive systems efforts include both synthesis and analysis.

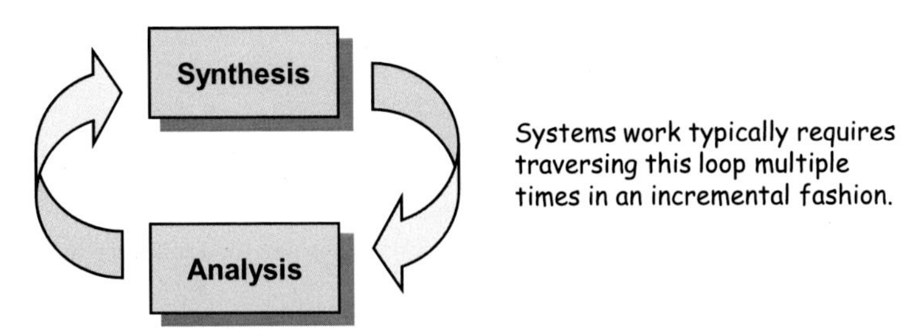

Figure 7.11 Combined synthesis and analysis.

Choice generation is essentially a synthesis activity, while *selection* is an analysis activity. Consistencies—and suggestions of a broad universality—definitely are emerging.

In this chapter, we have demonstrated that the evolution model has very wide applicability. It applies across spatial and temporal scales. It applies across system domains (natural systems and artificial systems). In the artificial systems domain, we have shown that the evolution model applies to system design, social system development, evolutionary engineering, the scientific method, and the development of scientific fields. The notion of broad universality is strengthened.

The evolution model is quite simple. The process is a "simple program" involving just two elements that follow "simple rules." The process is repeated again and again. These evolution model features very closely match the concepts that have been developed and articulated by Wolfram (2002) in his book *A New Kind of Science*. Wolfram hypothesizes and demonstrates that the mechanism of simple programs with simple rules, repeated over and over, can yield extremely complex system behavior. It is the underlying mechanism "that allows nature seemingly so effortlessly to produce so much that appears to us so complex." This hypothesis "implies a radical rethinking of how processes in nature and elsewhere work." It seems to apply "to systems throughout the natural world and elsewhere."

And how can the resulting extremely complex system behavior be characterized? Benoit Mandelbrot has said that "bottomless wonders spring from simple rules . . . repeated without end."[11] The bottomless wonders that he was talking about are *fractals*. In my work (reported in detail later in the book), I have found that the dynamics of complex systems—generated by simple programs with simple rules repeated over and over—are characterized by fractal behavior.[12]

So, the evolution model seems to be not only a universal development model, but it is also consistent with and part of even larger emerging universal principles. We have Wolfram's innovative hypothesis on the underlying mechanism of complex system behavior, Mandelbrot's innovative insights on fractal behavior, and my humble systems and engineering knowledge of complex system dynamics along with my hypothesis on the behavioral characteristics of these complex system dynamics. It seems that

[11] Mandelbrot (2010a) concluded his talk on *Fractals and the Art of Roughness* at the TED2010 conference with this statement.

[12] I have already noted the presence of fractal dynamics in complex systems several times in the book.

we may have the necessary ingredients for proposing a unified set of important universal dynamics principles. This is exciting!

Here is the path forward. We will cover the essentials of relevant existing complex systems theory next in Part III of the book. Chapters 8 and 9 address network theory. Chapter 10 addresses nonlinear dynamics theory. Wolfram's work and Mandelbrot's work are described in some detail in Chapters 11 and 12, respectively. We will then cover my hypothesis work on the behavioral characteristics of complex system (ecosystem) dynamics—including hypothesis generation, modeling, testing, and results—in Parts IV and V of the book. We will bring it all together in the conclusions of Part VI.

PART III

Complex Systems Theory: Networks, Nonlinear Dynamics, Cellular Automata, and Fractals (Roughness)

In this part, we conduct an extensive review of the pertinent extant complex systems theory. Network theory, nonlinear dynamics theory, cellular automata investigations, and fractals (roughness theory) are addressed. In some areas, I provide additional commentary based on my systems, engineering, and ecological perspectives. The material here serves as a valuable and necessary resource for our work. Application and, in some cases, extensions of the theory contribute to a "synthesis of ideas" that is pursued in the subsequent parts of the book.

CHAPTER 8

Network Theory: The Structure of Complex Networks

The current interest in networks is part of a broader movement towards research on complex systems.

Steven Strogatz (2001)

It is increasingly recognized that . . . real networks are governed by robust organizing principles.

Albert and Barabási (2002)

The scientific community has, by drawing on ideas from a broad variety of disciplines, made an excellent start on . . . the characterization and modeling of network structure.

Mark Newman (2003)

Networks are the architectural basis of complex systems. Knowledge of the structure and dynamics of these networks is critical. Most network investigation to date in the scientific community has focused on the structure of complex networks. In many scientific pursuits, structural issues seem to get addressed first. We'll follow that pattern. This chapter focuses on the *structure* of complex networks. The next chapter focuses on the *dynamics* of complex networks. Here, in this chapter, the fundamental concepts and properties that characterize network structure are defined and described. A primary underlying goal of this book is to increase our understanding of real-world network-based systems (particularly ecological systems). Next, therefore, the chapter takes a comprehensive look at empirical data from real-world networks, with an emphasis on properties that are common to many of them. The chapter then takes a theoretical perspective, and considers mathematical developments relevant to complex system networks and their structural properties.

8.1 NETWORK STRUCTURE CONCEPTS, PROPERTIES, AND CHARACTERISTICS

A *network* is a set of elements called *nodes* (or vertices), with connections between them called *links* (or edges). Many authors agree that complex

Understanding Complex Ecosystem Dynamics
http://dx.doi.org/10.1016/B978-0-12-802031-9.00008-5

systems very often take the form of networks. Newman (2003): "Systems taking the form of networks (also called 'graphs' in much of the mathematical literature) abound in the world. Examples include the Internet, the World Wide Web, . . . neural networks, metabolic networks, food webs, distribution networks such as blood vessels . . . ." Albert and Barabási (2002): "Complex networks describe a wide range of systems in nature and society." Strogatz (2001): "The study of networks pervades all of science, from neurobiology to statistical physics" and "In the longer run, network thinking will become essential to all branches of science."

There is a spectrum of network types. Section 8.1.1 discusses that spectrum, and the position of most real-world networks within the spectrum. A set of important network properties is described in Section 8.1.2. Network structural characteristics, based on those properties, are examined in Sections 8.1.3 (small-world characteristics) and 8.1.4 (scale-free characteristics).

8.1.1 Random and Nonrandom Networks

There is a spectrum of network (graph) types, from random to well ordered. A random graph is a network in which elements are connected at random. An example of a very well-ordered network is a regular lattice—a network in which all elements have the same number of connections and are arranged in a highly periodical manner. So-called scale-free networks lie between these two extremes.

Random graphs have been extensively analyzed, and the analyses have yielded some important benchmarks and insights applicable to networks in general. Newman (2003) says, "Perhaps the simplest useful model of a network is the random graph . . . . In this model, undirected edges are placed at random between a fixed number n of vertices to create a network in which each of the $n(n-1)/2$ possible edges is independently present with some probability p, and the number of edges connected to each vertex—the degree of the vertex—is distributed according to a binomial distribution, or a Poisson distribution in the limit of large n. The random graph has been well studied by mathematicians, and many results, both approximate and exact, have been proved rigorously." Albert and Barabási (2002) say, "Since the 1950s large scale networks with no apparent design principles have been described as random graphs, proposed as the simplest and most straightforward realization of a complex network. Random graphs were first studied by the Hungarian mathematicians Paul Erdös and Alfréd Rényi" (see Erdös and Rényi, 1960).

Despite their usefulness, however, do random graphs actually represent real-world networks? Apparently, they do not. Albert and Barabási (2002) ask, "Are the real networks behind such diverse complex systems as the cell or the Internet fundamentally random?" They answer, "Our intuition clearly indicates that complex systems must display some organizing principles." Newman (2003) adds, "Most of the interesting features of real-world networks that have attracted the attention of researchers in the last few years . . . concern the ways in which networks are *not* like random graphs." It turns out that many complex networks are "scale-free" networks.

Figure 8.1 provides a schematic illustration of the random and scale-free network architectures. The following description of this figure is due mostly to Strogatz (2001). On the left is a random graph constructed by placing n nodes on a plane, then joining pairs of them together at random until m links are used. Nodes may be chosen more than once, or not at all. The resulting wiring diagram (not shown) would be a snarl of crisscrossed lines. For clarity, Strogatz has segregated the different connected components, colored them, and eliminated as many spurious crossings as possible. The main topological features are (a) the presence of a single giant component, as expected (see Section 8.3) for a random graph with $m > n/2$ (here $n = 4200$, $m = 4193$), and (b) the absence of any dominant hubs. The degree k, or number of connections to/from a node, is Poisson distributed across the nodes. Most nodes have between one and four connections, and all have between zero and six. On the right-hand side of Figure 8.1 is a so-called scale-free network, grown by attaching new nodes to previously existing nodes. The probability of attachment is proportional to the degree of the existing node; thus, richly

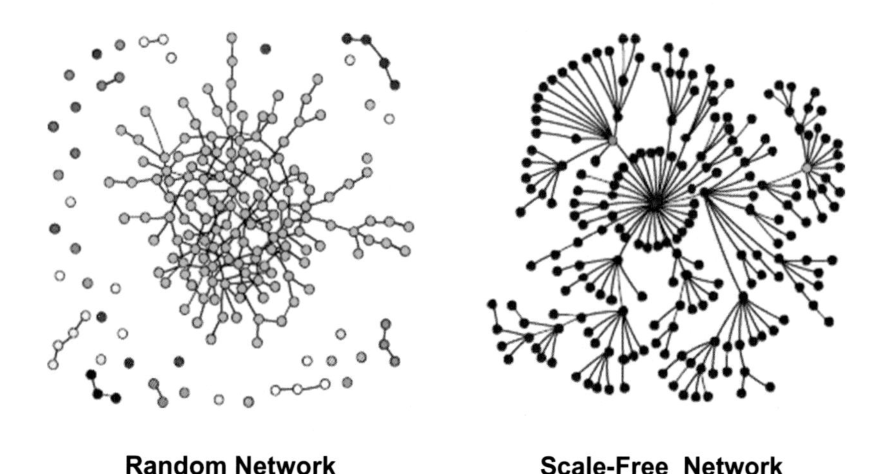

Figure 8.1 Schematic illustration of two important network architectures. *(Adapted with permission from Macmillan Publishers Ltd: NATURE (http://www.nature.com/), Strogatz, S. H., Exploring Complex Networks, Volume 410, Issue 6825, March 8, 2001, pp. 268–276. Copyright (2001)).*

connected nodes tend to get richer, leading to the formation of hubs and a skewed degree distribution with a heavy tail (i.e., a power-law distribution). Here, the three nodes with the most links have $k=433$, $k=412$, and $k=411$ links, respectively. For this network, $n=4200$ nodes, $m=4199$ links. Network visualization was done using the Pajek program for large network analysis (http://vlado.fmf.uni-lj.si/pub/networks/pajek/). Many of the terms used in this description will be explained further as we proceed in the chapter.

8.1.2 Some Important Network Properties

Newman (2003) observes, "The study of networks, in the form of mathematical graph theory, is one of the fundamental pillars of discrete mathematics". ... "Recent years, however, have witnessed a substantial new movement in network research, with the focus shifting away from the analysis of single small graphs and the properties of individual vertices or edges within such graphs to consideration of large-scale statistical properties of graphs." According to Newman, these large-scale statistical properties that describe the structure of networked systems include path length, clustering, and node degree distribution. Albert and Barabási (2002) agree that we should "focus on three robust measures of a network's topology: average path length, clustering coefficient, and degree distribution."

Let me add some additional perspective on why these three properties may be particularly important. In ecological systems, clustering (which can be viewed as modularization) facilitates local processing of stimuli from the environment. The existence of short path lengths facilitates global connection across the network. Local processing and global connection are essential features of ecological systems (and other complex systems). Degree distribution is of interest because it turns out that a large percentage of real-world networks have the same general degree distribution: the scale-free (or scale-invariant) degree distribution. Note that, in the network literature, typically a network is considered "scale-free" if its degree distribution follows a power law. My research indicates that other important network parameters—many of which are associated with network dynamics—also have scale-invariant/power-law distributions. We'll discuss that quite extensively in Part IV and beyond in this book.

The next three subsections address path length, clustering, and node degree, respectively. To help with the network jargon, Box 8.1—*A Short Glossary of Network Terms* provides a summary of definitions of important network terms for easy reference.

BOX 8.1 A Short Glossary of Network Terms

From Newman (2003):

Vertex (pl. vertices): A fundamental element of a network, also called a node or a site (physics).

Edge: The line connecting two vertices, also called a link or a bond (physics).

Directed/undirected: An edge is directed if it runs in only one direction and undirected if it runs in both directions. Directed edges can be represented as arrows indicating their orientation. A graph is directed (a *digraph*) if all of its edges are directed. Directed graphs can be either cyclic, meaning they contain closed loops of edges, or acyclic, meaning they do not.

Degree: The number of edges connected to a vertex. A directed graph vertex has both an in-degree and an out-degree, which are the numbers of incoming and outgoing edges, respectively.

Geodesic path: A geodesic path is the shortest path through the network from one vertex to another. Note that there may be and often is more than one geodesic path between two vertices.

Network diameter: The diameter of a network is the length (in number of edges) of the longest geodesic path between any two vertices.

From Csermely (2006):

Random graph: A random graph is a network in which vertices are connected at random.

Degree: The degree of a network vertex corresponds to the number of links to/from this element.

Regular lattice: A network in which all vertices have the same degree, and are arranged in a highly periodical manner.

Path length: The path length is the number of links that are traversed when traveling between two network vertices.

Shortest path length: The shortest route between two vertices.

Characteristic path length: The average of all the shortest path lengths in the network.

Network diameter: The maximum number of links in the shortest path between any pair of network vertices.

Clustering coefficient: The probability that two neighbors of a given vertex are also neighbors of each other.

8.1.2.1 Path Length

Csermely (2006) defines *path length* as the number of links traversed when traveling between two network elements (nodes). The *shortest path length* is the shortest route (smallest number of links) between two elements. The *characteristic path length* is the average of all the shortest path lengths in the network. The *network diameter* is the maximum number of links in the shortest path between any pair of network elements.

Newman (2003) expresses the terminology slightly differently. He refers to characteristic path length as *mean geodesic distance*. "Consider an undirected network, and let us define ℓ to be the mean geodesic (i.e., shortest) distance between vertex pairs in a network:

$$\ell = \frac{1}{\frac{1}{2}n(n+1)} \sum_{i \geq j} d_{ij}$$

where d_{ij} is the geodesic distance from vertex i to vertex j [and $n =$ the number of vertices/nodes]." Only pairs that actually have a connecting path are included in the sum. The geodesic path is the shortest path through the network from one vertex to another. (Note that there may be and often is more than one geodesic path between two vertices.) The network diameter is the length (in number of edges) of the longest geodesic path between any two vertices.

8.1.2.2 Clustering

The notion of *clustering* in networks seems to have originated in sociology and social networking. The pertinent network parameter is *clustering coefficient*.

Newman (2003) provides two definitions for clustering/clustering coefficient:

Definition 1 In many networks it is found that if vertex A is connected to vertex B and vertex B to vertex C, then there is a heightened probability that vertex A will also be connected to vertex C. In the language of social networks, the friend of your friend is likely also to be your friend. In terms of network topology, transitivity (clustering) means the presence of a heightened number of triangles in the network—sets of three vertices, each of which is connected to each of the others. It can be quantified by defining a clustering coefficient C thus:

$$C = \frac{3 \times \text{number of triangles in the network}}{\text{number of connected triples of vertices}}$$

where a "connected triple" means a single vertex with edges running to an unordered pair of others. See Figure 8.2. As illustrated, this network has one triangle and eight connected triples, and therefore has a clustering coefficient of $3 \times 1/8 = 3/8$.

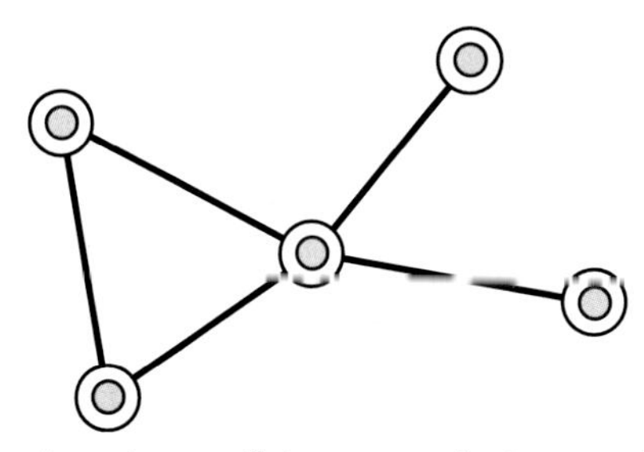

Figure 8.2 Network for clustering coefficient example. *(Reprinted from Newman, M. E. J., The Structure and Function of Complex Networks, SIAM Review, Volume 45, Number 2, May 2003, pp. 167–256. Copyright ©2003 Society for Industrial and Applied Mathematics. Reprinted with permission. All rights reserved).*

In effect, C measures the fraction of triples that have their third edge filled in to complete the triangle. The factor of three in the numerator accounts for the fact that each triangle contributes to three triples and ensures that C lies in the range $0 \leq C \leq 1$. In simple terms, C is the mean probability that two vertices that are network neighbors of the same other vertex will themselves be neighbors. Newman notes that, in the sociology literature, the previous definition is referred to as the "fraction of transitive triples."

Definition 2 An alternative approach defines a local value of clustering coefficient as:

$$C_i = \frac{\text{number of triangles connected to vertex } i}{\text{number of triples centered on vertex } i}$$

The clustering coefficient for the whole network is the average:

$$C = \frac{1}{n} \sum_i C_i$$

where n = the number of nodes in the network.

This definition effectively reverses the order of the operations of taking the ratio of triangles to triples and of averaging over vertices. Here one calculates the mean of the ratio, rather than the ratio of the means.

Albert and Barabási (2002) provide a similar definition for clustering/clustering coefficient. A common property of social networks is that cliques form, representing circles of friends or acquaintances in which every member knows every other member. This inherent tendency to cluster is quantified by the clustering coefficient, a concept that has its roots in sociology, appearing under the name "fraction of transitive triples." Let us focus first on a selected node i in the network, having k_i edges that connect it to k_i other nodes. If the nearest neighbors of the original node were part of a clique, there would be $k_i(k_i - 1)/2$ edges between them. The ratio between the number E_i of edges that actually exist between these k_i nodes and the total number $k_i(k_i - 1)/2$ gives the value of the clustering coefficient of node i:

$$C_i = \frac{2E_i}{k_i(k_i - 1)}$$

The clustering coefficient of the whole network is the average of all individual C_i values.

Csermely (2006) defines the clustering coefficient of an element (node) simply as:

$$C = \frac{g}{G}$$

where g is the number of actual links between all neighbors of an element and G is the number of links that could possibly exist between all neighbors of an element. (Actually, that's the same as the Albert and Barabási definition.)

With respect to clustering, there is a clear deviation between the behavior of real networks and random graphs. Newman (2003) says that, for a real network, "in general, regardless of which definition of the clustering coefficient is used, the values tend to be considerably higher than for a random graph with a similar number of vertices and edges." Albert and Barabási (2002) say that "in a random graph, since the edges are distributed randomly, the clustering coefficient is $C = p$ [where p is the node-to-node connection probability]. However, in most, if not all, real networks the clustering coefficient is typically much larger than it is in a comparable random network (i.e., having the same number of nodes and edges as the real network)." We will see this effect when we look at some numbers for real networks in Section 8.2.

Networks with clusters (modules) and sparse links between and among modules (minimal coupling) are very familiar to me from my systems engineering work. These are the network architectures of functionally partitioned systems.

8.1.2.3 Node Degree

The degree of a network node is the number of links originating/terminating on the node. Random networks have Poisson degree distributions. Scale-free networks have power-law degree distributions. Newman (2003) notes that for directed graphs each vertex has both an in-degree and an out-degree, and the degree distribution therefore becomes a function $P(i, j)$ of two variables, representing the fraction of vertices that simultaneously have in-degree i and out-degree j.

From Albert and Barabási (2002): Not all nodes in a network have the same number of edges (same node degree). The spread in the node degrees is characterized by a distribution function $P(k)$, which gives the probability that a randomly selected node has exactly k edges. Because in a random graph the edges are placed randomly, the majority of nodes have approximately the same degree, close to the average degree $<k>$ of the network. The degree distribution of a random graph is a Poisson distribution with a peak at $P(<k>)$. One of the interesting developments in our understanding of complex networks was the discovery that for most real large networks, the degree distribution significantly deviates from a Poisson distribution. In particular, for a large number of networks, including the World Wide Web, the Internet, and metabolic networks, the degree distribution has a power-law tail:

$$P(k) \propto k^{-\gamma}$$

where γ is the scaling exponent.
(The above relationship is for non-directed graphs, i.e., graphs that do not distinguish between in-degree and out-degree.)

From Newman (2003): The degree of a vertex in a network is the number of edges connected to that vertex. We define $P(k)$ to be the fraction of vertices in the network that have degree k. Equivalently, $P(k)$ is the probability that a vertex chosen uniformly at random has degree k. A plot of $P(k)$ for any given network can be formed by making a histogram of the degrees of vertices. This histogram is the degree distribution for the network. In a random graph of the type studied by Erdös and Rényi, each edge is present or absent with equal probability; hence, the degree distribution is binomial, or Poisson in the limit of large graph size. Real-world networks are mostly found to be very unlike the random graph in their degree distributions. Far from having a Poisson distribution, the degrees of the vertices in most networks are highly right-skewed, meaning that their distribution has a long right tail of values that are far above the mean.

Newman also notes that an alternative way of presenting degree data is to plot the node degree cumulative distribution function:

$$P_c(k) = \sum_{k'=k}^{\infty} P(k')$$

which is the probability that the degree is greater than or equal to k. If $P(k)$ follows a power law with scaling exponent γ, then $P_c(k)$ also follows a power law, but with scaling exponent $\gamma - 1$. (One is the mathematical derivative of the other.)

In summary, with respect to node degree, random graphs have a characteristic scale—it is the mean of the Poisson distribution. Nonrandom networks that have "long tail" degree distributions are known as scale-free, which is meant to suggest that all scales are present (free of any characteristic scale) and their likelihood of occurrence (probability distribution) follows a power law.

As noted previously, Box 8.1 provides a summary of definitions of important network terms discussed in Sections 8.1.1 and 8.1.2.

The path length, clustering, and node degree properties we have discussed have importance in certain real-world network structural characteristics. Section 8.1.3 addresses small-world characteristics. Section 8.1.4 addresses scale-free characteristics and provides more detail on scale-free degree distributions.

8.1.3 Small-World Characteristics

Real networks have been observed to have the "small-world" characteristics of *high clustering* (for local processing) and *short path lengths* that provide relatively easy node-to-node accessibility (for global propagation). In ecological systems, these capabilities facilitate effective handling of stimuli from the environment.

Both Csermely (2006) and Strogatz (2001) depict small-world networks as lying between the extremes of well-ordered networks and random networks. From Csermely (2006): Small-world networks lie between lattice-type networks and random networks. They have a high clustering coefficient, close to that of a lattice network and much higher than a random network. They have a small characteristic path length, close to that of a random network and much smaller than a lattice network. From Strogatz (2001): Although well-ordered networks and random graphs are both useful idealizations, many real networks lie somewhere between the extremes of order and randomness—between regular lattices and random graphs. Such networks are so-called "small world" networks, with short paths between any two nodes, just as in the giant component of a random graph. Yet the network is much more highly clustered than a random graph, in the sense that if A is linked to B and B is linked to C, there is a greatly increased probability that A will also be linked to C.

Watts and Strogatz (1998) and Strogatz (2001) conjectured that these two properties—short paths and high clustering—would hold for many natural and technological networks. Furthermore, they conjectured that dynamical systems coupled in this way would display enhanced signal propagation speed compared with regular lattices of the same size. The reasoning is that the short paths could provide high-speed communication channels between distant parts of the system, thereby facilitating any dynamical process that requires global coordination and information flow.

Here's a view from Albert and Barabási (2002): "The small-world property appears to characterize most complex networks." The small-world concept in simple terms describes the fact that despite their often large size, in most networks there is a relatively short path between any two nodes. The distance between two nodes is defined as the number of edges along the shortest path connecting them. The most popular manifestation of small worlds is the "six degrees of separation" concept, uncovered by social psychologist Stanley Milgram (1967), who concluded that there is a path of acquaintances with a typical length of about six between most pairs of people in the United States. Albert and Barabási also point out that, as Erdös and Rényi have demonstrated, the typical distance between any two nodes in a random graph scales as the logarithm of the number of nodes. Random graphs, therefore, have short path lengths as well (but not high clustering).

Three observations from Newman (2003) follow:

Milgram's experiments involved letters, passed from person to person, which were able to reach a designated target individual in only a small number of steps. These experiments were one of the first direct demonstrations of the small-world effect: the fact that most pairs of vertices in most networks seem connected by a short path through the network.

The short path length aspect of the small-world effect is mathematically obvious. If the number of vertices within a distance r of a typical central vertex grows exponentially with r—and this is true of many networks, including the random graph—then the value of ℓ, the average shortest path length, will increase only as $\log n$. In recent years, the term small-world effect has thus taken on a more precise meaning: networks are said to show the small-world effect if the value of ℓ scales logarithmically or slower with network size for fixed mean degree. This behavior has been observed in various real networks.

The small-world effect has obvious implications for the dynamics of processes taking place on networks. For example, if one considers the spread of information, or indeed anything else, across a network, the small-world effect implies that the spread will be fast on most real-world networks.

8.1.3.1 Commentary

The small-world network structural properties of high clustering and short path lengths clearly have important implications for network dynamics. They facilitate dynamic local-to-global processing in networks. More detail on theoretical aspects of small-world networks is given in Section 8.3.3 of this chapter. We will revisit the small-world effect when we synthesize a view of ecological network dynamics in Chapter 15.

Here's an interesting point about the relationship between the small-world characteristic and the scale-free characteristic (covered in the next subsection): it seems that small-world networks are not necessarily scale-free, but real networks that are scale-free are also small-world. As we shall see, the empirical data in Section 8.2 of this chapter supports that premise.

8.1.4 Scale-Free Characteristics

Networks with "long tail" degree distributions are known as scale-free networks. The "scale-free" label is meant to suggest that all scales are present (free of any one particular scale) and the likelihood of occurrence (probability distribution) of any given node degree value follows a power law.

So, node degree is statistically scale-free if its probability distribution follows a power law, which can be written as:

$$P(k) = Ck^{-\gamma}$$

where C is a constant, k is the node degree, and γ is the scaling exponent. Taking the logarithm of both sides of the previous equation, we have

$$\log P = \log C - \gamma \log k$$

The logarithm of the probability is a linear function of the logarithm of node degree. It plots as a straight line in a log-log representation. The slope of the line is the negative of the scaling exponent. Figure 8.3 illustrates both forms of the power-law distribution.

The previous power-law distribution holds over a limited domain (e.g., the probability cannot go to ∞ as k goes to zero). Most natural scale-free distributions lose their power-law shape after a few orders of magnitude.

Csermely (2006) states that "scale-free networks have been reported in all areas of biology, human relations and constructs appearing in every moment of our everyday lives." He also says that "networks have surprisingly common features in spite of their vastly different constituents and linkage systems."

Some thoughts and results from Strogatz (2001) and Broder et al. (2000):

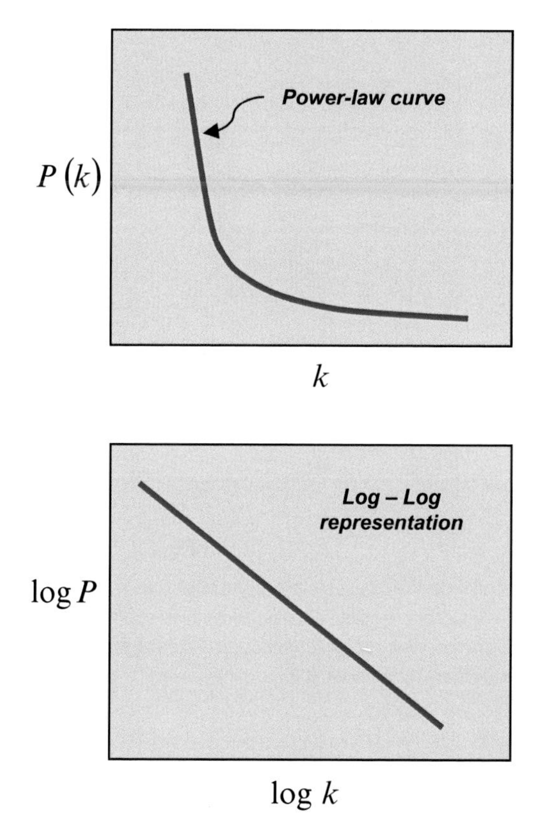

Figure 8.3 Scale-free/power-law illustration.

Strogatz says the simplest random graph models predict a bell-shaped Poisson degree distribution. But, in any real network, some nodes are more highly connected than others—the degree distribution is highly skewed and decays much more slowly than a Poisson. For instance, the distribution decays as a power law for the Internet backbone, metabolic reaction networks, the telephone call graph, and the World Wide Web. Remarkably, the degree distribution scaling exponent is in the range 2.1–2.4 for all of these cases. Strogatz also notes that the earliest work in scale-free networks came from Herbert Simon (1955).

Figure 8.4 provides a log-log plot of the in-degree distribution for the World Wide Web. The network nodes are web pages and the links are the directed URL hyperlinks from one page to another. The plot shows the number of web pages that have a given in-degree (number of incoming links).

Strogatz (2001) generated the plot in Figure 8.4, but he credits Broder et al. (2000) as the source for the data and the analysis. Broder et al. explain that

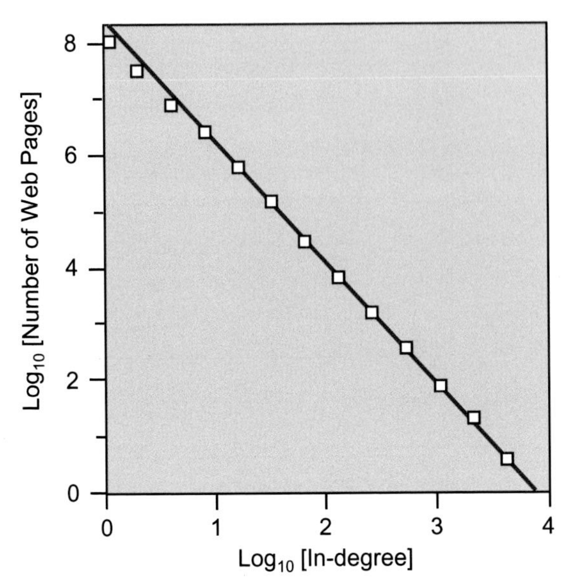

Figure 8.4 The World Wide Web is scale-free. *(Adapted, with permission of Elsevier, from Broder, A. et al, Graph Structure in the Web, Computer Networks, Volume 33, Issue 1–6, June 2000, pp. 309–320 (http://www.sciencedirect.com/science/journal/computer_networks). Copyright © 2000 Published by Elsevier B.V.).*

the raw data come from "web crawl" experiments of approximately 200 million web pages and 1.5 billion links. The raw data have been logarithmically binned. Logarithmic binning is a procedure for averaging data that fall in certain ranges of values (bins). The equal logarithmic bin widths are specified as:

$$\text{Bin width} = \log(\text{upper edge of bin}) - \log(\text{lower edge of bin})$$
$$\text{Bin width} = \log(x_{i+1}) - \log(x_i)$$

The straight line plotted in Figure 8.4 is a "best fit" linear regression. The result indicates that the World Wide Web in-degree distribution follows a power law with a scaling exponent of approximately 2.1.

Newman (2003) offers an important insight:

"The term 'scale-free' refers to any functional form $f(x)$ that remains unchanged to within a multiplicative factor under a rescaling of the independent variable x. In effect this means power-law forms, since these are the only solutions to $f(ax) = bf(x)$, and hence 'power-law' and 'scale-free' are, for our purposes, synonymous."

Csermely (2006) discusses several other occurrences of power-law behavior:

- The allometric scaling laws in biology are empirical power laws that describe the behavior and illustrate the similarity of various complex systems, such as cells, organs, organisms, and others, over a wide range of masses. The allometric scaling laws can be written as:

$$P(m) = cm^{\alpha}$$

where c is a constant, m is the mass of the organism or organelle, and α is the scaling exponent.

The concept is very general, and the expression describes the mass dependence of metabolic rate ($\alpha = 3/4$), lifespan ($\alpha = 1/4$), growth rate, heart rate ($\alpha = -1/4$), tree height, mass of cerebral gray matter, and so on.

- Distribution of wealth in society follows a power law. This phenomenon is described by the Pareto law (1897), also known as the 80-20 rule; for example, 80% of the wealth is possessed by 20% of the people. (The 80-20 wealth proportions need a twenty-first-century update.) The Pareto law applies more widely than just wealth.
- One could argue that the quality of artistic achievement (music, dance, visual art, literature . . .) exhibits power-law behavior. The probability of occurrence of lower quality events is substantially greater than the probability of high-quality events.
- The Gutenberg-Richter law states that both the occurrence and the magnitude of earthquakes follow a power law. Rain also exhibits power-law behavior in occurrence and magnitude.

In Csermely's words (but referring to Kauffman's work (2002)), power-law behavior seems "tightly linked to the self-organization of matter in the Universe. The scale-free distribution is related to the emergence and maintenance of life."

Some caution with respect to power-law behavior, however, is in order. Such behavior is common but does not apply to everything. Regarding node degree, Newman (2003) points out that the degree distributions of the electric power grid and railway networks do not have a power-law form, but an exponential form. Some network degree distributions follow a power law, but only in limited regions. Note also that while a particular form may be seen in the degree distribution for the network as a whole, specific subnetworks within the network can have other forms. The World Wide Web, for instance, shows a power-law degree distribution overall, but unimodal distributions within domains.

8.2 EMPIRICAL VIEW OF REAL-WORLD NETWORKS

In Section 8.2, we take a look at empirical data from real networks, before taking a further look at the theoretical mathematics of networks in

Section 8.3. Newman (2003) says, "Recent work on the mathematics of networks has been driven largely by observations of the properties of actual networks and attempts to model them, so network data are the obvious starting point. . . . It also makes sense to examine . . . data from different kinds of networks. One of the principal thrusts of recent work in this area . . . has been the comparative study of networks from different branches of science, with emphasis on properties that are common to many of them." So, let's now look at some empirical data from networks in different disciplines.

Albert and Barabási (2002) say, "The study of most complex networks has been initiated by a desire to understand various real systems, ranging from communication networks to ecological webs." We'll pay particular attention to "three robust measures of a network's topology: average path length, clustering coefficient, and degree distribution." The corresponding empirical data are summarized in Tables 8.1 and 8.2 (adapted from Albert and Barabási), which follow.

Table 8.1 provides some published basic statistics for several real networks in various disciplines. For each network, the network size (number of nodes), the characteristic path length ℓ, and the clustering coefficient C

Table 8.1 Empirical Data for Several Real Networks

Network	Size (# Nodes)	ℓ	ℓ_{random}	C	C_{random}
World Wide Web	153,127	3.1	3.35	0.1078	0.00023
Los Alamos Lab co-authorship	52,909	5.9	4.79	0.43	1.8×10^{-4}
Medical co-authorship	1,520,251	4.6	4.91	0.066	1.1×10^{-5}
Physics co-authorship	56,627	4.0	2.12	0.726	0.003
Neuroscience co-authorship	209,293	6.0	5.01	0.76	5.5×10^{-5}
E. coli substrate graph	282	2.9	3.04	0.32	0.026
E. coli reaction graph	315	2.62	1.98	0.59	0.09
Ythan estuary food web	134	2.43	2.26	0.22	0.06
Silwood Park food web	154	3.40	3.23	0.15	0.03
C. elegans	282	2.65	2.25	0.28	0.05
Electric power grid	4941	18.7	12.4	0.08	0.005

Table 8.2 More Empirical Data for Several Real Networks

Network	Type	Size (# Nodes)	γ	γ_{out}	γ_{in}	ℓ_{real}	ℓ_{random}
World Wide Web	Directed	325,729	-	2.45	2.1	11.2	8.32
World Wide Web	Directed	2×10^8	-	2.72	2.1	16	8.85
Internet	Undirected	3888	2.48	-	-	12.15	8.75
Internet	Undirected	150,000	2.4	-	-	11	12.8
Physics co-authorship	Undirected	56,627	1.2	-	-	4	2.12
Neuroscience co-authorship	Undirected	209,293	2.1	-	-	6	5.01
Metabolic, *E. coli*	Directed	778	-	2.0-2.4	2.0-2.4	3.2	3.32
Protein, *S. cerevisiae*	Undirected	1870	2.4	-	-	-	-
Ythan estuary food web	(Assumed) undirected	134	1.05	-	-	2.43	2.26
Silwood Park food web	(Assumed) undirected	154	1.13	-	-	3.4	3.23

Adapted, with permission, from Albert, R. and Barabási, A-L., *Reviews of Modern Physics*, Volume 74, January 2002, pp. 47–97. Copyright (2002) by the American Physical Society (http://link.aps.org/).

are provided. (See Section 8.1.2 for the definitions used here; we use the Albert and Barabási definition for clustering coefficient.) For comparison, the characteristic path length ℓ_{random} and clustering coefficient C_{random} of a random graph of the same size and average degree are also included (see Section 8.3.1 for these definitions).

In Section 8.1 of this chapter, we noted that many real-world natural and technological networks have been observed to be "small-world" networks with short path lengths and high clustering coefficient. For such real networks, we said that the characteristic path length would be close to that of a random network and that the clustering coefficient would be much higher than in a comparable random network (i.e., having the same number of nodes and edges as the real network). These effects can be clearly seen in Table 8.1.

Table 8.2 provides more empirical data for several scale-free/power-law real networks. For each network, the network type (directed or undirected), the network size (number of nodes), and the scaling exponent(s) are provided. Because undirected networks do not distinguish out and in connections, only one exponent value is indicated for each of those networks. For each of the directed networks, the out-degree scaling exponent (γ_{out}) and the in-degree scaling exponent (γ_{in}) are listed separately. Note that the distribution of outgoing edges and the distribution of incoming edges are given by

$$P_{out}(k) \propto k^{-\gamma_{out}} \quad \text{and} \quad P_{in}(k) \propto k^{-\gamma_{in}}$$

The columns ℓ_{real} and ℓ_{random} compare the characteristic path lengths of the real networks with the predictions of random-graph theory (see Sections 8.1.2 and 8.3.1 for definitions).

In Section 8.1 of this chapter, we made the point that small-world networks are not necessarily scale-free, but it seems that real networks that are scale-free are also small-world. Tables 8.1 and 8.2, taken together, support that conjecture. Consider the four networks that are common to the two tables: the physics coauthorship network, the neuroscience coauthorship network, the Ythan estuary food web, and the Silwood Park food web. They are all scale-free and also small-world. The electric power grid network from Table 8.1, on the other hand, is small-world but is known not to be scale-free.

Here's some additional explanation of the real networks of Tables 8.1 and 8.2. The World Wide Web is a logical network of web pages linked together by hyperlinks from one page to another. The web should not be confused with the Internet—a physical network of computers and routers linked together by optical fiber and other data connections (wired and wireless). Many biological and ecological systems are complex networks. The *E. coli* metabolic network statistics in Table 8.2 come from a paper by Jeong et al. (2000). Albert and Barabási (2002) relate that Jeong et al. studied the metabolism of 43 organisms representing all domains of life, reconstructing them in networks in which the nodes are the substrates and the edges represent the predominantly directed chemical reactions in which these substrates can participate. The distributions of the outgoing and incoming edges have been found to follow power laws for all organisms, with the degree exponents varying between 2.0 and 2.4. The average path length was found to be approximately the same in all organisms, with a value near 3.3.

The food web network statistics in Tables 8.1 and 8.2 come from a paper by Montoya and Solé (2000), but of course many authors have studied food webs. Albert and Barabási (2002) explain that food webs are used regularly by ecologists to quantify the interaction between various species. In a food web, the nodes are species and the edges represent predator-prey relationships between them. In a recent study, Williams et al. (2000) investigated the topology of the seven most documented and largest food webs, namely, those of Skipwith Pond, Little Rock Lake, Bridge Brook Lake, Chesapeake Bay, Ythan Estuary, Coachella Valley, and St. Martin Island. While these webs differ widely in the number of species and their average degree, they all indicate that species in these habitats are approximately three or fewer edges from each other. This result was supported by the independent investigations of Camacho et al. (2001) and Montoya and Solé. Their investigations showed that food webs are highly clustered as well. The degree distribution was first addressed by Montoya and Solé, focusing on the food webs of Ythan Estuary, Silwood Park, and Little Rock Lake, considering these networks as being nondirected. Although the size of these webs is small (the largest of them has 186 nodes), they appear to share the properties of their larger counterparts. In particular, Montoya and Solé concluded that the degree distribution is consistent with a power law with an unusually small exponent of $\gamma \cong 1.1$. The small size of these webs does leave room, however, for some ambiguity in degree distribution. While the well-documented existence of key species that play an important role in food webs points toward the existence of hubs (a common feature of scale-free networks), an unambiguous determination of degree distribution is more likely with larger data sets.

In my view, there may be an issue with the available empirical data for food webs. Network representations of food webs are often highly aggregated. Individuals are aggregated into species. Similar species may be further aggregated, e.g., into "trophic species" containing all taxa that share the same set of predators and prey (according to Strogatz (2001)). The aggregation process may include oversimplifications that hide important aspects of food web interactions.

8.3 THEORETICAL VIEW OF MODEL NETWORKS

In the previous section we examined empirical data from networks in various disciplines, with an emphasis on properties that are common to many of them. In this section, we consider mathematical developments relevant to those networks and properties.

Earlier in this chapter (Section 8.1.1), it was noted that random graphs are important because they can yield useful benchmarks and insights applicable to networks of many types. Furthermore, they are the simplest and most straightforward realization of a complex network. As a result, the random graph has been well studied by mathematicians, and many results, both approximate and exact, have been proved rigorously. In fact, much of the mathematical analysis on the structure of complex networks has been done for random graphs.

We will review the available and pertinent theoretical work here. The mathematics assumes undirected networks. Section 8.3.1 covers Poisson random graphs. Section 8.3.2 addresses generalized random graphs, with a focus on graphs with scale-free degree distribution. Finally, Section 8.3.3 examines theoretical aspects of the small-world network model of Watts and Strogatz.

Some comments from several investigators follow. Solé and Bascompte (2006): "The random graph is a null model that can be used as a first approximation to an arbitrary complex network." Albert and Barabási (2002): "A particularly rich source of ideas has been the study of random graphs. Networks with a complex topology and unknown organizing principles often appear random; thus random-graph theory is regularly used in the study of complex networks." Newman (2003): "Much of our basic intuition about the way networks behave comes from the study of the random graph. In particular, the presence of the phase transition and the existence of a giant component [more on this later, mostly in Chapter 9] are ideas that underlie much of the work [describing real-world networks]." Strogatz (2001): Using "graph theory to explore the structure of complex networks [is] an approach that has recently led to some encouraging progress, especially when combined with the tools of statistical mechanics and computer simulations."

8.3.1 Poisson Random Graphs

Paraphrasing Newman (2003): The first serious attempt (around 1950) at constructing a model for large, random networks was the "random net" of Rapoport and collaborators. The model was independently rediscovered a decade later by Erdös and Rényi. They studied it exhaustively and rigorously and gave it the name "random graph," by which it is most often known today.

8.3.1.1 Poisson Degree Distribution

Consider an undirected network with n vertices. The maximum possible number of edges m in the network is

$$m_{\max} = \frac{n(n-1)}{2}$$

Connect each pair of vertices (or not) with probability p (or $1-p$). The resulting random graph has $m < m_{\max}$ edges for $p < 1$.

Newman (2003) says that many properties of this random graph are exactly solvable in the limit of large graph size. Typically, the limit of large n is taken holding the mean degree constant, i.e.,

$$\langle k \rangle = p(n-1)$$

In that case, the model clearly has a Poisson degree distribution, because the presence or absence of edges is independent. The probability, therefore, of a vertex having degree k is

$$P(k) = \binom{n}{k} p^k (1-p)^{n-k} \cong \frac{\langle k \rangle^k \exp\{-\langle k \rangle\}}{k!}$$

with the approximate equality on the right becoming exact in the limit of large n and fixed $\langle k \rangle$. (In the limit, the binomial distribution on the left becomes the Poisson distribution on the right.)

The Poisson degree distribution has mean $\langle k \rangle$ and standard deviation $\sqrt{\langle k \rangle}$. The distribution, therefore, has a narrow peak and most nodes have the same degree—the same scale (*not* scale-free).

Albert and Barabási (2002) comment, "Despite the fact that the position of the edges is random, a typical random graph is rather homogeneous, the majority of the nodes having the same number of edges."

8.3.1.2 Other Random Graph Properties

Let's look at path length and clustering coefficient of random graphs. (The equations in the following are from Albert and Barabási (2002), with some modifications for consistency.)

Earlier, in Section 8.1, we defined *shortest path length* as the smallest number of links traversed when traveling between two network nodes; *network diameter* as the largest of all the shortest path lengths in the network; and *characteristic path length* as the average of all the shortest path lengths in the network. Random graphs tend to have small diameters and small characteristic path lengths. To a first approximation, the characteristic path length of a random graph is

$$\ell_{\text{random}} \cong \frac{\ln(n)}{\ln(\langle k \rangle)}$$

where n is the number of nodes and $\langle k \rangle$ is the mean degree. This is the expression for ℓ_{random} used in Tables 8.1 and 8.2 in Section 8.2.

Unlike real-world networks, random graphs have low clustering coefficients. Consider a node in a random graph and its nearest neighbors. The probability that two of these neighbors are connected is equal to the probability (p) that two randomly selected nodes are connected. Consequently, the clustering coefficient of a random graph is

$$C_{\text{random}} = p = \frac{\langle k \rangle}{n-1}$$

Because mean degree is held constant, C_{random} tends to zero in the limit of large system size. The previous expression for C_{random} is the one used for Table 8.1 in Section 8.2.

8.3.2 Generalized Random Graphs and Scale-Free Networks

In this section, we briefly address generalized random graphs, with a focus on graphs with scale-free degree distribution. Generalized random graphs can be and have been used for theoretical exploration of real complex networks (although we do not cover the specifics here).

Newman (2003) says, "The random graph, while illuminating, is inadequate to describe some important properties of real-world networks, and so has been extended in a variety of ways. In particular, the random graph's Poisson degree distribution is quite unlike the highly skewed distributions of... [real networks]. Extensions of the model to allow for other degree distributions lead to the class of models known as 'generalized random graphs.'"

From Albert and Barabási (2002): The degree distribution of real networks often follows a power law—it is scale-free. Because random graphs do not capture the scale-free character of real networks, we need a different model to describe real systems. "One approach is to generalize random graphs by constructing a model that has the degree distribution as an input but is random in all other respects. In other words, the edges connect randomly selected nodes, with the constraint that the degree distribution is restricted to a power law."

The remainder of this section follows Newman (2003): "Random graphs can be extended in a variety of ways to make them more realistic. The property of real graphs that is simplest to incorporate is the property of non-Poisson degree distributions, which leads us to the so-called configuration model." Let's take a look at the configuration model. A degree distribution $P(k)$ is specified such that $P(k)$ is the fraction of vertices in the network having degree k. A *degree sequence* is chosen. This is a set of n values of the degrees k_i of vertices $i = 1, \ldots, n$. We can think of this as giving each vertex i in the graph k_i "stubs" or "spokes" sticking out of it, which are the ends of edges-to-be. Then, pairs of stubs are chosen at random from the network and connected together. "It is straightforward to demonstrate that this process generates graphs with the given degree sequence." The *configuration model* is defined as the ensemble of graphs so produced.

Newman derives properties of the configuration model network. The mean degree of the network is, as before,

$$\langle k \rangle = \sum_k kP(k)$$

The phase transition—at which a giant component first appears—occurs at

$$\sum_k k(k-2)P(k) = 0$$

An expression for the clustering coefficient of the configuration model is

$$C = \frac{\langle k \rangle}{n} \left[\frac{\langle k^2 \rangle - \langle k \rangle}{\langle k \rangle^2} \right]^2$$

which is essentially the value $C = \langle k \rangle / (n-1)$ for the Poisson random graph times an extra factor. The clustering coefficient of a generalized random graph, therefore, also tends to zero in the limit of large system size. It is much smaller than the clustering coefficient of a real network.

Newman concludes, "The generalized random graph models ... effectively address one of the principal shortcomings of early network models such as the Poisson random graph: their unrealistic degree distribution. However, they have a serious shortcoming in that they fail to capture the common phenomenon of transitivity [clustering]."

8.3.3 The Small-World Network Model

The small-world model discussed here yields networks that have short characteristic path lengths and high clustering coefficients, but are not scale-free. These networks, therefore, exhibit only a limited subset (path length and clustering coefficient) of the characteristics of real networks. We'll look at a development of the small-world model from Newman (2003) and then a few comments from Albert and Barabási (2002). These three authors draw on the work of Watts and Strogatz (1998), the originators of the small-world model.

Newman's development: The small-world model starts with a network built on a low-dimensional regular lattice and then "rewires" a small fraction of the edges to create a low density of "shortcuts" that join remote parts of the lattice to one another. The process is illustrated in Figure 8.5. The one-dimensional lattice on the left has connections between all vertex pairs separated by x or fewer lattice spacings, with $x=3$ in this case. As shown on the right in Figure 8.5, the network is then rewired by choosing at random a fraction p of the edges and moving one end of each edge to a new location, also chosen uniformly at random. The rewiring process allows the small-world model to interpolate between a regular lattice (which has high clustering) and something that is similar to a random graph (which has low average path length).

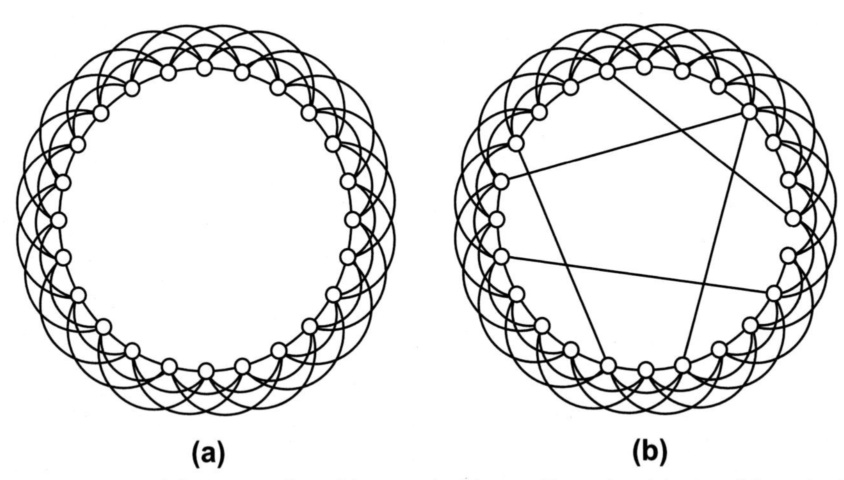

Figure 8.5 Modeling a small-world network, (a) one-dimensional lattice, (b) rewired lattice. *(Reprinted from Newman, M. E. J., The Structure and Function of Complex Networks, SIAM Review, Volume 45, Number 2, May 2003, pp. 167–256. Copyright ©2003 Society for Industrial and Applied Mathematics. Reprinted with permission. All rights reserved).*

Watts and Strogatz (1998) showed, by numerical simulation, that there exists a sizable region for which the model has both low path lengths and high clustering. Newman (2003) generated Figure 8.6.

Figure 8.6 shows the mean vertex-vertex distance ℓ and the clustering coefficient C in the small-world model of Watts and Strogatz as a function of the rewiring probability p. For convenience, both C and ℓ are divided by their maximum values, which they assume when $p=0$. Between the extremes $p=0$ and $p=1$, there is a region in which clustering is high and mean vertex-vertex distance (path length) is simultaneously low.

Newman says that the primary use of the small-world model has been as a substrate for the investigation of various processes taking place on graphs such as percolation, diffusion processes, and epidemic processes. We'll discuss these processes when we address the dynamics of complex networks in Chapter 9.

Here are comments from Albert and Barabási (2002) on the small-world model:

Path length: In the small-world model, every shortcut, created at random, is likely to connect widely separated parts of the graph, and thus has a

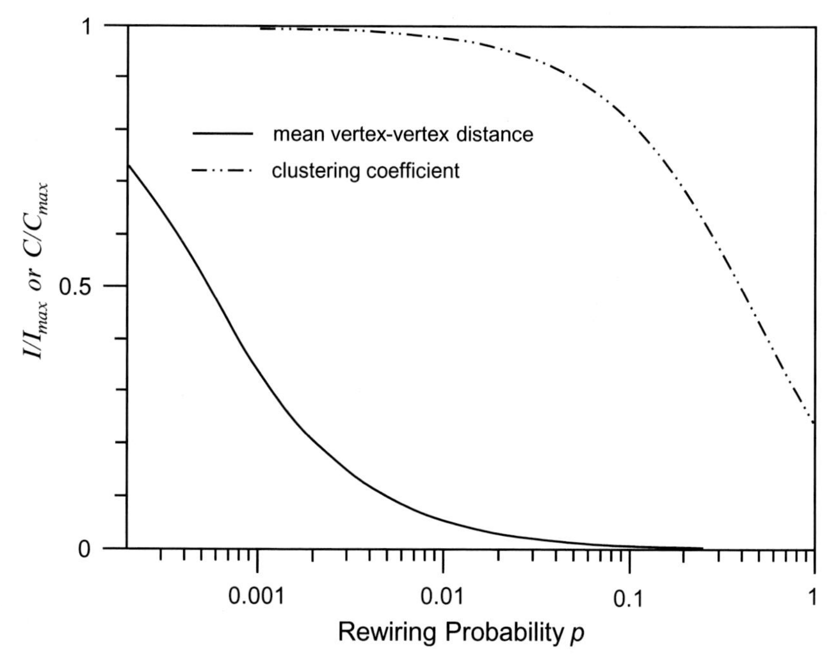

Figure 8.6 Small-world model—path length and clustering coefficient. *(Reprinted from Newman, M. E. J., The Structure and Function of Complex Networks, SIAM Review, Volume 45, Number 2, May 2003, pp. 167–256. Copyright ©2003 Society for Industrial and Applied Mathematics. Reprinted with permission. All rights reserved).*

significant impact on the characteristic path length of the entire graph. Even a relatively low fraction of shortcuts is sufficient to drastically decrease the average path length, yet locally the network remains highly ordered.

Degree distribution: In the small-world model, the shape of the degree distribution is similar to that of a random graph. The topology of the network is relatively homogeneous, with all nodes having approximately the same number of edges. The small-world model network is *not* scale-free.

Albert and Barabási (2002) also offer a summary of work done on the *structure* of complex networks: "We have learned through empirical studies, models, and analytic approaches that real networks are far from being random, but display generic organizing principles shared by rather different systems."

But there is much more work to do. The next chapter addresses the crucially important topic of the *dynamics* of complex networks.

CHAPTER 9

Network Theory: The Dynamics of Complex Networks

The previous chapter addressed network structure. This chapter focuses on extant network dynamics theory. It explores the complex dynamics of structurally complex networks. The overall goal is a more complete understanding of real-world networks, and ecological networks in particular. First, the chapter addresses the concepts of scaling, power laws, and fractals, which are prominent in both structural and dynamical behavior of complex networks. After discussing the relationships among these three concepts, some introductory comments on spatial and temporal fractals are provided. (Details on fractals are provided in later chapters.) Next, the dynamics of network phase transitions are covered. The transition from an unconnected to a connected network, the phenomenon of network percolation, and other important aspects of network phase transitions are described and discussed. Network dynamics are then explored via examples of processes that take place on networks. These include epidemic processes, network failure/attack processes, macroevolution processes, and network growth processes.

9.1 INTRODUCTION

The scientific community (and, more generally, Western culture) has a tendency to focus on things, on structure. Mark Newman (2003) says that the scientific community has made an excellent start on characterizing the structure of networked systems. Studies on network dynamics, on the other hand, "are still in their infancy." ... "Progress on this front has been slower than progress on understanding network structure."

There is much important work to be done in the area of dynamics of complex networks. David Bohm (1983) says that to fully understand structure, we must understand the dynamics, that is, "the process in which it forms, maintains itself, and ultimately dissolves." Newman (2003) says that perhaps the most important direction for future network study is "the behavior of processes taking place on networks." ... "This, in a sense, is our ultimate goal in this field: to understand the behavior and function of

Understanding Complex Ecosystem Dynamics
http://dx.doi.org/10.1016/B978-0-12-802031-9.00009-7

the networked systems we see around us. If we can gain such understanding, it will give us new insight into a vast array of complex and previously poorly understood phenomena." Albert and Barabási (2002)—echoing Strogatz (2001)—say, "In general, when it comes to understanding the dynamics of networks ... we are only at the beginning of a promising journey."

We continue our own journey toward increased understanding of complex network dynamics by reviewing the extant literature and theory in this chapter.

9.2 SCALING, POWER LAWS, AND FRACTALS

9.2.1 Relationships

Scaling, power laws, and fractals are prominent in both the structure and dynamics of complex networks. We established the relationship between scale-free and power-law behavior in Chapter 8. There it was noted, for example, that "the term 'scale-free' refers to any functional form $f(x)$ that remains unchanged to within a multiplicative factor under a rescaling of the independent variable x. In effect this means power-law forms, since these are the only solutions to $f(ax) = bf(x)$, and hence 'power-law' and 'scale-free' are, for our purposes, synonymous." This argument (and the quote) is from Newman (2003).

I also noted in Chapter 8 that, in the extant network literature, the term "scale-free" usually refers to networks with a scale-invariant/power-law *node degree* distribution. Other important network parameters—many in the dynamics realm—can also have scale-invariant/power-law distributions. Henceforth in this book, I'll endeavor to use the more general term "scale-invariant" when referring to any network parameter that has a power-law distribution. The term "scale-free" will still appear at times, because it is pervasive in the extant network literature.

Solé et al. (1999) provide an argument (similar to Newman's) that establishes the further relationship between scale-invariant/power-law behavior and fractal behavior. Here's the argument: The term *fractal* is essentially defined as *self-similar on all scales*. The common feature of self-similar behavior is the presence of scaling laws (power laws). Consider the distribution $N(s)$ of some quantity s. This distribution can be viewed as the number of objects (or events) associated with a particular value of quantity s vs. s. Let's assume here that s stands for object size. The distribution $N(s)$ then plots the number of events having size s vs. s. $N(s)$ is said to follow a power law if

$$N(s) = Cs^{-\gamma}$$

Here, C is a constant and $\gamma > 0$ is a given exponent, often called the scaling exponent. The reason why these laws are characteristic of fractals is that they are the only functions displaying invariance under scale change. If we look at a larger or smaller scale, that is, if we take

$$s' = as$$

where a is a multiplicative factor, it is not difficult to see that

$$N(s') = cN(s) \quad \text{where } c \text{ is another constant}$$

or, in other words, a change of scale does not modify the basic statistical behavior. The behavior is self-similar at all scales. The behavior is fractal.

9.2.2 Fractals in Space and Time

A fractal displays self-similarity at all scales. Fractals are everywhere—in ecological systems and other complex systems. It is important to note that fractals in nature are usually so in the statistical sense; that is, their features are *statistically*, rather than *exactly*, the same when observed and measured at different scales. We will discuss fractals in detail in Chapters 12 and 15. For now, here are some general comments from two well-known sources, and then a comment on spatial and temporal fractals from me.

> *"The patterns displayed by many natural systems do not allow for a simple description using Euclidean geometry: they present scale-invariance; that is, no characteristic length measure can be obtained from them. Therefore, when observed at different resolutions, they display the same pattern. This is the case of river networks and mountains, tree branching and blood vessels or forest spatial structures. Even at the molecular level, fractals can be observed: if we analyze the linear distribution of nucleotides in a DNA chain, a self-similar pattern can also be detected."*
>
> *"Fluctuations in ecological systems are known to involve a wide range of spatial and temporal scales, often displaying self-similar (fractal) properties."*
>
> *"Fractals are widespread in nature and have features that look the same when there is a change in scale: they are called 'self-similar.' In biology, self-similar patterns are known to occur [in space] at many levels. But fractals are also present in time: the fluctuations of a given quantity can appear the same when observed at different temporal resolutions. This is the case for heartbeat intervals, epidemics in small islands, breeding bird populations or the fossil record."*
>
> **Solé et al. (1999)**

> *"A striking, widespread feature of many complex systems is that some of their properties are reproduced at different scales in such a way that we perceive the same patterns when looking at different subparts of the same system. This property, known as scale-invariance, is widespread in many systems under nonequilibrium conditions. This is the case for ecological systems, where flows of energy enter into*

the system and are dissipated at different, interconnected scales. The origin of such fractal patterns, named after the pioneering work by Benoit Mandelbrot, is a fundamental problem in many areas of science. Actually, empirical evidence has been mounting in support of the unexpected possibility that many different systems arising in disparate disciplines such as physics, biology, and economy may share some intriguingly similar scale invariant features."

Solé and Bascompte (2006)

Note that the terms *spatial fractal* and *temporal fractal* are used by many investigators and authors. In my view, this categorization is not adequate. Spatial implies physical structure. That's fine. What is called a temporal fractal, however, is typically associated with the dynamical behavior of a *process* that occurs over time *and* space. Therefore, instead of the terms spatial fractal and temporal fractal, I prefer the terms *structure fractal* and *process fractal*. These terms are more accurate and they are consistent with the systems triad discussed in Part I and Part II of this book. I will use these terms going forward.

9.3 THE DYNAMICS OF NETWORK PHASE TRANSITIONS

In this section, we consider the dynamics of network phase transitions. We cover unconnected-to-connected transitions, network percolation, and so-called "self-organized criticality," as well as some additional aspects of network phase transitions.

9.3.1 Network Transition from Unconnected to Connected

As the number of edges in a random graph increases, the graph rather abruptly transitions from unconnected to connected—from separate components to a giant component. Strogatz, Albert and Barabási, and Newman each provide their view of the phenomenon.

Strogatz (2001) describes how the topology of a random graph changes as a function of the number of edges, m. When m is small, the graph is likely to be fragmented into many small clusters of nodes, called components. As m increases, the components grow, at first by linking to isolated nodes and later by coalescing with other components. A phase transition occurs at

$$m = \frac{n}{2} \quad \text{where } m = \text{the number of edges and } n = \text{the number of nodes}$$

when many clusters cross-link spontaneously to form a single *giant component*. For $m \geq n/2$, this giant component contains on the order of n nodes (more precisely, its size scales linearly with n, as $n \rightarrow \infty$), while its closest rival contains only about $\log n$ nodes. Furthermore, all nodes in the giant component are connected to each other by short paths. The maximum number of "degrees of separation" between any two nodes grows slowly, like log n.

Albert and Barabási (2002) focus more on connection probabilities and offer the following view. Consider a network of n nodes and m edges. The node degree variable is k. If every pair of nodes is connected with probability p, then the number of edges m and node degree k are random variables with the following expected values:

$$E(m) = p\left[\frac{n(n-1)}{2}\right]$$
$$E(k) = \langle k \rangle = p(n-1) \text{ where } \langle k \rangle \text{ denotes mean degree}$$

We will use these expected value expressions in a moment. For now, you might want to note that the maximum number of edges equals $n(n-1)/2$ when p equals 1.

In the mathematical literature, the construction of a random graph is often called an evolution. Starting with a set of n isolated vertices, the graph develops by the successive addition of random edges. The graphs obtained at different stages of this process correspond to larger and larger connection probabilities p. For example, Figure 9.1 starts with $n=10$ isolated nodes and $p=0$ (upper diagram). The lower portion of the figure shows two different stages in the graph's development, corresponding to $p=0.1$ and

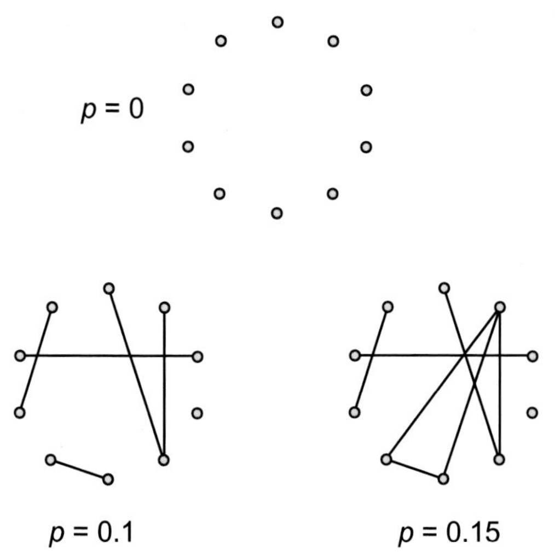

Figure 9.1 Illustration of the graph evolution process. *(Reprinted, with permission, from Albert, R. and Barabási, A-L., Reviews of Modern Physics, Volume 74, January 2002, pp. 47–97. Copyright (2002) by the American Physical Society (http://link.aps.org/)).*

$p=0.15$. Notice the emergence of trees and cycles in the graph, and a connected cluster that unites half of the nodes at $p=0.15$.

A main goal of random-graph theory is to determine at what connection probability p a particular property of a graph will most likely arise. According to Albert and Barabási, perhaps the greatest discovery of Erdös and Rényi was that many important properties of random graphs appear quite suddenly. That is, at a given probability either almost every graph has some property Q (e.g., every pair of nodes is connected by a path of consecutive edges) or, conversely, almost no graph has it. The transition from a property's being very unlikely to its being very likely is usually swift. For many such properties there is a critical probability $p_c(n)$. Physicists trained in critical phenomena will recognize in $p_c(n)$ the critical probability familiar in percolation.

Albert and Barabási (2002) show further that the critical point occurs, for large n, when the expected number of edges equals $n/2$. It follows that:

$$\begin{aligned}&\text{At the critical point:}\\&E(m)=p\left[\frac{n(n-1)}{2}\right]=\frac{n}{2}\ \text{ and }\ p(n-1)=1\\&\text{therefore}\\&E(k)=\langle k\rangle=p(n-1)=1\ \text{ and }\ \langle k\rangle=1\end{aligned}$$

When the mean degree $\langle k\rangle<1$, there are isolated clusters. When $\langle k\rangle\geq 1$, a giant component appears.

We conclude with Newman's (2003) view. The expected structure of the random graph varies with the value of p. The edges join vertices together to form components, that is, subsets of vertices that are connected by paths through the network. Perhaps the most important property of the random graph is that it achieves a phase transition—from a low-density, low-p state in which there are few edges and all components are small, to a high-density, high-p state in which almost all (on the order of n) vertices are joined together in a single *giant component*.

Figure 9.2 depicts network growth and transition behavior from an interesting perspective. Variable $\langle s\rangle$ is the mean component size, S is the fraction of the graph occupied by the giant component, and $\langle k\rangle$ is the mean degree. The mean component size (solid line) excluding the giant component if there is one and the giant component size (dotted line) are plotted for the Poisson random graph. The phase transition occurs at $\langle k\rangle=1$. This is also the point at which $\langle s\rangle$ diverges, a behavior that will be recognized by

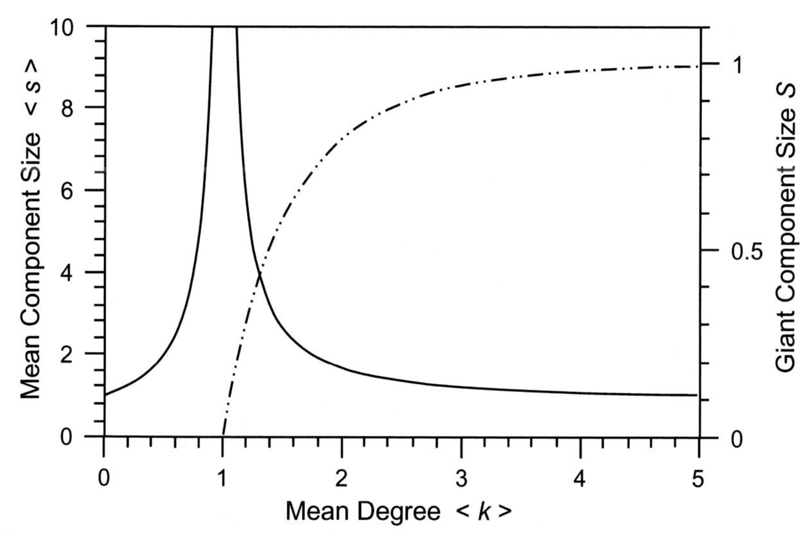

Figure 9.2 Network growth and transition behavior. *(Reprinted from Newman, M. E. J., The Structure and Function of Complex Networks, SIAM Review, Volume 45, Number 2, May 2003, pp. 167–256. Copyright © 2003 Society for Industrial and Applied Mathematics. Reprinted with permission. All rights reserved).*

those familiar with the theory of phase transitions. S plays the role of the order parameter in this transition and $\langle s \rangle$ the role of the order-parameter fluctuations.

In Figure 9.2, you can see the dramatic growth of mean component size as the transition is approached, and then the collapse as the giant component appears. Adding more edges after the transition occurs changes the topology very little.

Here's a summary of the particulars of the phase transition to a connected network. This applies to undirected random graphs with large n (say $n \geq 100$).

$$E\{m\} = p\left[\frac{n(n-1)}{2}\right] \quad \text{and} \quad E\{k\} = \langle k \rangle = p(n-1)$$

where

$n =$ the number of network nodes
$p =$ the connection probability of any pair of nodes
$m =$ the number of network edges
$k =$ the node degree $=$ the number of edges connected to a node
$E\{m\}$ is the expected value of m
$E\{k\} = \langle k \rangle$ is the expected value of k (the mean degree)

It can be shown that at the critical phase transition point, the expected number of edges equals $n/2$. Therefore, at the critical point:

$$E\{m\} = p\left[\frac{n(n-1)}{2}\right] = \frac{n}{2}$$

$$p\,(n-1) = 1$$

It follows that

$$\langle k\rangle_c = 1 \text{ and } p_c = \frac{1}{(n-1)}$$

where

$\langle k\rangle_c =$ the critical mean degree

$p_c =$ the critical probability

9.3.2 Percolation in Networks

The abrupt phase transition from an unconnected network to a connected network is central to the percolation phenomenon. The critical probability derived in the previous subsection applies to undirected random graphs. Percolation, however, was originally studied in physics and statistical mechanics rather than graph theory. Traditional percolation theory is usually based on regular lattice structures and not on random graphs. For general lattices, the transition point (the critical probability p_c) is much more difficult to determine. In a few cases, the critical probability may be calculated explicitly. For example, for the square lattice in two dimensions, $p_c = 0.5$.

Albert and Barabási (2002) look at percolation theory and consider a regular square two-dimensional lattice whose edges are present with probability p and absent with probability $1-p$. Percolation theory studies the emergence of paths that percolate through the lattice (starting at one side and ending at the opposite side). For small p, only a few edges are present and only small clusters of nodes connected by edges can form. At larger values of p—specifically at critical probability p_c, called the *percolation threshold*—a percolating cluster of nodes connected by edges appears. See Figure 9.3. For this illustration of percolation, nodes are placed on a 25 × 25 square lattice, and two nodes are connected by an edge with probability p. For $p = 0.315$ (on the left), which is below the percolation threshold of $p_c = 0.5$, the connected nodes form isolated clusters. For $p = 0.525$ (on the right), which is above the percolation threshold, the lattice network percolates.

Solé and Bascompte (2006) and Solé et al. (1999) discuss a forest fire model (a cellular-automata model). It is specifically about the spread of forest fires, but is generally illustrative of percolation phenomena in natural systems. See Figure 9.4. The model uses a 150 × 150 grid of cells in which each cell is

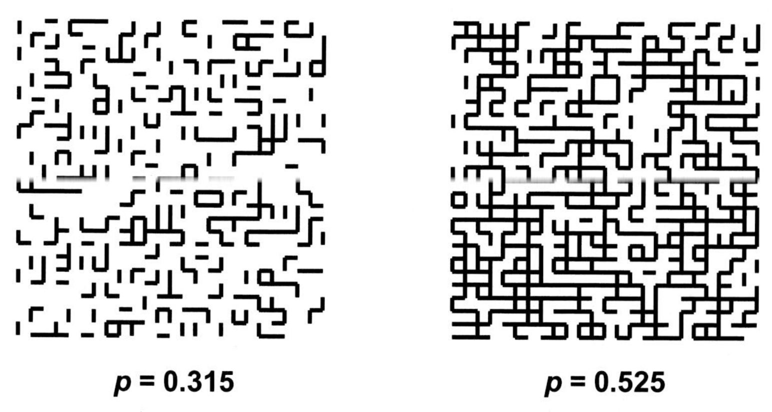

Figure 9.3 Percolation in a two-dimensional lattice. *(Reprinted, with permission, from Albert, R. and Barabási, A-L., Statistical Mechanics of Complex Networks, Reviews of Modern Physics, Volume 74, January 2002, pp. 47–97. Copyright (2002) by the American Physical Society (http://link.aps.org/)).*

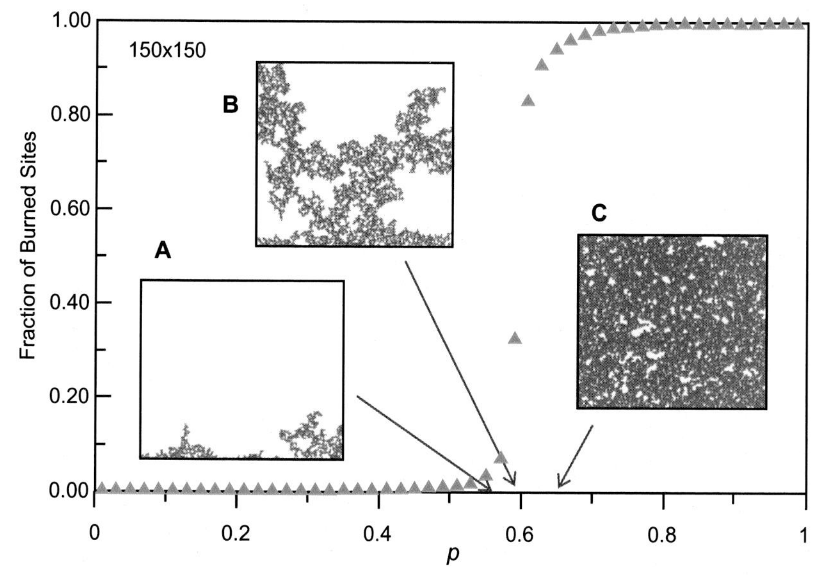

Figure 9.4 Percolation in a forest fire model. *(Reprinted by permission of Princeton University Press. Ricard V. Solé and Jordi Bascompte. Self-Organization in Complex Ecosystems. © 2006 Princeton University Press).*

occupied (by a tree) with probability p. A simple burning algorithm is applied. Initially a small number of trees starts burning; then any tree with a burning neighbor also starts to burn (the neighbors are the eight nearest trees), and the fire propagates. Percolation occurs at a critical point p_c. For $p < p_c$, few trees are burning. For $p > p_c$, almost all trees are burning. The progression of the forest fire is visually depicted by insets A, B, and C in Figure 9.4.

We can summarize some basic aspects of percolation. Percolation is sudden—a threshold transition phenomenon—separating two well-defined phases. In the subcritical phase ($p < p_c$), there are short-range (local) interactions. In the supercritical phase ($p > p_c$), there are long-range (global) interactions. Processes with global correlations arise from local interactions.

Related observations (quotes are from Solé and Bascompte, 2006):

Percolation is not persistent (and is reversible). "For a given ecosystem, where high diversity needs a minimal spatial habitat, random habitat loss can lead to percolation [in reverse], which means breaking available space into many patches and triggering ecosystem collapse."

Percolation occurs "at a so-called critical point. Critical points are associated with the presence of a narrow transition domain separating two well-defined phases, and we generally speak of critical phenomena to refer to such transitions."

Newman (2003) provides a formulation of *bond percolation*. Consider a set of existing nodes with potential links (or edges or *bonds*) that are initially inactive. As the network formation process proceeds, edges or bonds become active over time until the critical threshold is reached and percolation occurs. Figure 9.5 illustrates the situation. The figure is a pre-percolation-threshold snapshot. The active edges are black and the inactive edges are gray. The nodes that are connected together by active edges form the clusters of interest. (There are two clusters in Figure 9.5.)

This bond percolation formulation is clearly consistent with the local-to-global interactions of real-world ecological networks that I mentioned in Chapters 6 and 8, and will revisit when I synthesize a view of ecological network dynamics in Chapter 15. In terms of percolation dynamics, an

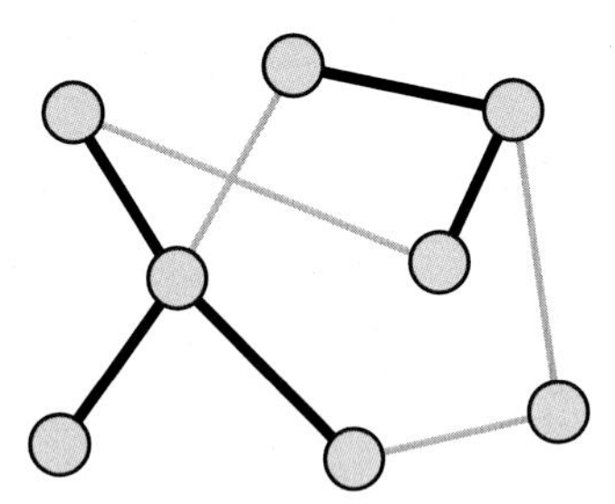

Figure 9.5 Bond percolation on a network. *(Reprinted from Newman, M. E. J., The Structure and Function of Complex Networks, SIAM Review, Volume 45, Number 2, May 2003, pp. 167–256. Copyright © 2003 Society for Industrial and Applied Mathematics. Reprinted with permission. All rights reserved).*

ecosystem scenario can be described as follows: At some point in time, a set of ecological network nodes is in place with network links (bonds) that are initially inactive. A stimulus (e.g., energy or biomass) arrives from the environment. A few bonds become active in an attempt to process the stimulus locally. Then, if necessary, more and more bonds become active to process globally in order to dissipate the stimulus. *The network percolates*. The bonds eventually become inactive once again—until the next stimulus arrives.

9.3.3 The Self-Organized Criticality Hypothesis

Percolation and other network phase transitions (from unconnected to connected) might be considered "self-organized criticality" occurrences, which are somehow associated with fractal behavior.

According to the literature, Per Bak et al. (1987) were the first to use the term *self-organized criticality* (SOC) and the first to propose that spatial and temporal fractals are identified with the dynamics of this critical state. In their own words, dynamical systems "naturally evolve into a self-organized critical point." The critical point "is an attractor reached by starting far from equilibrium." ... "The 'physics of fractals' could be that they are the minimally stable states originating from dynamical processes which stop precisely at the critical point." ... "We believe that the new concept of self-organized criticality ... might be *the* underlying concept for temporal and spatial scaling in a wide class of dissipative systems."

Here are some additional relevant statements from other investigators:

> *"The basic idea in Bak's theory is that some far-from-equilibrium systems formed by many interacting units can spontaneously drive themselves into the critical point, thus spontaneously leading to spatial and temporal fluctuations of all sizes." ... "The presence of fractal patterns reveals the nonequilibrium conditions under which structures at different levels are created and how large-scale patterns are generated from local interactions."*
>
> **Solé and Bascompte (2006)**

> *"In physics, fractal structures in space and time were known to emerge in the proximity of some types of phase transition." ... "Self-similarity is a defining characteristic of the critical state. The surprise came when physicists realized that very different systems behaved exactly the same when close to critical transition points. Surprisingly, extremely simple models of these systems provided an exact description. This is a consequence of the so-called 'universality.' Universality [in this context] means that systems sharing a small number of basic features behave identically at the critical point."*
>
> **Solé et al. (1999)**

"SOC is remarkable because of its 'holistic' character, and because it emerges in such a wide range of physical systems."

Halley (1996)

I view this self-organized criticality hypothesis with some skepticism. I certainly accept the notion of self-organization and the notion of criticality, and I also accept the fact that some form of fractal behavior can be observed at critical points. Saying (or at least implying) that self-organized criticality is the *source* of fractal behavior, however, seems inappropriate. In my view, self-organized criticality is not a persistent phenomenon; it comes and goes. As a complex system traverses its dynamics trajectory, it reaches the critical point only when required by the real environment and the current status of the system. In my research, I have found that many dynamical processes on networks clearly exhibit *process fractal* behavior, with the critical point being one of many possible outcomes encountered during the process. So the question then becomes, Is fractal behavior a characteristic of self-organized criticality or is self-organized criticality a characteristic of fractal behavior? I think it is the latter. The results of my work support that view. (See also Chapter 14, Section 14.3.)

9.3.4 Other Topological Phase Transitions in Networks

The network phase transitions we have talked about thus far are extremely important. They are a key part of the operational dynamics of well-functioning complex network systems. Additional topological phase transitions, however, can also occur, due to a lack of resources or other environmental stresses. These phase transitions are addressed in this subsection. We start with a theoretical perspective (the source is Derényi et al., 2004) and then move to a qualitative interpretation (the source is Csermely, 2006).

According to Derényi et al., to provide a phenomenological theory for various interesting network topological phase transitions, a statistical mechanical approach should be employed. By assigning energies to the different network topologies and defining appropriate order parameters, topological phase transitions can be identified as singular changes in network energy and (thus) network connectivity.

Derényi et al. use a thermodynamic formalism and a lattice base structure to describe the network changes, and focus on rearrangements of links between nodes. The authors begin with a given number of interacting nodes in a given environment. The probability of establishing a new connection or ceasing an existing connection between two nodes depends on two parameters: the fluctuation level of the system (which the authors call "noise") and the level of the "energy" (desirability) associated with the connection configuration. In this analysis, temperature is used to represent "noise." Network connectivity transitions occur as a function of temperature and are determined both analytically and by numerical simulations. In the series of transitions, scale-free networks (i.e., networks with a scale-invariant node degree distribution) are observed between random-like networks and star networks dominated by a large hub(s).

Derényi et al. describe their procedure and results: "The basic event of the rearrangement is the relocation of a randomly selected edge (link) to a new position either by 'diffusion' (keeping one end of the edge fixed and connecting the other one with a new node) or by removing the given edge and connecting two randomly selected nodes." Then, the energy difference between the original and new configurations is calculated and the relocation is carried out following a defined algorithm based on energy difference and temperature. "Our Monte Carlo simulations demonstrate that as we cool down the system, the dispersed random graph first assembles to a configuration with a few large stars (sharing most of their neighbors), and then at lower temperatures it reorganizes into [separate subgraphs]." . . . "A remarkable feature of the Monte Carlo dynamics is that . . . a scale-free graph (with a degree distribution $P(k) \approx k^{-\gamma}$ and $\gamma \cong 3$) appears at some point of the evolution of the graph from the random configuration towards the star."

Csermely (2006) provides a useful qualitative view of the Derényi et al. (2004) results via a chart similar to Figure 9.6. The abscissa variables used in this figure are proxies for the Derényi et al. "network temperature." (Network temperature would vary from high on the left to low on the right.)

Let's focus first on the *resources* and *stress* (abscissa) and *complexity* (ordinate) variables. Figure 9.6 indicates that ample resources and low stress result in a moderate complexity random network. More limited resources and increased stress yield the highest complexity configuration—a scale-free network. Scarce resources and high stress yield a complex star network. This is a "winner takes all" configuration with one or very few mega-hubs dominating. Very scarce resources and very high stress result in the lowest complexity case considered—disconnected subnetworks. The giant component disappears and the network disintegrates. In an ecological context, the overall chart essentially shows the ecological impact of decreasing availability of resources and increasing environmental stresses.

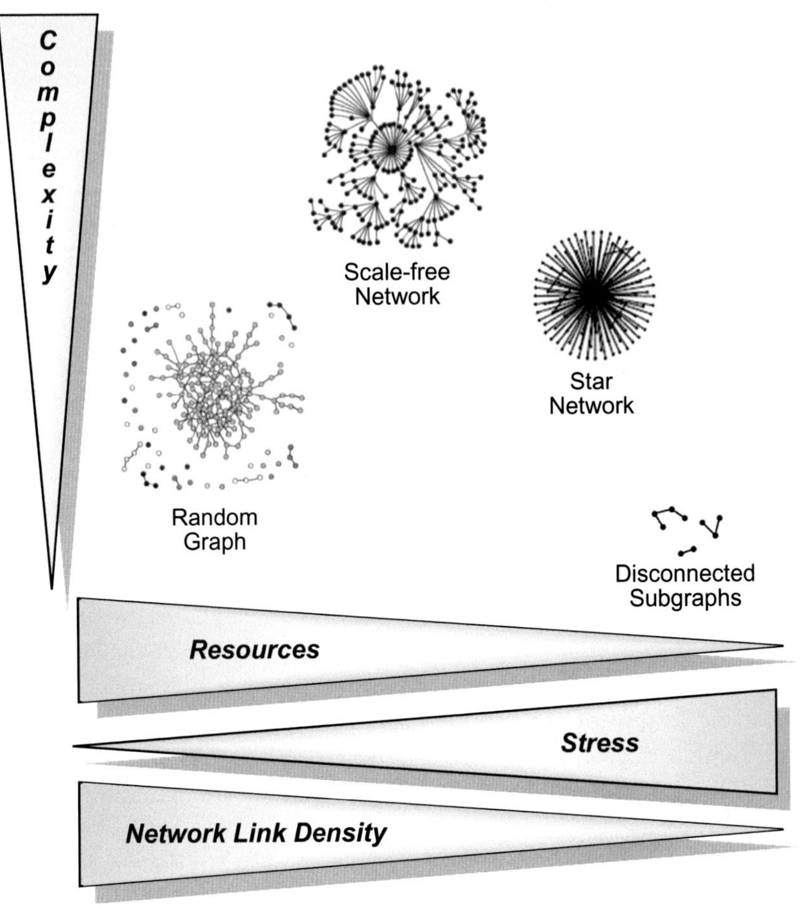

Figure 9.6 Topological phase transitions of networks. *(Adapted from Csermely, P., Weak Links, Springer, Berlin, Germany, 2006, p. 75. © Springer-Verlag 2006).*

Now let's address the *network link density* aspect of Figure 9.6. I define this variable as:

$$\text{Network link density} = \frac{L}{L_{\max}}$$

where

L = number of actual links

$L_{\max}$ = maximum possible number of links

In the figure, note that high resources (and temperature and energy) correspond to high network link density, while lower resources/temperature/energy correspond to lower density. Note also that a "cost" variable (not shown on the chart) decreases from left to right with link density. There is a "parsimony principle" at work here. As the network attempts to survive, less and less "wiring" cost can be supported as resources decrease and stress increases.

Csermely notes that these "topological phase transitions of networks seem to be a fairly general phenomenon."

(Using this topological phase transition context, perhaps one could do an analysis of habitat loss → food web fragmentation → biodiversity loss.)

9.4 NETWORK DYNAMICS: PROCESSES TAKING PLACE ON NETWORKS

In this section, several dynamic processes that can take place on complex networks are discussed. These are specific processes that have been addressed in the scientific literature. They are epidemic processes on networks, network failure processes, macroevolution processes, and so-called preferential attachment processes.

Recall that the ecological network dynamics framework described in Chapter 5 includes operational and developmental tiers and core processes that implement self-organization, regulation/adaptation, and propagation. The core processes are fundamental and are involved in the other processes taking place on complex networks. In this context, *epidemic processes* are (often deleterious) operational processes, having to do with the propagation of some "currency"—in this case, disease. (Trophic (food web) dynamics also involve operational propagation processes. The currency is biomass.) *Network failure processes* are adaptive developmental processes, which impact operational processes. (Developmental processes and operational processes almost always impact each other.) *Macroevolution processes* are developmental processes that incorporate adaptation and propagation. In macroevolution time frames, adaptation can be viewed as the system response to longer term system perturbations, and it involves developmental changes to the network of species. Propagation transports the impacts of the changes from node-to-node. Self-organization processes underlie all of these complex system processes.

9.4.1 Epidemic Processes on Networks

One important reason for studying system networks is to understand the processes by which matter and energy are propagated over the networks. Here we consider the spread of diseases per Newman (2003).

"The simplest model of the spread of a disease over a network is the SIR model of epidemic disease. This model . . . divides the population into three classes: susceptible (S), meaning they don't have the disease of interest but can catch it if exposed to someone who does, infective (I) meaning they have the disease and can pass it on, and recovered (R), meaning they have recovered from the disease and have permanent immunity, so that they can never get it again or pass it on. (Some authors consider the R to stand for 'removed,' a general term that encompasses also the possibility that people may die of the disease and remove themselves from the infective pool in that fashion. Others consider the R to mean 'refractory,' which is the common term among those who study the closely related area of reaction diffusion processes.)" The SIR model applies to epidemic diseases such as influenza, which sweeps through the population rapidly and infects a significant fraction of individuals in a short outbreak.

The basic SIR model can be generalized in a straightforward manner to become a network process model—to model an epidemic taking place on a network. "The important observation that allows us to make progress . . . is that the model can be mapped exactly onto bond percolation on the same network." (This last quote suggests, more generally, that energy and biomass propagation on ecological networks might be modeled with a bond percolation scenario.)

"The extraction of predictions about epidemics from the percolation model is simple: the distribution of percolation clusters (i.e., components connected by occupied edges) corresponds to the distribution of the sizes of disease outbreaks that start with a randomly chosen initial carrier; the percolation transition corresponds to the 'epidemic threshold' of epidemiology, above which an epidemic outbreak is possible; and the size of the giant component above this transition corresponds to the size of the epidemic." A solution of the bond percolation model—with uniform edge occupation probability T (which would also be the disease transmission probability)—is given by Callaway et al. (2000).

> *"One of the most important conclusions of this work is for the case of networks with power-law degree distributions, for which . . . there is no nonzero epidemic threshold so long as the exponent of the power law is less than 3. Since most power-law networks satisfy this condition, we expect diseases always to propagate in these networks, regardless of transmission probability between individuals. . . ."*
>
> ***Newman (2003)***

9.4.2 Network Failure Processes

As we have already established (see Chapter 8), many real-world network systems have scale-invariant/power-law node degree distributions (they are "scale-free" networks). Strogatz (2001) notes (summarizing Albert et al., 2000) that "scale-free networks are resistant to random failures because a few hubs dominate their topology. Any node that fails probably has small degree (like most nodes) and so is expendable. The flip side is that such networks are vulnerable to . . . [failures of] the hubs."

Newman (2003) provides interesting results regarding network vulnerability to node removal (failure). Here's the setup: The network that is considered has the form of a generalized random graph with a scale-invariant degree distribution and a scaling exponent of γ, that is,

$$P(k) \propto k^{-\gamma}$$
$$\text{where} \quad k = \text{node degree}$$

Nodes are removed from the network in decreasing order of their degrees. (The highest degree nodes—the hubs—are removed first.) The results are depicted in Figure 9.7. The chart shows the fraction of nodes that must be removed from the network to destroy the giant component. Newman explains that "the maximum fraction is less than three percent, and for most values of γ the fraction is significantly less than this. This appears to imply that networks . . . that have power-law degree distributions are highly susceptible to such [hub failures or] attacks."

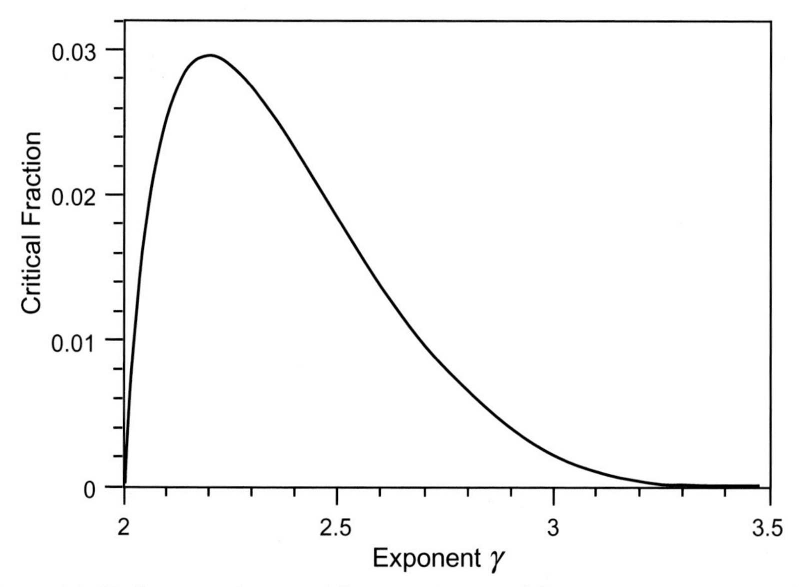

Figure 9.7 Node removal—critical fraction. *(Reprinted from Newman, M. E. J., The Structure and Function of Complex Networks, SIAM Review, Volume 45, Number 2, May 2003, pp. 167–256. Copyright © 2003 Society for Industrial and Applied Mathematics. Reprinted with permission. All rights reserved).*

Comment: The removal of nodes (and therefore edges) can have a very significant impact on the processes and functions of the network. The subject of cascading failures begins to get at this. The failure of one node can prevent another node(s) from performing its function and causing it to fail, which can prevent another node(s) from performing its function and causing it to fail, and so on. Important related subjects are (a) species extinction and secondary extinction and (b) keystone species.

Albert and Barabási (2002) provide a detailed analysis of their view of network failure/attack processes and effects. Some of their analysis and discussion is covered in the next several paragraphs.

"Many complex systems display a surprising degree of tolerance for errors." The error tolerance of some lower level organisms can be attributed to the robustness of their underlying metabolic and/or genetic networks. "Complex communication networks display a high degree of robustness: while key components regularly malfunction, local failures rarely lead to loss of the global information-carrying ability of the network." Network node-and-edge topology plays an important role in network robustness. The scale-free node degree topology plays an important role.

Consider network edge removal. Note that the unconnected-to-connected network phase transition discussion from Section **9.3.1** applies here (in reverse). Albert and Barabási say, "The first results regarding network reliability when subjected to edge removal came from random-graph theory. Consider an arbitrary connected graph H_n of n nodes, and assume that a fraction p of the edges have been removed. What is the probability that the resulting subgraph is connected, and how does it depend on the removal probability p? For a broad class of starting graphs H_n there exists a threshold probability $p_c(n)$ such that if $p < p_c(n)$ the subgraph is connected, but if $p > p_c(n)$ it is disconnected."

Consider network node removal. "As the removal of a node implies the malfunctioning of all of its edges as well, node removal inflicts more damage than edge removal." . . . "We shall call a network error tolerant (or robust) if it contains a giant cluster comprised of most of the nodes even after a fraction of its nodes are removed." Results indicate that "scale-free networks are more robust than random networks against random node failures, but are more vulnerable when the most connected nodes are targeted" in intentional attacks.

Albert and Barabási discuss their numerical simulation efforts. They consider random graphs and scale-free networks that have the same number of nodes and edges. Two types of node removal are addressed: (1) removal of randomly selected nodes and (2) removal of the most highly connected nodes. The "numerical simulations indicate that scale-free networks display a topological robustness against random node failures. The origin of this error tolerance lies in their heterogeneous topology: low-degree nodes are far more abundant than nodes with high degree, so random node selection will more likely affect the nodes that play a marginal role in the overall network topology. But the same heterogeneity makes scale-free networks fragile to intentional attacks, since the removal of the highly connected nodes has a dramatic disruptive effect on the network." The results present "a simple but compelling picture: scale-free networks display a high degree of robustness against random errors, coupled with a susceptibility to attacks. This double feature is the result of the heterogeneity of the network topology, encoded by the power-law degree distribution."

Melanie Mitchell (2009) makes some interesting observations about brain networks. Brain networks are scale-free, small-world networks. They are resilient with respect to the most likely failures, that is, random node failures. Brain networks may also represent an optimal compromise between two modes of brain behavior: efficient local processing and efficient global processing.

Finally, here are a few words about the robustness of real ecological networks. As a result of human actions or environmental changes, species can be removed from food webs. Solé and Montoya (2001) studied the response of food webs to the removal of species (nodes). Albert and Barabási (2002) comment on the work of Solé and Montoya: "The authors measured the relative size S of the largest cluster ... and the fraction of species becoming isolated due to the removal of other species on whom their survival depended (secondary extinctions). The results indicate that random species removal causes the fraction of species contained in the largest cluster to decrease linearly. At the same time ... the secondary extinction rates remain very low (smaller than 0.1) even when a high fraction of the nodes is removed. The estimate of ... the critical fraction [f_c] at which the network fragments gives f_c^{fail} values around 0.95 for all networks, indicating that these networks are error tolerant. However, when the most connected (keystone) species are successively removed, S decays quickly and becomes zero at $f_c^{\text{attack}} \approx 0.2$ .... The secondary extinctions [secondary extinction rates] increase dramatically, reaching 1 at relatively low values of f ($f \approx 0.16$ for the Silwood Park web)." This is consistent with expected scale-free network behavior.

9.4.3 Macroevolution Processes

We will discuss two issues in this subsection: macroevolution in a network context and macroevolution modeling.

9.4.3.1 Macroevolution as a Network Phenomenon

*Micro*evolution studies typically focus on a single species—or, in coevolution, a small number of species. In *macro*evolution studies, on the other hand, the focus is on large communities of species over long time periods. Particular attention must be paid to the interactions among these species. Such interactions reflect network properties, which are not reducible to single species or to a small number of coevolving species. Macroevolution, it would seem, is a network phenomenon.

Solé and Bascompte (2006) certainly support the network view. "We might ask if an ecological-based network theory of species changes can provide some understanding of large-scale evolution." They suggest that simple network models can capture long-term general behavior, independent of specific species details. The dynamics of macroevolution are "not species-dependent, but network dependent." Solé et al. (1999) also discuss ecosystem macroevolution network modeling. They say that model behavior can be "highly nonlinear and might generate large extinction events." (We will discuss details of a macroevolution model momentarily.) Solé et al. note that "a particularly important consequence of these models is the decoupling between micro- and macroevolutionary processes." They assert that a decoupling hypothesis is needed to explain the very different micro and macro behaviors. "If network properties are important, a change in one species could propagate through the system in a highly nonlinear, unpredictable way. The consequences of such propagation are unlikely to be explained through the adaptation or selection processes that can be applied to single species or simple two-species systems." ... "The highly unpredictable network dynamics could provide the natural source of decoupling between micro- and macroevolutionary dynamics."

Solé et al. note further that, in macroevolution network models, "Power laws are observed ... in the distribution of extinction events and lifetimes ...." Fractal behavior occurs in these large-scale, long-term network processes. We will see that as the discussion proceeds.

9.4.3.2 Macroevolution Matrix Modeling

We are concerned here with macro effects (networks of species) rather than micro effects (individual species). Macroevolution modeling work has been done by many noteworthy investigators, including Newman, Solé, Kauffman, Bak, and others. We discuss one particular family of macroevolution extinction/speciation matrix models in the following paragraphs (see Manrubia and Paczuski, 1998; Solé and Goodwin, 2000; Solé et al., 1998).

The model addresses an ecological system network of N species (nodes) and their links. The network (digraph) is represented by an NxN link matrix and an Nx1 node state vector. Link matrix and node state vector initial conditions are random values. Link activity is represented by changes to the matrix, based on a set of matrix rules. Node activity results from link activity. The N nodes are binary:

State = 1 if species is viable.

State = 0 if species is extinct.

Node/link activity can propagate locally as well as globally, yielding both small extinction events and large extinction events. The dynamics of these changes are not species dependent but *network dependent*. Extinct species are replaced by new species (diversification).

Here is a set of matrix rules that can be used in the model (Solé and Goodwin, 2000):

- *Random changes in the connectivity matrix*. At each time step, one input connection is selected for each species and assigned a new, random value. This rule introduces changes into the system. These changes could be due to species interactions or to environmental changes of some sort. In this sense, the two possible sources of change (intrinsic and extrinsic) are incorporated.
- *Extinction*. At each time step, the sum of the inputs for each species is computed. This sum defines the condition for extinction. If it is negative, the species is extinct (state $S_i = 0$) and all its connections are removed. Otherwise, it remains alive ($S_i = 1$).
- *Diversification*. As species disappear as a consequence of the previous rule, their empty sites are refilled by diversification. Each extinct species is replaced by a randomly chosen survivor. The connections of the survivor are simply copied into the empty site.

Solé and Goodwin (2000) provide graphical results for this version of the macroevolution matrix model. The extinction event time series contains a wide range of small, moderate, and (very) large extinction events. The dynamics are clearly punctuated. The extinction event distribution indicates power-law/fractal behavior.

Solé et al. (1998) note that several features of the fossil record—including the extinction event time series and the frequency spectrum representations—indicate "fractal behavior, characterized by the presence of power-law distributions." They state further that the set of macroevolution matrix models can reproduce "all the reported statistical features" of the fossil record. "These models suggest that [network-based] multispecies interactions are the relevant

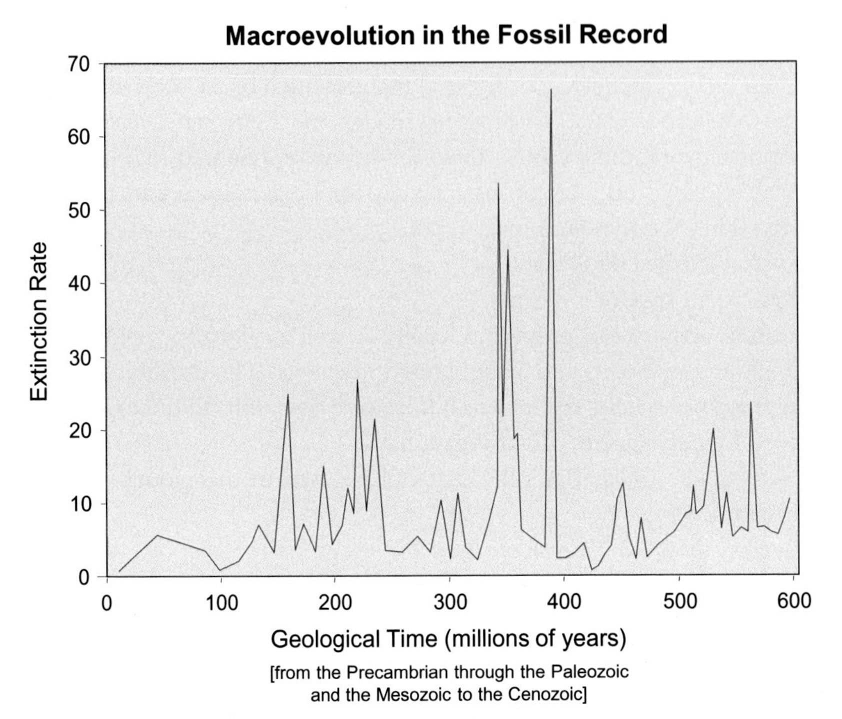

Figure 9.8 Fossil record extinction rate time series.

ingredient" that drives the complex behavior. (I fully agree that the network interactions drive the behavior.) Solé et al. (1998) also suggest that the behavior is a consequence of self-organized criticality. (I do not agree with the self-organized criticality claim, as I discussed in Section **9.3.3** of this chapter and will also discuss in Chapter 14, Section 14.3.)

Here is an example of fossil record results. Figure 9.8 displays an extinction rate time series generated from data in the fossil record. This figure provides the time series of extinction rates (mostly) during the Phanerozoic eon—from the Precambrian through the Paleozoic and the Mesozoic to the Cenozoic. Michael Benton (1993, 1995) has comprehensively described and analyzed the fossil record. A website that provides a full set of fossil record data is available online.[1] I obtained the data to generate Figure 9.8 from that website.

What about species diversity? Benton (1995) notes that diversification occurred rapidly during the past 600 million years, and at exponential rates

[1] The Fossil Record 2 at http://www.fossilrecord.net/fossilrecord/summaries.html.

through most of the Mesozoic and the Cenozoic (the past 250 million years). He suggests, "There is no evidence in the fossil record of a limit to the ultimate diversity of life on Earth." Benton also acknowledges that this rapid diversification was interrupted by mass extinctions, followed by high-rate diversification right after the extinctions. He says, "The numbers of new families that appeared through geological time varied greatly. High measures of origination ... indicate bursts of diversification into new habitats, and many of the high frequencies follow after mass extinction events, when empty ecospace was filled."

Solé et al. (1998) model and analyze species diversity. Using a version of the macroevolution matrix model,[2] they have generated results that generally match the fossil record diversity data. Figure 9.9 displays some of the model results. Solé et al. note that the model "reproduces the diversification curves displayed by the FR [fossil record] data, with a transient increase in diversity punctuated by some

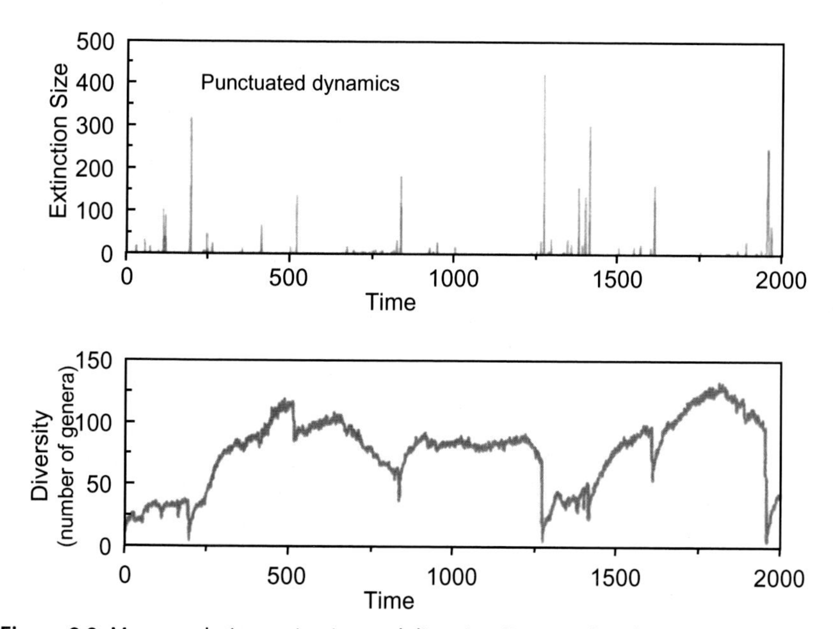

Figure 9.9 Macroevolution extinction and diversity. *(Reprinted, with permission of World Scientific Publishing, from Long-range Correlations in the Fossil Record and the Fractal Nature of Macroevolution, R. V. Solé, S. C. Manrubia, J. Pérez-Mercader, M. Benton, and P. Bak, Advances in Complex Systems, Volume 01, Issue Number 02 and 03, Copyright © 1998 World Scientific Publishing).*

[2] See Manrubia and Paczuski (1998).

drops associated with large extinction events." They also state, "We have a highly nonlinear dynamics leading to large extinction events and strong changes in diversity at the (very) large time scale. This result should prevent us from conclusions about specific trends in the fossil record."

There is an interesting question that might be asked at this point: how many species are likely to exist in the ecosphere? Is the answer as many as possible (i.e., as many as the environment will support), or is there some limiting principle due to the "highly nonlinear" network dynamics at work here? The answer may be the latter. Although there have been trends (up and down) over time, Figure 9.9 seems to indicate a remarkable constancy of diversity (number of genera) in the ecosphere over many hundreds of millions of years.

9.4.4 Network Growth via Preferential Attachment

In this subsection, we'll briefly review material from the literature on particular network growth models. The primary source is Newman (2003). Comments from Albert and Barabási (2002) are also provided.

First, consider material from Newman (this paragraph and the next three). He says that "the best studied class of network growth models by far, and the class on which we concentrate, is the class of models aimed at explaining the origin of the highly skewed degree distributions. Indeed these models are some of the best studied in the whole of the networks literature, having been the subject of an extraordinary number of papers in the last few years." Newman is talking about preferential attachment models that yield networks with scale-invariant degree distributions.

Here's some history: "The physicist-turned-historian-of-science, Derek de Solla Price, described in 1965 probably the first example of what would now be called a scale-free network; he studied the network of citations between scientific papers and found that both in- and out-degrees (number of times a paper has been cited and number of other papers a paper cites) have power-law distributions." What is the growth mechanism of such a network? Consistent with earlier work by Herbert Simon, Price's "idea was that the rate at which a paper gets new citations should be proportional to the number that it already has. This is easy to justify in a qualitative way. The probability that one comes across a particular paper whilst reading the literature will presumably increase with the number of other papers that cite it, and hence the probability that you cite it yourself in a paper that you write will increase similarly. The same argument can be applied to other networks also, such as the Web." Price called this effect *cumulative advantage* and showed analytically that the resulting citation network degree distribution has a power-law tail—it is scale-free. "The mechanism of cumulative advantage proposed by Price is now widely accepted as the probable explanation for the power-law degree distribution observed not only in citation networks but in a wide variety of other networks . . . including the World Wide Web, collaboration networks, and possibly the Internet and other technological networks . . . ." Barabási and Albert rediscovered cumulative advantage some decades later and gave it the new name *preferential attachment*. The Barabási-Albert model, however, assumes an undirected network. They sacrifice some of the realism of Price's model in favor of simplicity.

"The model of Barabási and Albert has attracted an exceptional amount of attention in the literature. In addition to analytic and numerical studies of the model itself, many authors have suggested extensions or modifications of the model that alter its behavior or make it a more realistic representation of processes taking place in real-world networks." One example of suggested modifications pertains to the World Wide Web. "Although it is rarely pointed out, it is clearly the case [for the World Wide Web] that a different mechanism must be responsible for the out-degree distribution from the one responsible for the in-degree distribution. We can justify preferential attachment for in-degree by saying that Web sites are easier to find if they have more links to them, and hence they get more new links because people find them. No such argument applies for out-degree. It is usually assumed that out-degree is subject to preferential attachment nonetheless. One can certainly argue that sites with many out-going links are more likely to add new ones in the future than sites with few, but it's far from clear that this must be the case."

Newman further notes, "There are some networks that appear to have power-law degree distributions but for which preferential attachment is clearly not an appropriate model. Good examples are biochemical interaction networks of various kinds. A number of studies have been performed, for instance, of the interaction networks of proteins in which the vertices are proteins and the edges represent reactions. These networks do change on very long time-scales because of biological evolution, but there is no reason to suppose that protein networks grow according to a simple cumulative advantage or preferential attachment process. Nonetheless, it appears that the degree distribution of these networks obeys a power law, at least roughly."

Next, consider comments from Albert and Barabási (2002). They say, "While it is straightforward to construct random graphs that have a power-law degree distribution, these constructions only postpone an important question: what is the mechanism responsible for the emergence of scale-free networks?" They conclude that the network assembly processes must be modeled. Albert and Barabási continue, "If we capture correctly the processes that assembled the networks that we see today, then we will obtain their topology correctly as well. Dynamics takes the driving role, topology being only a byproduct of this modeling philosophy." Albert and Barabási take credit: "Two ingredients, growth and preferential attachment, inspired the introduction of the Barabási-Albert model, which led for the first time to a network with a power-law degree distribution."

In my view, preferential attachment is real in many complex networks, both natural and human-generated. In ecological networks, for example, preferences in trophic interactions are clear and widely documented in food web studies. In social networks, person-to-person (node-to-node) interaction preferences are evident. In the World Wide Web, Google websites may be some of the premier beneficiaries of preferential attachment. The Albert and Barabási proposed preferential attachment mechanism ("the rich get richer"; the addition of new links to a node is proportional to the number of the node's existing links), however, is likely an overly simplified generalization. It works well for some complex networks but not for others, as pointed out by Newman and other authors. Surely, additional factors can be involved in preferential attachment. Factors that come to mind are interaction type, node availability, physical distance to a node, and chance.

9.5 CONCLUDING REMARKS

We began this chapter with comments from several respected investigators. Reiterating some of Newman's (2003) remarks is an appropriate way to conclude:

> *"In looking forward to future developments in this area [complex network dynamics] it is clear that there is much to be done. The study of complex networks is still in its infancy."*
>
> *Perhaps the most important direction for future study is "the behavior of processes taking place on networks." . . . "This, in a sense, is our ultimate goal in this field: to understand the behavior and function of the networked systems we see around us. If we can gain such understanding, it will give us new insight into a vast array of complex and previously poorly understood phenomena."*

My work heads precisely in the direction that Newman suggests.

CHAPTER 10

Fundamentals of Nonlinear Dynamics

10.1 INTRODUCTION

Because this book is about the dynamics of complex systems and because those dynamics are primarily nonlinear, it is quite appropriate to review existing fundamental theory of nonlinear dynamics. This chapter provides an introductory understanding of the subject. We explore basic systems—single-node systems and very small networks consisting of only a few nodes. (I have found that the extant nonlinear dynamics theory typically does not address highly complex systems.) Equipped with basic theoretical knowledge from this chapter, we will be in a better position to begin to understand the dynamical behavior of complex ecosystems (and highly complex systems in general).

Much of the material for the current chapter comes from two sources: Solé and Bascompte (2006) and Strogatz (2001), with some additional contributions from other authors as noted. I provide distillation, synthesis, and organization, as well as my perspectives. In their book, *Self-Organization in Complex Ecosystems*, Solé and Bascompte "focus on very simple models that, despite their simplicity, encapsulate fundamental properties of how ecosystems work. They are based on the nonlinear interactions observed in nature and predict the existence of thresholds and discontinuities that can challenge the usual linear way of thinking." Their book "presents theoretical evidence of the potential of nonlinear ecological interactions to generate nonrandom, self-organized patterns at all levels."

Here's a map of the chapter. Section 10.2 introduces fundamental concepts of nonlinear dynamical systems: rate equations, state space, and attractors. These concepts are applied to a simple example in Section 10.3 (to help ease us into our nonlinear dynamics investigations). In Section 10.4, we get into a comprehensive description of dynamics attractors, with interesting examples and illustrations. The phenomenon of bifurcation plays an important role in understanding nonlinear dynamics. Section 10.5, therefore, addresses bifurcation in some detail. Dependence on initial conditions and resulting unpredictability are hallmarks of so-called "deterministic chaos" systems. Section 10.6 explains why

Understanding Complex Ecosystem Dynamics
http://dx.doi.org/10.1016/B978-0-12-802031-9.00010-3

this is so. In Section 10.7, we discuss the fact that much system behavior seems to be general—universal—and independent of the details of specific systems.

10.2 FUNDAMENTAL CONCEPTS

The fundamental concepts of nonlinear dynamical systems are introduced here. Rate equations, state space, and attractors are defined in Sections 10.2.1, 10.2.2, and 10.2.3, respectively. These concepts are the basis for investigations in the succeeding sections.

10.2.1 Rate Equations

Dynamical systems are often modeled by differential equations (see Strogatz, 2001):

$\frac{d\mathbf{x}}{dt} = \mathbf{v}(\mathbf{x})$ where
$\mathbf{x}(t) = (x_1(t), x_2(t), \ldots, x_n(t))$ is a vector of state variables
t is time and
$\mathbf{v}(\mathbf{x}) = (v_1(\mathbf{x}), v_2(\mathbf{x}), \ldots, v_n(\mathbf{x}))$ is a vector of functions that encode the dynamics

For example, in a chemical reaction, the state variables represent concentrations, and the differential equations represent the kinetic rate laws, which usually involve nonlinear functions of the concentrations. In ecological stock-and-flow models, the state variables represent the ecological network compartment stocks.[1] The differential equations encode the input/output flow dynamics (inflow minus outflow).

Here's a formulation often used in stock-and-flow modeling:

$$x(t+\Delta t) = x(t) + r(*)\Delta t$$

$$x(t+\Delta t) - x(t) = r(*)\Delta t \quad \text{or} \quad \frac{dx}{dt} = r(*)$$

where
$x =$ a state variable of interest (a compartment stock)
$t =$ time
$r =$ rate of change of x and
$r =$ a constant or $r(x)$ or $r(t)$ or $r(t, x)$

When r is a constant, x is a linear function of time:

$$x(t) = x(0) + rt$$

[1] Compartment stock is the biomass/energy contents of a given ecological network compartment.

When r is proportional to x, then x is an exponential function of time:

$$r = r(x) = ax$$
and
$$x(t) = x(0)e^{at}$$

When r is a function of both t and x, we get the general form of an ordinary differential equation of first order:

$$\frac{dx}{dt} = r(t, x)$$

Typically, this equation cannot be solved mathematically. (It usually can be approximately solved, however, by numerical methods and computer simulation.) We very often deal with nonlinear relationships in the rate equations.

10.2.2 State Space

Strogatz (2001) observes that, although nonlinear rate equations are typically impossible to solve analytically, "one can gain qualitative insight by imagining an abstract n-dimensional state space with axes $x_1, \ldots, x_n$. As the system evolves, $\mathbf{x}(t)$ flows through state space, guided by the 'velocity' field $d\mathbf{x}/dt = \mathbf{v}(\mathbf{x})$ like a speck carried along in a steady, viscous fluid."

To obtain a workable visual state space depiction, two or three important state variables are usually chosen as coordinates and the rest are held constant. The dynamic "trajectory" of the system is then plotted parametrically with time. The resulting plot allows us to observe the state or states that the system trajectory is "attracted to" over time.

We'll look at several state space depictions (some authors call it phase space) and the associated system "attractors" as the chapter proceeds.

10.2.3 Attractors

So, a system trajectory in state space moves toward an attractor. According to the extant theory, there are three types of attractors:

- A single state (a *point attractor*)
- A relatively small set of repeating states (a *limit cycle* or *periodic attractor*)
- A large set of nonrepeating states (a *strange attractor*)

We will ease into this. We will consider a very simple system (that has a point attractor) in Section 10.3, before getting into a more comprehensive discussion of types of attractors in Section 10.4.

10.3 STARTING SIMPLE

Let's look at the fundamental concepts—rate equations, state space, and attractors—for a very basic system: a simple pendulum.

A perfect pendulum is a deterministic system described by only two variables: its position and its velocity. See inset (a) of Figure 10.1. It can be shown that two deterministic rate equations describe the pendulum's motion, defining how the position (angle φ) and the velocity (v) change through time:

$$\frac{d\varphi}{dt} = v$$
$$\frac{dv}{dt} = -\mu v - \frac{g}{L}\sin\varphi$$

The motion takes place under the action of a gravitational field (gravitational constant g) acting upon a mass m (although the value of m doesn't affect the motion). The effects of a frictional force proportional to the pendulum's speed have been added (μ is the proportionality factor). Consistent with our experience, over time

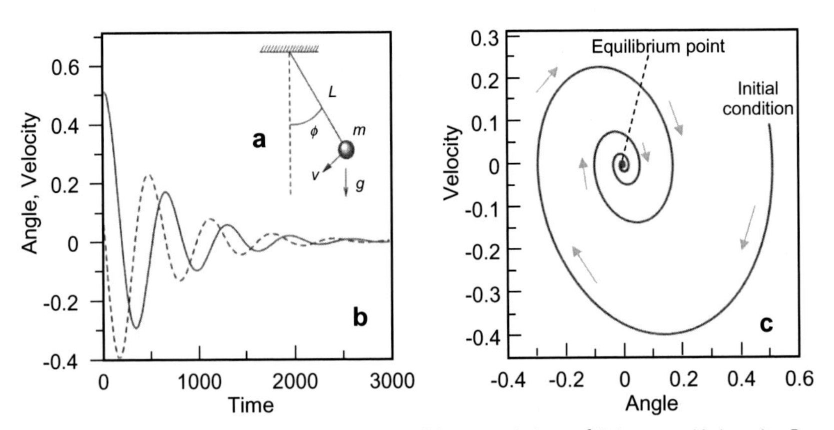

Figure 10.1 Pendulum dynamics. *(Reprinted by permission of Princeton University Press from Ricard V. Solé and Jordi Bascompte. Self-Organization in Complex Ecosystems. © 2006 Princeton University Press).*

the system approaches a state of rest ($\varphi = 0,\ v = 0$). The dynamical evolution in time is displayed in part b of Figure 10.1. Both the position (continuous line) and the velocity (dashed line) exhibit damped oscillations as the system approaches equilibrium. This state will be reached starting from any possible initial condition. It is a global attractor—and, in this case, a point attractor—for the system. Part c of Figure 10.1 provides the state space (φ, v) view. It shows the system trajectory converging to its single-point attractor ($\varphi^* = 0,\ \ v^* = 0$).

10.4 TYPES OF ATTRACTORS

Solé and Bascompte (2006) state, "A popular view assumes a balance of nature." For example, "according to this view, species fluctuate until reaching a stationary abundance, a specific number given by the energetic constraints of their habitat." Everything is in balance. Solé and Bascompte then go on to describe why this is not true. Balance is not the view that describes complex systems; the notion of trajectories and attractors in nonlinear dynamical systems is the more correct view.

As previously described, a system trajectory in state space moves toward an attractor. There are three types of attractors: a single state (a *point attractor*), a relatively small set of repeating states (a *limit cycle*), and a large set of nonrepeating states (a *strange attractor*). Let's continue now with a more comprehensive treatment of attractors.

In Section 10.2, some material and quotes from Strogatz (2001) were used to help describe rate equations and state space. We continue with his mathematical development here to help describe attractors: "Suppose $\mathbf{x}(t)$ eventually comes to rest at some point $\mathbf{x}^*$. Then the velocity must be zero there, so we call $\mathbf{x}^*$ a fixed point. . . . If all small disturbances away from $\mathbf{x}^*$ damp out, $\mathbf{x}^*$ is called a stable fixed point—it acts as an attractor [a point attractor] for states in its vicinity. Another long-term possibility is that $\mathbf{x}(t)$ flows towards a closed loop and eventually circulates around it forever. Such a loop is called a limit cycle. It represents a self-sustained oscillation of the physical system. A third possibility is that $\mathbf{x}(t)$ might settle onto a strange attractor, a set of states on which it wanders forever, never stopping or repeating."

Section 10.3 provided an example of a very simple system with a point attractor. Here, in Section 10.4, more examples—of point attractors, a limit cycle, and a strange attractor—are provided.

Consider a Lotka-Volterra predator/prey model with predator satiation. The defining differential equations are:

$$\frac{dR}{dt} = \mu R\left(1 - \frac{R}{K}\right) - \Phi(R, C)C$$

$$\frac{dC}{dt} = \beta\Phi(R, C)C - \delta C$$

where

R is the resource (prey population)

C is the consumer (predator population)

$\Phi(R, C)$ describes the functional response involving predator satiation

μ is the population growth rate

K is the population carrying capacity

β and δ are additional scaling parameters

(The form of the equations is instructive but, I think, the derivation details are not necessary in the context of this chapter.)

Figure 10.2 displays three different examples of the nonlinear dynamics of this Lotka-Volterra model. The figure is from Solé and Bascompte (2006).

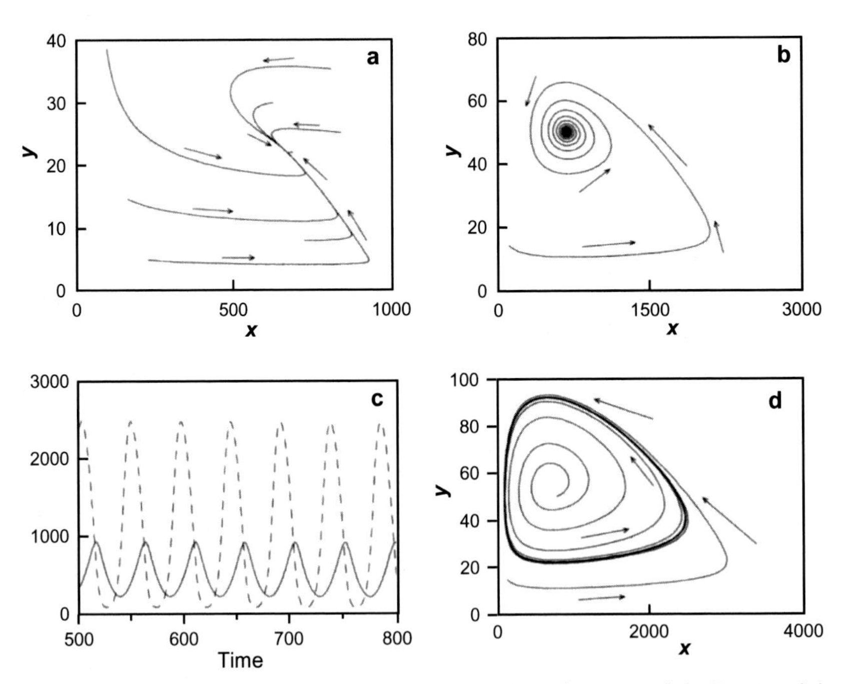

Figure 10.2 Nonlinear dynamics of a simple Lotka-Volterra model. *(Reprinted by permission of Princeton University Press from Ricard V. Solé and Jordi Bascompte. Self-Organization in Complex Ecosystems. © 2006 Princeton University Press).*

It is based on work by Roughgarden (1979). Parts a, b, and d of the figure are state space depictions. Each represents a different value of carrying capacity, K, which yields differing system dynamics. Parts a and b have point attractor dynamics and part d has limit cycle dynamics. Part c of the figure displays the temporal series of predator (y, continuous line) and prey (x, dashed line) densities that correspond to the dynamics of part d. (Predator densities in c have been rescaled by a factor of ten.)

In the state space plots, coordinates x and y represent prey density and predator density, respectively. At each time step, the system is defined as a point in state space—a particular density of prey and predator. The sequence of points defines the system trajectory and, after a large enough number of iterations, the system enters its attractor—the solution that is reached independent of the initial conditions. The arrows in parts a, b, and d of Figure 10.2 indicate movement along the trajectories in state space.

Solé and Bascompte (2006) observe that the qualitative change that takes place as a system transitions from point attractor dynamics to limit cycle dynamics is called a Hopf bifurcation. In general, qualitative changes in system dynamics are called bifurcations. The parameter values at which bifurcations occur are called bifurcation points. There are several types of bifurcations, and they play an important role in understanding biocomplexity. The presence of bifurcation points provides one of the clearest illustrations of nonlinearity: small changes in key (bifurcation) parameters can lead to totally different types of system behaviors. We'll discuss this further in Section 10.5.

In 1963, meteorologist Edward Lorenz formulated a weather model with three coupled, nonlinear differential equations in three climatic variables (say x, y, and z). The form of the equations is:

$$\frac{dx}{dt} = a(y - x)$$
$$\frac{dy}{dt} = x(b - z) - y$$
$$\frac{dz}{dt} = xy - cz$$

The three variables (x, y, and z) can be interpreted as the vertical component of air speed, and the vertical and horizontal variations in air temperature. Parameters a, b, and c are related to viscosity and thermic diffusion. According to Solé and Bascompte, Lorenz was surprised to find that "despite the system's being completely deterministic, it shows aperiodic dynamics, the type of dynamics that one would expect for a noisy system."

Figure 10.3 illustrates the system dynamics of the weather model. The upper diagram provides the state space depiction. It is a two-dimensional

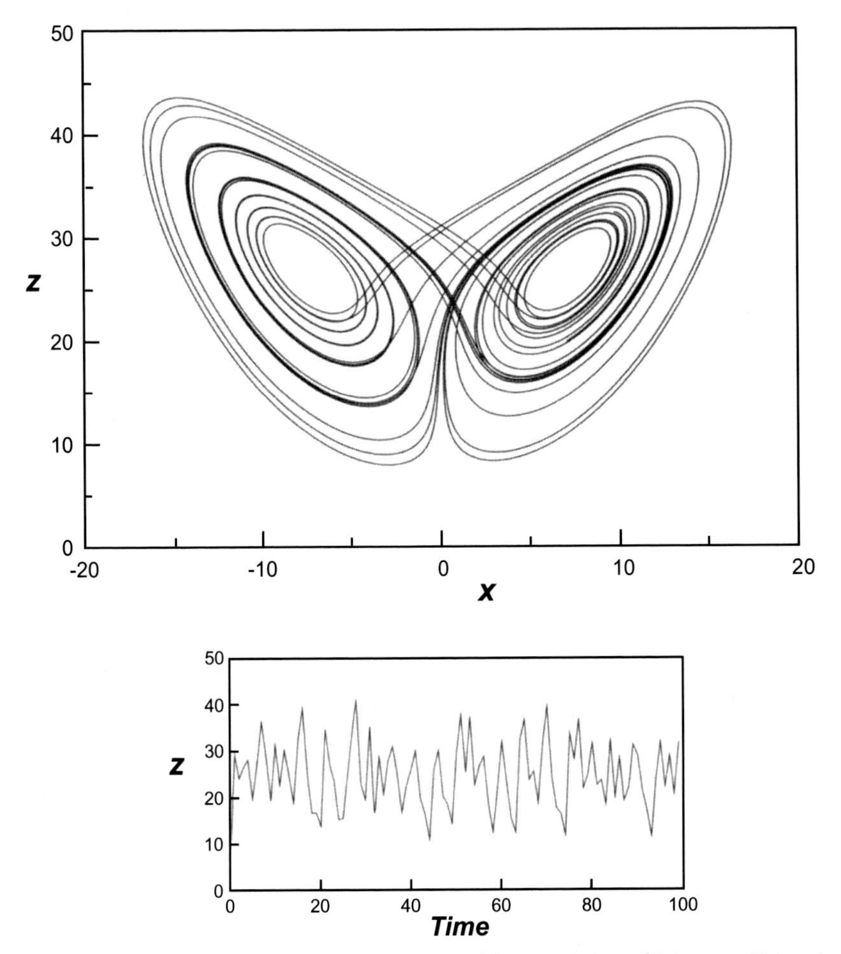

Figure 10.3 Lorenz's strange attractor. *(Reprinted by permission of Princeton University Press from Ricard V. Solé and Jordi Bascompte. Self-Organization in Complex Ecosystems. © 2006 Princeton University Press).*

projection on the *z*-*x* plane of the three-dimensional attractor of Lorenz's system. (In three dimensions, the Lorenz attractor trajectories never cross.) The lower diagram displays the temporal series for variable *z*.

Instead of the random-like behavior that can be observed in the time series, the object in state space is stunning. This is something certainly more complex than a limit cycle. The term *strange attractor* has been coined to describe this type of attractor. Solé and Bascompte: "Since the system is deterministic, the attractor has a well-defined shape, and all [system] trajectories will ultimately become trapped there. But two different trajectories will never cross (repeat) each other...." For systems such as this, there seems to be some order within chaotic-like behavior.

10.5 BIFURCATION AND COMPLEX NONLINEAR BEHAVIOR

We have seen that simple, nonlinear deterministic equations can generate complex system dynamics. Solé and Bascompte: "An especially important feature of most complex systems is the presence of multiple attractors. ... A remarkable feature of this multiplicity is that, under some conditions, as a given parameter smoothly changes, an abrupt transition in a system's state can occur." Such abrupt qualitative changes in system dynamics are called bifurcations. The parameter values at which bifurcations occur are called bifurcation points. Let's take a closer look at this phenomenon by considering an analysis (by Robert May) and a corresponding *bifurcation diagram*. We will see that small changes in bifurcation parameters can lead to different attractor types and totally different types of system behaviors.

The following development is based on Solé and Bascompte (2006). In the 1970s, ecology theoretician Robert May (1974, 1976) made an outstanding contribution to the study of system dynamics. He started with a discrete-time version of the single-species logistic equation:

$$x_{t+1} = F_\mu(x_t) = \mu x_t(1 - x_t)$$

where

x_t = the normalized population size at generation t

μ = the growth rate

This is a *deterministic* model. There are no random or probabilistic terms. If the population value at any given generation is known, one can iterate and estimate population values at successive generations. Observe that

$$\mu \leq 1;\quad x_{t+1} \to 0 \quad \text{(extinction)}$$
$$\mu > 1;\quad x_{t+1} \to x^* = 1 - \frac{1}{\mu}$$

Analysis shows that attractor x^* is stable for $\mu \in (1,3)$ and then becomes periodic and eventually random-like for $\mu > 3$. The dynamics of the system are very dependent on μ.

The system dynamics and the dependence on μ are clearly illustrated by the time series plots and the *bifurcation diagram* (in the center) of Figure 10.4. The ordinate in all cases is the normalized population size. Time series A

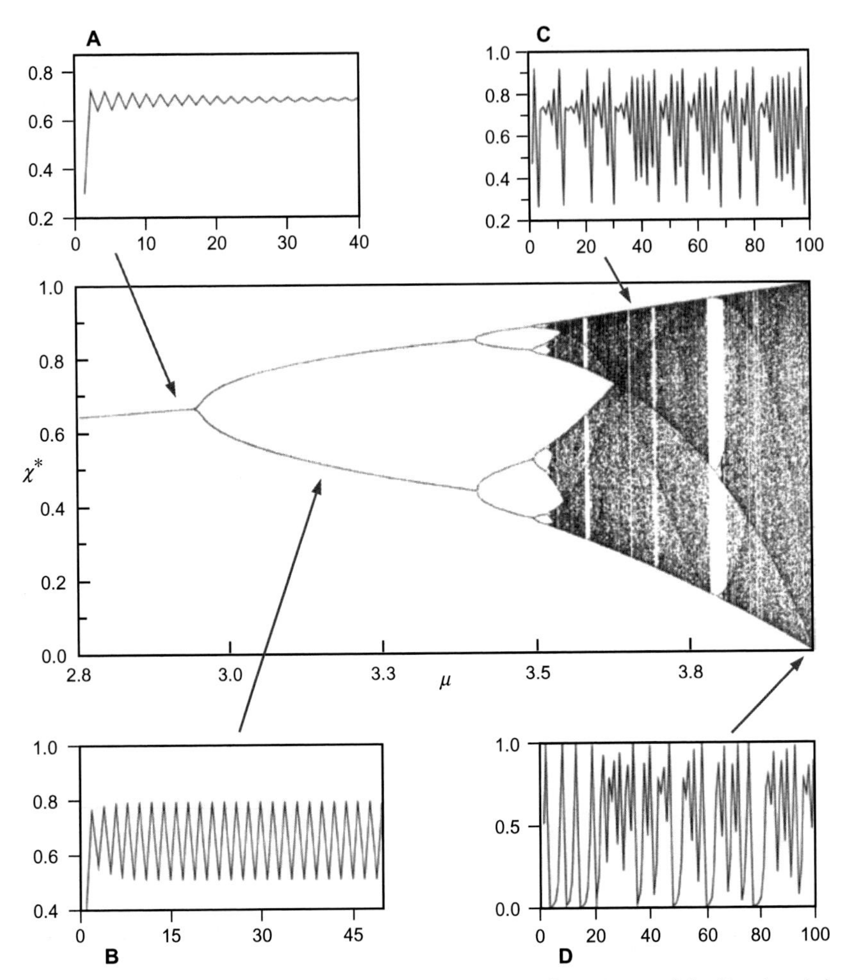

Figure 10.4 System dynamics of the discrete-time logistic model. *(Reprinted by permission of Princeton University Press from Ricard V. Solé and Jordi Bascompte. Self-Organization in Complex Ecosystems. © 2006 Princeton University Press).*

corresponds to $\mu = 2.95$. As shown, normalized population grows through time until it reaches a stable value of $x^* = 0.661$. In time series B, $\mu = 3.2$. Here the population does not converge to a time-invariant value, but oscillates in time between two values. One year the population has a low value, the next year a high value, and so on. In this case, there is a period-two cycle. Here's what has happened: as the value of μ is increased, a critical value $\mu_{c_1} \cong 3$ is encountered and a bifurcation takes place. At this critical point, the previously stable solution x^* becomes unstable. A small perturbation drives the system far from x^*. The next critical point on the bifurcation diagram occurs at $\mu_{c_2} \cong 3.45$. Here, a period-four cycle commences. Next, on this period doubling route, a period-eight cycle is encountered at $\mu_{c_3} \cong 3.5$. Then something strange happens to the system dynamics. Time series C has $\mu = 3.7$ and time series D has $\mu = 4.0$. For these values of μ, a completely different and unexpected kind of dynamics is observed. The population oscillates in a seemingly random way, without periodicity. The dynamics never repeat, no matter how long we iterate the model.

The domains of each of the system attractor types are clear from the bifurcation diagram in Figure 10.4. The *point attractor* domain is $\mu < 3$. The *limit cycle* domain is approximately $3 \leq \mu \leq 3.5$. The *strange attractor* domain is $\mu > 3.5$.

Here's how the bifurcation diagram is generated. Select a value of μ. Iterate the system dynamics equation enough times to get beyond transients. Then iterate another N (maybe 100) times and plot each of the resulting N values of x. If the system is in the point attractor domain, all x values will be the same and look like one point. If the system is in the period-two cycle domain, x will alternate between two values and will look like two points, and so on. Then select another value of μ and continue .... In the strange attractor domain, lots of points are plotted for each value of μ.

Solé and Bascompte claim that, in the bifurcation diagram (again, see Figure 10.4), the periodic windows (white stripes) inside the strange attractor domain are nested, and the nesting is fractal with self-similar structure at different scales. Inside such a periodic window, there is another period doubling route to complex dynamics with other periodic windows within another strange attractor domain, and so on.

The bifurcation development in this section is for a discrete-time case. Solé and Bascompte say that the general results also apply to the continuous-time case, with three or more interacting components.

Although they don't look it, the dynamics depicted in the bifurcation diagram are, in fact, deterministic. The strange attractor domain of complex nonlinear behavior is often referred to as "deterministic chaos" (in *chaos theory* literature).

10.5.1 Comments

When there is a relationship between a bifurcation parameter and an environmental variable, that variable might strongly influence system behavior. Human-caused changes to environmental variables, therefore, can yield unexpected results by suddenly changing the dynamics of an ecological system.

Simple nonlinear systems can generate complex dynamics. Perhaps a message here is that relatively simple nonlinear models can help us understand the complex dynamics of real ecological systems. In many cases, simple models also seem to describe dynamics properties that are general—independent of the details of specific systems. (See Section 10.7.) Solé and Bascompte seek to "use simple models in the hope that they capture the essential, while omitting the irrelevant."

10.6 UNPREDICTABILITY

We might expect that if we know the equations that drive a deterministic system, and can measure an initial condition, then future values should be predictable. Solé and Bascompte (2006) note however that nonlinear deterministic systems, such as the May model (discrete-time logistic model) from the previous section, are very sensitive to initial conditions in the chaotic domain. There is always an error when measuring/estimating a given system variable (such as population size), and the initial error between the real value and the estimate will not remain small, as it would in the linear case. Due to the nonlinearities in the system, the error will grow exponentially with time. There is a time horizon beyond which prediction has nothing to do with reality. Accurate prediction of future values is not possible.

Dependence on initial conditions and resulting unpredictability are hallmarks of deterministic chaos systems. This is very well illustrated in Figure 10.5 for a discrete-time logistic population model. The figure displays the two time series of two runs of the logistic model with a growth rate of $\mu = 4$ (within the chaotic strange attractor domain). The continuous-line time series has a normalized population initial condition of $x_0 = 0.10$. The dashed-line time series has an initial condition of $x_0 = 0.10001$. The two initial condition values could correspond to a population estimate and the real value, respectively. There is an initial error of only 0.001%. At the start the two time series coincide, but they soon begin to diverge and end up completely uncorrelated. Long-term prediction is not possible.

Solé and Bascompte say that the behavior of such systems "can be predicted at short time windows (they are deterministic), but long-term prediction is impossible despite knowing the dynamics underlying [the] system and having a good (as good as one wants) estimation of the initial condition." Strogatz (2001) agrees that long-term prediction is impossible because of the exponential amplification of small uncertainties or measurement errors.

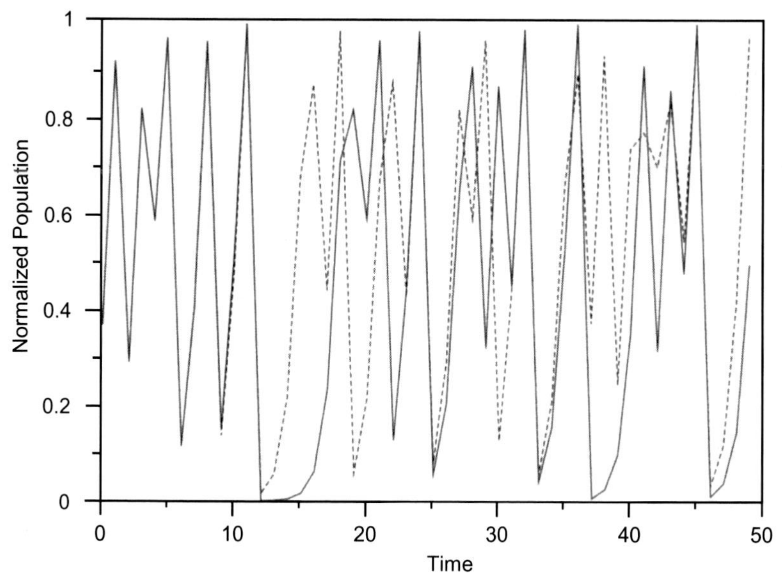

Figure 10.5 Unpredictability of deterministic chaos systems. *(Reprinted by permission of Princeton University Press from Ricard V. Solé and Jordi Bascompte. Self-Organization in Complex Ecosystems. © 2006 Princeton University Press).*

Dependence on initial conditions and other dynamical behaviors are also conveyed by the so-called Lyapunov exponent—a measure of how fast nearby system trajectories diverge in state space. For the discrete-time logistic model, Solé and Bascompte derive the following expression for the Lyapunov exponent:

$$\lambda \approx \frac{1}{n}\sum_{i=0}^{n-1} \log\left|\frac{\partial F_{\mu}(x_i)}{\partial x_i}\right|$$

where

$\lambda =$ the Lyapunov exponent

$n =$ the number of iterations of the logistic equation

$x_i =$ normalized population size at iteration i

$F_{\mu}(x_i) = \mu x_i(1 - x_i)$

$\mu =$ the growth rate

Figure 10.6 plots this expression for the Lyapunov exponent versus the population growth rate (μ).

There is a direct correspondence between the Lyapunov diagram of Figure 10.6 and the bifurcation diagram in Figure 10.4. For small values of μ, the negative Lyapunov exponent indicates point attractor domain

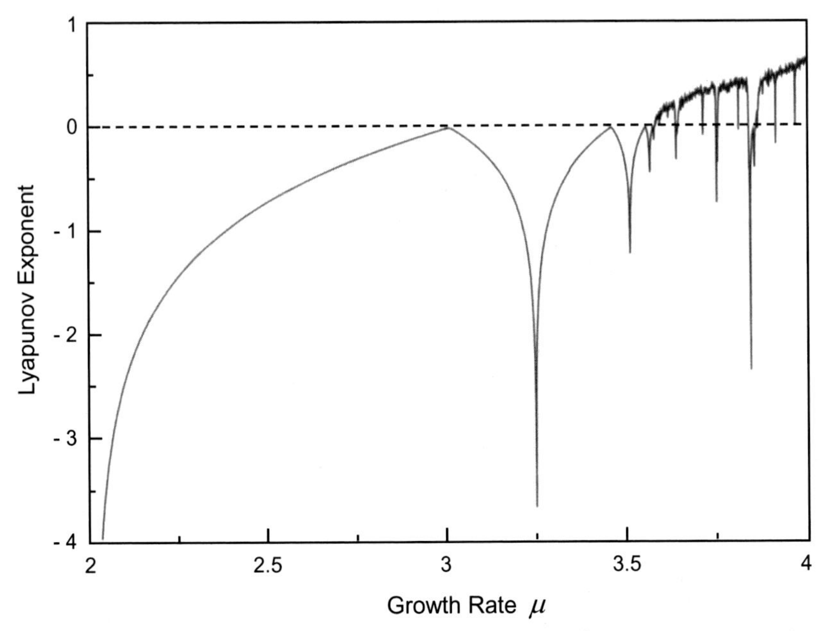

Figure 10.6 Lyapunov dynamics. *(Reprinted by permission of Princeton University Press from Ricard V. Solé and Jordi Bascompte. Self-Organization in Complex Ecosystems. © 2006 Princeton University Press).*

behavior. At $\mu \cong 3$, the exponent reaches zero, which signals entry into a period-two limit cycle. At $\mu \cong 3.45$ and $\mu \cong 3.5$, respectively, the exponent also reaches zero to indicate entry into a period-four limit cycle and then a period-eight limit cycle. For $\mu > 3.5$, the Lyapunov exponent is generally positive and corresponds to the strange attractor domain, where system trajectories with arbitrarily small differences in initial conditions will diverge exponentially with time. Observe that, within this domain, the Lyapunov exponent spikes negative several times. These occurrences correspond to the "white stripe" periodic windows (mentioned in our previous discussion) of the bifurcation diagram.

10.7 UNIVERSALITY

Much system behavior seems to be general—universal—and independent of the details of a given specific system. Solé and Bascompte (2006) say that "ecological networks . . . share a number of (sometimes surprising) universal properties that can [sometimes] be explained by means of simple mathematical models."

There seems to be universality in the behavior depicted by Lyapunov dynamics and bifurcation diagrams. This was first noted by Mitchel Feigenbaum (1978), a physicist at Los Alamos National Laboratory. Consider a bifurcation diagram. Each new bifurcation k takes place at μ_k, where μ is the bifurcation parameter (growth rate, in this case). For example, bifurcation 1 takes place at μ_1 (a period-two cycle begins), bifurcation 2 takes place at μ_2 (a period-four cycle begins), and so on. The bifurcation diagram is said to illustrate a "period-doubling" route to complex nonlinear behavior. Feigenbaum showed that for large k:

$$\delta = \lim_{k\to\infty} \frac{\mu_k - \mu_{k-1}}{\mu_{k+1} - \mu_k} = 4.6692\ldots$$

This is a general relationship for the "distance" δ between successive bifurcations. There is common, universal structure in bifurcation diagrams. Solé and Bascompte say, "Any nonlinear, one-dimensional map that has a single hump or maximum [such as the discrete-time logistic model], despite differences in details, has the same constant." They further claim that multi-peak continuous-time systems can be characterized by discrete-time models—so the previous results apply to aspects of continuous-time systems as well.

Solé and Bascompte: "The implications of universality are that some basic mechanisms are general and so independent of details of specific systems. ... Thus, despite the omission of many details from these simplistic models, we may expect that the relevant features are maintained. Simple [nonlinear] models may be, after all, relevant when dealing with complex ecological systems."

So, we might expect relatively simple models to be very productive and we might expect to encounter dynamics mechanisms and characteristics that are universal. These are very significant insights.

CHAPTER 11

Cellular Automata Investigations and Emerging Complex System Principles

11.1 INTRODUCTION

In Chapter 7, *Evolution and Universal Development Concepts*, I made the case that the evolution process model can be considered a universal development model. This model is essentially a "simple program" that operates with "simple rules." The program and rules are repeated again and again. Stephen Wolfram (2002), in his book *A New Kind of Science*, developed and articulated the simple program/simple rules ideas and principles. In my view, these ideas and principles are clearly reflected in the evolution/universal development model of Chapter 7. Wolfram hypothesized and demonstrated that the repeated simple program/simple rules mechanism can yield extremely complex system behavior. He says it is the underlying mechanism "that allows nature seemingly so effortlessly to produce so much that appears to us so complex." This hypothesis "implies a radical rethinking of how processes in nature and elsewhere work." It seems to apply "to systems throughout the natural world and elsewhere."

Wolfram conducted his innovative work mostly in the context of cellular automata investigations. This chapter is about Wolfram's work. The primary source is Wolfram's *A New Kind of Science* (also known as NKS).[1] I also include some contributions from others, as well as my own observations and perspectives.

Here's a list of the topics covered in the chapter:

Some Cellular Automata History

Explicit Experimentation via Cellular Automata

Simple Programs/Simple Rules and Highly Complex Behavior

Cellular Automata Classes

[1] The quotes in this chapter are from Wolfram (2002), except as noted otherwise.

Understanding Complex Ecosystem Dynamics
http://dx.doi.org/10.1016/B978-0-12-802031-9.00011-5

All Kinds of Systems
A Computational View of Systems
The Principle of Computational Equivalence
Summary of My Perspectives on these Complex System Principles

The chapter begins with some cellular automata history and background, and then provides a description of Wolfram's cellular automata "explicit experimentation" work. The experimentation work shows that simple programs with simple rules, repeated over and over, can yield highly complex behavior. Wolfram has identified four classes of cellular automata. Those classes and their characteristics are discussed. In my view, there is a direct correspondence between cellular automata classes and the attractors of nonlinear dynamics theory (Chapter 10). I describe that correspondence. Another of Wolfram's important insights is that the behavior of cellular automata is indicative of the behavior of systems in general. To corroborate and demonstrate that, Wolfram considered a spectrum of simple-program systems, from human-made systems to systems in nature. He found behavioral consistency across these system types. That work is discussed. The latter part of the chapter addresses Wolfram's computational view of systems. The topics addressed include computation as a framework for system principles, the concept of computational universality, and the identification of computationally universal cellular automata. Wolfram's Principle of Computational Equivalence is then described and discussed. The chapter concludes with a summary of my perspectives on the emerging complex system principles.

11.2 SOME CELLULAR AUTOMATA HISTORY

It is my understanding that John von Neumann was a principal creator of and a key figure in the development of cellular automata theory. He was a brilliant mathematician, a computer and computer science pioneer, and a charter member of the Institute for Advanced Study in Princeton (along with, for example, Einstein and Gödel).

Stephen Wolfram is also a brilliant scientist and cellular automata investigator. Wolfram became an undergraduate at Oxford University at age 17. He moved to the California Institute of Technology (Caltech) at age 18. He received his PhD from Caltech at age 20 and joined the Caltech faculty. Soon after, he was awarded a MacArthur Foundation "genius" grant. A few years later, he became a faculty member at the Institute for Advanced Study,

Princeton.[2] Currently, he is president and CEO of Wolfram Research, which he founded in 1987.

Wolfram, in *A New Kind of Science*, provides the following account of cellular automata history. In the 1950s, "several different kinds of systems equivalent to cellular automata were independently introduced." John von Neumann, in 1952–53, designed and constructed elaborate two-dimensional cellular automata, including a cellular automaton "specifically set up to emulate the operations of components of an electronic computer and various mechanical devices." Around 1960, "Stanislaw Ulam and others used computers at Los Alamos to produce a handful of examples of what they called recursively defined geometrical objects—essentially the result of evolving generalized 2D cellular automata from single black cells." In the mid-1950s to mid-1960s, cellular automata-like rules were used in digital image processing. By 1970, while experimenting with different two-dimensional cellular automata rules, John Conway came up with a simple set of rules he called The Game of Life "that exhibit a range of complex behavior." "By the late 1970s, despite all these different directions, research on systems equivalent to cellular automata had largely petered out." Wolfram began his work on cellular automata in the 1980s.

11.3 EXPLICIT EXPERIMENTATION VIA CELLULAR AUTOMATA

Wolfram believed that experimenting with cellular automata would be a profitable way of studying and understanding system behavior. He wanted to "set up a sequence of possible simple programs, and then run them and see how they behave." He chose to focus on elementary cellular automata, that is, one-dimensional, two-state, three-cell-neighborhood cellular automata. Wolfram reasoned that the behavior of these simple cellular automata could be indicative of the behavior of cellular automata (and real-world systems) in general. The elementary cellular automata universe was small enough to permit exhaustive exploration of the entire set. For every member of the set, he could implement the corresponding simple program with simple rules, run it, and see how it behaved. Wolfram referred to his approach as *explicit experimentation*.

Wolfram developed a rule-numbering scheme for the members of the elementary cellular automata set. The number of possible configurations of

[2] This account of Wolfram's early career is from Melanie Mitchell (2009), who wrote about Wolfram and his work in her book *Complexity: A Guided Tour*.

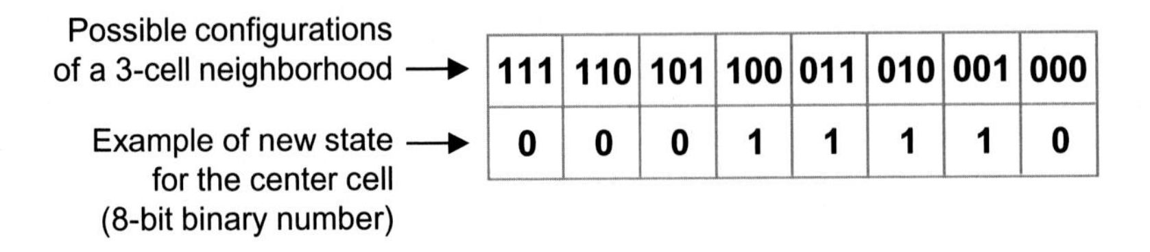

Decimal equivalent of the 8-bit binary number:

$$0 + 0 + 0 + 2^4 + 2^3 + 2^2 + 2^1 + 0 = 16 + 8 + 4 + 2 = 30$$

This, therefore, is Rule 30

Figure 11.1 Rule-numbering scheme.

a three-cell neighborhood (target cell plus the two adjacent cells) of two-state cells is $2^3 = 8$. The number of possible rules for the eight configurations is $2^8 = 256$. The outcome of a given rule is interpreted as an 8-bit binary number. The rule number (0-255) is the decimal equivalent of that 8-bit binary number. The rule-numbering scheme is illustrated in Figure 11.1 for Rule 30.

For each rule, Wolfram implemented and ran the corresponding simple program to demonstrate rule behavior. He illustrated the behavior using space-time diagrams. Time steps proceed from top to bottom in the diagram. The top row corresponds to the system initial conditions. In one version of the diagrams, the top is a single cell in the "black" state. In another version, the top row is an array of cells, often with random initial states. At each successive time step, the diagram displays the spatial evolution of the states of the cellular automaton cells, according to the rule. (Many examples of cellular automata space-time diagrams can be found at http://www.wolframscience.com/nksonline.)

11.4 SIMPLE PROGRAMS/SIMPLE RULES AND HIGHLY COMPLEX BEHAVIOR

The elementary cellular automaton space-time diagrams seemed to yield an epiphany for Wolfram, who was quoted as saying, "The Rule 30 automaton is the most surprising thing I've ever seen in science . . . . It took me several years to absorb how important this was. But in the end, I realized that this one picture contains the clue to what's perhaps the most long-standing mystery in all of science: where, in the end, the complexity of the natural world comes from."[3]

[3] Quoted in Malone, M.S., "God, Stephen Wolfram, and Everything Else," Forbes ASAP, Nov. 27, 2000 (per Melanie Mitchell, 2009).

In *A New Kind of Science*, Wolfram says that the Rule 30 cellular automaton is an "example of the fundamental phenomenon that even with simple underlying rules and simple initial conditions, it is possible to produce behavior of great complexity." He further posits that this fundamental phenomenon "is ultimately responsible for most of the complexity that we see in nature."

In his book, note that Wolfram sometimes suggests an equivalence between complexity and randomness.[4] "Even the most sophisticated mathematical and statistical methods of analysis" indicate that the complex behavior of the Rule 30 cellular automaton is perfectly random. He says later, however, that the Rule 110 cellular automaton displays even more complex behavior; it is "a remarkable mixture of regularity and irregularity."

In Section 11.3, I said Wolfram thought that the behavior of elementary cellular automata would be indicative of the behavior of systems in general. It turns out that is likely correct. With the benefit of his experimental results on elementary cellular automata, as well as further experimentation, Wolfram observed that if we make the cellular automata more "complicated," for example, three states instead of two states for a cell, we do not seem to get more complex results. It seems that all the "essential ingredients" for complexity exist in elementary cellular automata. "Using more complicated rules may be convenient if one wants, say, to reproduce the details of particular natural systems, but it does not add fundamentally new features." ... "Even though their underlying rules are extremely simple, certain cellular automata can nevertheless produce behavior of great complexity."

"To suppose that our universe is in essence just a simple program is certainly a bold hypothesis." If we are looking for an "ultimate model of the universe," the simple programs/simple rules mechanism may be an important part of the model. It would be "a model of nature that was not in any sense an approximation or idealization. Instead, it would be a complete and precise representation of the actual operation of the universe." It might be "the ultimate validation of the idea that human thought can comprehend the construction of the universe."[5]

Benoit Mandelbrot was also aware (perhaps earlier than Wolfram) of the mechanism of simple programs/simple rules repeated over and over. Mandelbrot talks about biological form and simplicity in his classic book

[4] Here is my view regarding randomness/complexity equivalence. While randomness is complex, it does not represent the maximal complexity of well-functioning, highly complex ecological systems (and living systems in general). I'll explain that in Part IV of the book.

[5] It might be a part—the mechanism part—of such comprehension.

on fractal geometry.[6] "Biological form being often very complicated, it may seem that the programs that encode this form must be very lengthy." ... "However, the complications in question are often highly repetitive in their structure." The generating rule can be systematic and simple. "The key is that the rule is applied again and again, in successive loops."

11.5 CELLULAR AUTOMATA CLASSES

Stephen Wolfram has identified four classes of cellular automata. "The number of fundamentally different types of [behavior] is very limited." There is cellular automata system behavior that becomes (1) single state, (2) periodic, or (3) random with some suggestions of order. But there is another type. "The greatest complexity lies between these extremes [essentially between (2) and (3)]—in systems that neither stabilize completely, nor exhibit close to uniform randomness forever." This is the fourth class. When cellular automata rules are sequenced in an orderly way and their behavior is examined, it becomes clear that class 4 behavior is positioned between class 2 and class 3 behavior. "For while class 4 is above class 3 in terms of apparent complexity, it is in a sense intermediate between class 2 and class 3 in terms of what one might think of as overall activity." In summary, there are just four basic classes of cellular automata:

Class 1—Single uniform final state
Class 2—Mixed final state that can repeat periodically
Class 3—Random with suggestions of order
Class 4—Mixture of order and irregularity with interacting/changing local structures

These classes apply to one-dimensional elementary cellular automata, two-dimensional cellular automata, and, it seems, to cellular automata in general.

Elementary cellular automata examples of the four classes are available at http://www.wolframscience.com/nksonline. In particular, you may want to take a look at the following space-time diagram examples in Chapter 6 of the online NKS book: Rule 250 (class 1), Rule 108 (class 2), Rule 30 (class 3), and Rule 110 (class 4).

Wolfram explains that, in class 4 cellular automata, "localized structures are produced which on their own are fairly simple, but these structures move around and interact with each other in very complicated ways."

[6] Benoit B. Mandelbrot (1982).

Wolfram also explains that both class 3 and class 4 cellular automata can contain "nested" spatial structure. He correctly uses the term *fractal* to describe this behavior, but his use of the term is limited to this relatively simple *structure fractal* case. Two points: (1) I agree that both class 3 and class 4 cellular automata can exhibit fractal behavior. (2) Fractal behavior is a much richer concept than just nesting. We will see the richness with respect to structure fractals in Chapter 12, and then with respect to *process fractals* in Chapter 15.

11.5.1 Information Handling and Communication

Cellular automata can be sensitive to initial conditions. Wolfram points out that sensitivity increases from class 1 (essentially no sensitivity) to class 2 (small, localized sensitivity) to class 3 (widespread sensitivity). Class 4 is intermediate between class 2 and class 3 with respect to sensitivity. Wolfram says that these behaviors relate to information handling:

Class 1—Information is forgotten and not communicated.

Class 2—Some information is retained locally, but not communicated globally.

Class 3—Global communication occurs across all parts of the system.

Class 4—Global communication is possible, but does not always occur.

These "differences in the handling of information are in some respects particularly fundamental." For class 2, Wolfram demonstrates that the absence of long-range communication leads to separate *piece part* features, which yield repetitive, periodic behavior. With respect to class 4 systems, Wolfram says, "A crucial feature of any class 4 system is that there must always be certain structures [patterns] that can persist forever in it." . . . "The existence of structures that move is a fundamental feature of class 4 systems." These structures "make it possible for information to be communicated from one part of a class 4 system to another." A "main point is just how diverse and complex the behavior of class 4 cellular automata can be—even when their underlying rules are very simple."

As I see it, highly complex ecological systems are class 4 systems. Earlier in this book (Chapters 6 and 8), I described local-to-global communication as an important characteristic of ecological systems. In these systems, global communication does not always occur, but it occurs when required by the real environment and the current state of the system. We have consistency with Wolfram's work.

11.5.2 Complexity and Simplicity

"There are many systems in nature that show highly complex behavior. But there are also many systems that show rather simple behavior—most often either complete uniformity, or repetition, or nesting." Wolfram further explains that higher class systems can exhibit lower class system behavior. There "exist special initial conditions for one cellular automaton that make it behave just like some other cellular automaton." Complex systems, therefore, can exhibit complex behavior and also simpler behavior:

Class 4 (maximal complexity) systems can exhibit class 4, 3, 2, and 1 behavior.

Class 3 systems can exhibit class 3, 2, and 1 behavior.

Class 2 systems can exhibit class 2 and 1 behavior.

Class 1 systems can exhibit only class 1 behavior.

Highly complex systems, including ecological systems, do not always (i.e., at every point in time) behave in highly complex ways.

11.5.3 Correspondence with Nonlinear Dynamics Attractors

In my view, there is a direct correspondence between cellular automata classes and the attractors of nonlinear dynamics theory.

As we discussed in Chapter 10, nonlinear dynamics theory defines three types of attractors for the dynamics of nonlinear systems: point attractors, limit cycle or periodic attractors, and strange attractors (or chaotic attractors). These attractor types were derived based on systems of relatively low complexity. It seems to me that the reason for working with lower complexity systems in nonlinear dynamics theory is that these are the systems that are amenable to mathematical and numerical analyses. The question then arises: Does this attractor classification scheme apply equally well to very complex systems such as ecological systems? My answer is no. The attractor set of conventional nonlinear dynamics theory seems incomplete. Cellular automata classes 1, 2, and 3 have dynamics that correspond to the three known classes of attractors (i.e., point attractors, periodic attractors, and chaotic attractors, respectively). Class 4 cellular automata, on the other hand, represent much more complex systems. Class 4 system dynamics are not well understood and they do not correspond to a conventional attractor type. Another question then arises: What is the dynamics attractor for these maximally complex systems? I propose a fourth type of nonlinear dynamics attractor and have coined a term for it: *edge-of-order attractor*. I will explain in Chapter 14.

My hypothesis[7] is that there is a one-to-one correspondence between cellular automata classes and nonlinear dynamics attractors as follows:

Class 1—Point attractor
Class 2—Periodic attractor
Class 3—Chaotic attractor
Class 4—Edge-of-order attractor

Moreover, I propose that the dynamics category of any system (not just cellular automata) may be referenced by either the class designation or the attractor designation.

My perspective is that natural systems can be any class. Complex ecological systems (and complex living systems in general) typically are class 4.

11.6 ALL KINDS OF SYSTEMS

11.6.1 Beyond Cellular Automata

We have seen that cellular automata with simple rules can "produce behavior of great complexity." Some may "assume that it must be a consequence of some rare and special feature of cellular automata, and must not occur in other kinds of systems." Are cellular automata rare and special? To answer that question, Wolfram examined and experimented with the following spectrum of simple-program systems:

- Turing Machines
- Substitution Systems (systems for which the number of elements/cells can grow or shrink over time)
- Sequential Substitution Systems
- Tag Systems
- Cyclic Tag Systems
- Register Machines (simple idealizations of present-day computers)
- Symbolic Systems

Wolfram found that cellular automata behavior is neither special nor rare. In his investigations of the listed system types, he found that in these other systems, when we take away, in turn, each of their cellular-automata-like features (e.g., rigid array, updating in parallel, etc.), we find "in the end none of these features actually matter much at all." We can still get "behavior of great complexity."

[7] I use the *hypothesis* terminology (here and elsewhere) in order to subject the claim to explicit corroboration/falsification procedures. Evidence—in either direction—by any reader or other interested party is welcome.

Wolfram also found further evidence of a threshold phenomenon. Going beyond a simple-rules threshold does not seem to produce more complexity. The threshold, which depends on the particular type of system, "is typically extremely low." ... "Once the threshold for complex behavior has been reached, what one usually finds is that adding complexity to the underlying rules does not lead to any perceptible increase at all in the overall complexity of the behavior that is produced." ... "The crucial ingredients that are needed for complex behavior are, it seems, already present in systems with very simple rules, and as a result, nothing fundamentally new typically happens when the rules are made more complex."

There is *behavioral universality*. "So this suggests that in fact the phenomenon of complexity is quite universal—and quite independent of the details of particular systems." ... "The typical types of behavior [per the classes] that occur are quite universal, and are almost completely independent of the details of underlying rules." How about higher dimension systems? After investigating higher dimension cellular automata, Turing machines, substitution systems, network systems, multi-way systems, and systems based on constraints, Wolfram concluded that, in general, system behavior "is not fundamentally much different in two or more dimensions than in one dimension." "The basic phenomenon of complexity does not seem to depend in any crucial way on the dimensionality of the system one looks at."

There seem to be "general principles that govern the behavior of a wide range of systems, independent of the precise details of each system." ... "Even if we do not know all the details of what is inside some specific system in nature, we can still potentially make fundamental statements about its overall behavior."

11.6.2 Natural Systems

What we have learned about cellular automata and other human-made simple-program systems applies as well to "actual phenomena in nature"—it applies to natural systems, to living systems. "Biological systems are often cited as supreme examples of complexity." The mechanism that produces the complexity appears to be simple programs/simple rules, repeated over and over. "Most aspects of the growth of plants and animals are in the end governed by remarkably simple rules." Wolfram sees "a deep correspondence between simple programs and systems in nature."

The notion of *behavioral universality* extends to natural systems. "When one looks at systems in nature, one of the striking things one notices is that even when systems have quite different underlying physical, biological or other components their overall patterns of behavior can often seem remarkably similar." ... "This suggests that a kind of universality exists in the types of behavior that can occur, independent of the details of underlying rules." ... "The crucial point is that I believe that this universality extends not only across [human-made] simple programs, but also to systems in nature." "And if this is the case, then it means that one can indeed expect to get insight into the behavior of natural systems by studying the behavior of simple programs. For it suggests that the basic mechanisms responsible for phenomena that we see in nature are somehow the same as those responsible for phenomena that we see in simple programs."

I note the important modeling implication here: cellular automata could be very effective vehicles for modeling the behavior of natural systems (e.g., ecological systems). See Chapter 16.

Wolfram provides some examples of natural system complex characteristics produced by simple programs/simple rules. Here's what he has to say about biological pigmentation patterns: All the different forms of mollusk shells are not "carefully crafted" by evolution/natural selection; they are the result of differing simple programs/simple rules—very similar to one-dimensional cellular automata. The pigmentation patterns of other animals often involve a simple-program activator/inhibitor process. "When a pattern forms, the color of each element will tend to be the same as the average color of nearby elements, and opposite to the average color of elements farther away." Such patterns can be produced by a two-dimensional cellular automaton with appropriate simple rules.

11.7 A COMPUTATIONAL VIEW OF SYSTEMS

In Section 11.7 and in Section 11.8, we consider Wolfram's computational view of systems.[8] Here, in Section 11.7, we discuss computation as a framework for system principles, the concept of computational universality, cellular automata and computational universality, and some overall implications of computational universality. Section 11.8 covers Wolfram's Principle of Computational Equivalence. As we proceed in these two sections, I will note my observations and views when it seems appropriate.

[8] In Section 11.6, we discussed system *behavioral universality*. Sections 11.7 and 11.8 address system *computational universality*, which is a related concept and, according to Wolfram's views and arguments, seems to include behavioral universality.

11.7.1 Computational Framework for System Principles

Stephen Wolfram develops a computation-based framework for thinking about simple programs/simple rules. "In many respects the single most important idea that underlies [a new kind of science] is the notion of computation." We can think about *system processes* in terms of the computations of simple programs. The system initial conditions comprise the input to a computation. The state of the system after some number of process time steps is the output of the computation. To illustrate, Wolfram provides some elementary examples. He shows that cellular automata system processes can compute whether a given number is even or odd, or compute the square of any number. He will show later that "cellular automata can in fact perform what are in effect arbitrarily sophisticated computations" and that this type of computation ability applies to "any system whatsoever." Using the *correspondence between system processes and computation*, we can formulate some general system principles.

11.7.2 The Concept of Computational Universality

A system is computationally universal if it can emulate the computational capability of any other system. Computers and computer languages are, in that sense, universal. "Any computer system or computer language can ultimately be made to perform exactly the same set of tasks." With respect to natural systems, Wolfram says that computational universality "occurs in a wide range of important systems that we see in nature" and that "universality is . . . quite crucial in finding general ways to characterize and understand the complexity we see in natural systems."

Given that Wolfram has established a direct correspondence between computation and system processes, we might conclude that a computationally universal system can emulate not only the computational capability but also the behavioral complexity of any other system. Wolfram's language, throughout his discussion of computational universality in his book, seems to support this conclusion. For example, in Wolfram's words, "If a system is universal, then it must effectively be capable of emulating any other system." A universal system can "emulate any type of behavior that can occur in any other system." "There is a close connection between universality and the appearance of complex behavior."

If we can show that a known simple programs/simple rules system (e.g., a cellular automaton) is universal, then that system would be able to emulate the behavior of any other system (artificial or natural). The very important implication would be that the mechanism of simple programs/simple rules repeated over and over can produce computational universality—and can produce the (arbitrarily high) behavioral complexity of "any other system."

Can we show that cellular automata can be universal?

11.7.3 Computational Universality and Cellular Automata

We indeed can identify cellular automata that are universal. In fact, we can identify elementary cellular automata (one-dimensional, two-state, nearest-neighbor, simple-rule cellular automata) that are universal.

Let's look at this by cellular automata class. Wolfram reasons that systems that propagate information only locally, or in limited ways over larger distances, are not universal. He concludes that class 1 and class 2 cellular automata, therefore, cannot be universal. Furthermore, systems with "uncontrolled transmission of information" (e.g., every change is propagated everywhere), as in at least some class 3 cellular automata, seem inconsistent with universal behavior. Universality seems to require not only propagation of information, but also controlled propagation of information. Although he notes that he is not sure, Wolfram says some class 3 cellular automata may turn out to be universal. The extremely complex class 4, rule 110 cellular automaton is universal. Wolfram provides a proof of this. Wolfram "strongly suspects" that all class 4 cellular automata "will turn out ... to be universal."

Wolfram also states, "Out of the 256 elementary rules one expects that perhaps as many as 27 will in fact be universal." Because there are only a few class 4 elementary cellular automata, those 27 would have to include some class 3 systems. Given that class 3 is less complex than class 4, a serious question results. How could a less-complex class 3 system emulate a more-complex class 4 system? In my view, it could not. In the *complexity and simplicity* subsection of Section 11.5, I stated that class 3 systems can exhibit class 3, 2, and 1 (but not class 4) behavior.

My hypothesis[9] is:

Only class 4 systems (cellular automata or other systems) can be universal.

Regarding the universal class 4, rule 110 cellular automaton, Wolfram says, "It is truly remarkable that a system with such simple underlying rules should be able to perform what are in effect computations of arbitrary

[9] This is a hypothesis. Stephen Wolfram and/or others may or may not agree with it. Any corroboration or falsification evidence is welcome from anyone.

sophistication, but that is what its universality implies." Very simple programs/simple rules can yield universality. Because a universal cellular automaton can emulate any other system, "nothing fundamental can ever be gained by using rules that are more complicated than those for the universal cellular automaton."

Wolfram provides experimental evidence that the system emulation capabilities of computationally universal systems applies across system types. Universal cellular automata can emulate other types of systems, and other types of universal systems can emulate cellular automata. Wolfram shows that universal cellular automata "can actually be made to emulate almost every single type of system" discussed in his book, including mobile automata, Turing machines, substitution systems, register machines, systems based on numbers, and even practical computers in their entirety. Wolfram also shows that many types of universal "systems can emulate cellular automata." These system types include mobile automata, Turing machines, some configurations of substitution systems, symbolic systems, cyclic tag systems, register machines, and systems based on numbers. Wolfram adds this historical note: a universal Turing machine was first constructed in 1936 (by Alan Turing), and was the first example of computational universality seen in any system.

11.7.4 Additional Implications of Computational Universality

Here are some additional comments from Wolfram on the implications of computational universality.

> Universality is a threshold phenomenon: There seems to be a "threshold of universality"—and once reached, nothing more can be gained. The threshold is relatively "easy" to reach. Simple programs/simple rules can get you there.
>
> Universal systems are common and widespread: Universality is a common phenomenon. Universal systems with simple rules that yield highly complex behavior are widespread. "So if we study computation at an abstract level, we can expect that the results ... will apply to a very wide range of actual systems" in nature and elsewhere.
>
> Universality seems insensitive to system details: "Essentially all of these various kinds of systems [discussed by Wolfram]—despite their great differences in underlying structure—can ultimately be made to emulate each other" ... and are "capable of exactly the same kinds of computations." "From a computational point of view a very wide variety of systems, with very different underlying structures, are at some level fundamentally equivalent."

We'll continue with Wolfram's study of computation and universality in the next section.

11.8 THE PRINCIPLE OF COMPUTATIONAL EQUIVALENCE[10]

The Principle of Computational Equivalence states that computationally universal systems have equivalent computational capabilities. Universality suggests that there is "an upper limit on the sophistication of computations that can be done" and all universal systems can "achieve this limit." Wolfram has established (and we have already discussed) the correspondence between system computation on the one hand and system processes and behavior on the other. So the Principle of Computational Equivalence becomes a set of principles about computation, processes, and behavior. Wolfram has referred to the set as a "new law of nature."

Stephen Wolfram further supports the computation/process/behavior correspondence in his discussions of the Principle of Computational Equivalence. He says, for example, that "It is possible to think of any process that follows definite rules as being a computation." This correspondence is very broad—it "applies to essentially any process of any kind, either natural or artificial." Similarly, "behavior can be viewed as corresponding to a computation." Complex behavior "in essentially any system ... can be thought of as corresponding to a computation of equivalent sophistication." The computations "generate the behavior of the system."

I am thus led to a process/behavior version of the Principle of Computational Equivalence. In his discussion, Wolfram says, "There is essentially just one highest level of computational sophistication, and this is achieved by almost all processes that do not seem obviously simple." In my view, a process version of the Principle of Computational Equivalence might be stated as follows: universal systems have equivalent process capabilities, which can produce the highest level of process sophistication. I use the phrase *process capabilities* because universal systems do not always *exhibit* their highest level of process sophistication. As I noted in Section 11.5, class 4 systems can exhibit class 4, 3, 2, and 1 behavior.

Perception and analysis might be regarded as behavioral processes. Wolfram suggests that the Principle of Computational Equivalence may apply to them as well. Humans appear to be class 4 universal systems. Other observed universal systems can exhibit extremely complex behavior (or, at

[10] As you read this section, please keep in mind that the Principle of Computational Equivalence is Wolfram's hypothesis. He offers evidence to corroborate validity, but of course cannot provide proof with certainty. As twentieth-century philosopher Karl Popper has said, no nontrivial inductive hypothesis can be proven true with certainty. Only falsification can be proven.

least, behavior that humans perceive to be extremely complex). Wolfram says, "Observers will tend to be computationally equivalent to the systems they observe—with the inevitable consequence that they will consider the behavior of such systems complex." In my view, a perception/analysis version of the Principle of Computational Equivalence, therefore, might be something like this: human perception and analysis capabilities are computationally equivalent to the behavioral complexity capabilities of other universal systems. (The notion of human perceptual complexity being equivalent to nonhuman system computational complexity is quite controversial, and is related to artificial intelligence claims and "mind/brain" considerations. I'll have some further comments on this in the last section of the chapter.)

Wolfram discusses a tantalizing conjecture regarding perception and analysis. Perhaps there is extraterrestrial intelligence in the universe. Perhaps we humans do not possess the perception and analysis sophistication necessary to detect and understand what could be meaningful complex messages from the extraterrestrials. To us, it just seems like noise. These extraterrestrial "systems" could have a higher level of perception and analysis capabilities and a higher level of behavioral complexity than humans do. If this is the case, it seems to me the stated Principle of Computational Equivalence would be violated. Either that or we would have to come up with a new "extra-universal" system designation. While this is an intriguing discussion, let's get back to our attempt to understand the planet on which we live.

11.8.1 Computational Properties of Complex Systems

Universal systems—systems with the highest level of computational capability—can exhibit the properties of *computational irreducibility*, *undecidability*, and *intractability*. We will discuss Wolfram's views regarding these properties in the following paragraphs.

Most great historical triumphs of theoretical science have involved finding a reduced description of system behavior. That description is usually a mathematical formula. But highly complex systems have no reduced description; rather, they exhibit *computational irreducibility*—"a very fundamental [and common] phenomenon." For computationally irreducible systems, we cannot predict system behavior except by computing all (or almost all) steps of the system evolution. Per the Principle of Computational Equivalence, we humans are *not* more computationally sophisticated than other highly complex systems in nature and elsewhere. Traditional theoretical science approaches, therefore, can work only for simpler systems.

So to determine the behavior of a highly complex system, essentially we have to "run it" and see what happens. Class 4 cellular automata exhibit computational irreducibility. Cellular automata can be a very effective vehicle for modeling other highly complex systems. "To capture the essential features even of systems with very complex behavior it can be sufficient to use models that have an extremely simple basic structure. Given these models the only way to find out what they do will usually be just to run them. But the point is that if the structure of the models is simple enough, and fits in well enough with what can be implemented efficiently on a practical computer, then it will often still be perfectly possible to find out many consequences of the model."

I note again the important and pertinent modeling implication here: cellular automata could be a very effective vehicle for modeling the behavior of highly complex ecological systems.

Computational irreducibility can yield *undecidability*. Undecidability is defined as the inability to determine/predict the ultimate outcome (after infinite time steps) of system behavior. There is "no finite computation that will guarantee to decide it." Highly complex systems can be undecidable. In some cases, even attempts to compute finite outcomes (after a finite number of time steps) of complex system behavior can be daunting. The required computations might be very extensive, time-consuming, and not practical. Such systems are termed *intractable*.

Wolfram expresses his views on how all of this relates to traditional mathematics and physics. "Like most other fields of human inquiry mathematics has tended to define itself to be concerned with just those questions that its methods can successfully address." "Mathematics has tended to . . . concentrate just on what are in effect limited patches of computational reducibility in the network of all possible theorems." Mathematics avoids "confrontation with undecidability or unprovability." "The field of mathematics as it exists today will come to be seen as a small and surprisingly uncharacteristic sample of what is actually possible." . . . "This is no different from what has happened, say, in physics, where the phenomena that have traditionally been studied are mostly just those ones that show enough computational reducibility to allow analysis by traditional methods of theoretical physics."

The Principle of Computational Equivalence "implies that all the wonders of our universe can in effect be captured by simple rules, yet it shows that there can be no way to know all the consequences of these rules, except in effect just to watch and see how they unfold."

11.9 SUMMARY OF MY PERSPECTIVES ON THESE COMPLEX SYSTEM PRINCIPLES

Stephen Wolfram's innovative cellular automata investigations and his resulting insights regarding complex system principles are extremely important contributions. Throughout this chapter, I have added my own observations and perspectives when I deemed it appropriate. I believe my views to be valid, but I welcome differing views and evidence. First I will summarize some of my perspectives, and then provide a few thoughts on the relationship between the Principle of Computational Equivalence and the "mind/brain" system.

Here's my perspective on randomness and complexity. While random behavior is complex, (a) it includes no event-to-event correlation; (b) it achieves no self-organized system functions; and (c) it has a white-noise frequency spectrum rather than the power-law spectrum widely observed in the behavior of living systems and certain other systems in nature. Randomness does not represent the maximal complexity of well-functioning highly complex ecological systems (and living systems in general). I'll explain that further in Part IV of the book.

In *A New Kind of Science*, use of the term *fractal* seems to be limited to spatial "nesting" behavior. While that is accurate for relatively simple structure fractals, fractal behavior is a much richer concept than that. We will see the richness in Chapter 12, and then again in Chapter 15 (where we discuss process fractals). Fractal behavior can occur in class 3 and class 4 systems. In Chapter 15, I focus on process fractal behavior in class 4 ecological systems.

Regarding system information handling and communication, Wolfram notes that class 4 system structures "make it possible for information to be communicated from one part of a class 4 system to another." He also notes that although class 4 global communication is possible, it does not always occur. As I see it, highly complex ecological systems are class 4 systems. In Chapter 6 of this book, I described local-to-global communication as an important characteristic of ecological systems. In these systems, global communication does not always occur, but it occurs when required by the real environment and the current state of the system. This is consistent with Wolfram's work.

In my view, there is a direct correspondence between cellular automata classes and the attractors of nonlinear dynamics theory. My hypothesis is that there is a one-to-one correspondence between cellular automata classes and nonlinear dynamics attractors as follows:

- Class 1—Point attractor
- Class 2—Periodic attractor
- Class 3—Chaotic attractor
- Class 4—Edge-of-order attractor

The dynamics category of any system (not just cellular automata) may be referenced by either the class designation or the attractor designation. My perspective is that natural systems can be any class. Complex ecological systems (and complex living systems in general) typically are class 4.

Wolfram says, "To capture the essential features even of systems with very complex behavior it can be sufficient to use models that have an extremely simple basic structure. Given these models the only way to find out what they do will usually be just to run them. But the point is that if the structure of the models is simple enough, and fits in well enough with what can be implemented efficiently on a practical computer, then it will often still be perfectly possible to find out many consequences of the model." Wolfram expects that universal cellular automata with simple rules would be able to emulate the behavior of any other system (artificial or natural) of any complexity, including the highest complexity. To determine the behavior of a given highly complex system, we could run an appropriate cellular automaton and see what happens. I again emphasize the important and relevant modeling implication here: cellular automata can be a very effective vehicle for modeling the behavior of highly complex ecological systems. See Part V of this book.

Regarding the occurrence of computational universality, my perspective is: only class 4 systems (cellular automata or other systems) can be universal.

11.9.1 The Principle of Computational Equivalence and the Mind/Brain System

Based on the systems triad (see Chapters 3 and 5), we know that processes operate over structure and that process-over-structure delivers system functions. The mind/brain system can be viewed in this context. The mind can be understood as a set of processes that operate over the structure of the brain. This process-over-structure flow delivers neurological functions. The Principle of Computational Equivalence states that universal systems have equivalent process capabilities, and therefore equivalent process-over-structure capabilities. Given that the mind/brain system is a universal system and that a digital computer is a universal system, it would follow that a digital computer could emulate the mind/brain system and its processes, functions, and responses. Many artificial intelligence investigators and some cognitive neuroscience researchers would agree with that statement. Other mind/brain researchers do not agree. For example, what about processes involving human emotions, including responses to beauty and the arts? Surely, many mind/brain process-structure-function outcomes are biological phenomena that are not fully understood. The controversy persists.

11.9.2 Wolfram's insights . . .

To me, the most significant of Wolfram's many important insights in *A New Kind of Science* is his contention that the mechanism of simple programs/simple rules repeated over and over produces the highly complex behavior of natural systems. In my view, this mechanism may be a key part of emerging principles of ecological system dynamics. I will address that in Part VI, Chapter 19.

CHAPTER 12

Fractals: The Theory of Roughness

12.1 INTRODUCTION

In Chapter 11, we saw that the mechanism of simple programs with simple rules, repeated over and over, can yield extremely complex system behavior. And how can the resulting complex system behavior be characterized? Benoit Mandelbrot has said that "bottomless wonders spring from simple rules . . . repeated without end."[1] The bottomless wonders he was talking about are *fractals*. Complex system structure and dynamics—generated by simple programs with simple rules repeated over and over—are very often characterized by fractal behavior.

In my view, there are two fundamental types of fractals that are of interest to us in this book: structure fractals and process fractals. *Structure fractals* are spatial and are usually treated as static (at least in the short term). *Process fractals* are dynamic and have both spatial and temporal aspects. Here in this chapter, we focus on the pioneering work of Benoit Mandelbrot (as well as the work of others) in the area of structure fractals. Later, in Parts IV and V of the book, I focus on process fractal behavior.

The chapter begins with a discussion of the concept of fractals, which, as Mandelbrot explains, is the basis of a theory of roughness. The chapter then addresses three important areas of fractal geometry, which is the spatial geometry of fractal objects. (1) Fractal dimension, mathematical expressions for calculating fractal dimension, and the relationship between fractal dimension and Euclidian topological dimension are described. (2) Both human-made and natural fractal objects are discussed and examples are provided. Fractal object self-similarity across spatial scales is observed and discussed. (3) Because networks are a very important part of the work covered in this book, relevant aspects of the fractal geometry of networks are addressed in considerable detail. Next, the chapter explores the Mandelbrot Set. It is fractal and self-similar across spatial scales. It demonstrates the phenomenon of

[1] Benoit Mandelbrot (2010a) concluded his talk on *Fractals and the Art of Roughness* at the TED2010 conference with this statement.

Understanding Complex Ecosystem Dynamics
http://dx.doi.org/10.1016/B978-0-12-802031-9.00012-7

extreme complexity generated from a simple rule, repeated again and again. The chapter concludes with a discussion of my perspectives on fractals.

12.2 FRACTALS AND ROUGHNESS

The term "fractal" was coined by Benoit Mandelbrot in the 1970s. It was taken from the Latin *fractus*, meaning broken, fractured, unsmooth. The corresponding verb *frangere* means to break.

Euclidean geometry is the geometry of the artificial. It deals with perfect rectangles, circles, triangles, spheres, cones, and so on. There is almost none of that in nature. In *The Fractal Geometry of Nature*, Mandelbrot (1982) says, "Clouds are not spheres, mountains are not cones, coastlines are not circles, and bark is not smooth, nor does lightning travel in a straight line." In a similar quote, Mandelbrot (2010b) says, "Many shapes of nature—for example, those shapes of mountains, clouds, broken stones, and trees—are far too complicated for Euclidian geometry. Mountains are not cones. Clouds are not spheres. Island coastlines are not circles. Rivers do not flow straight. Therefore, we must go beyond Euclid if we want to extend science to those aspects of nature." Compared with Euclidean concepts, Mandelbrot (1982) says, "Nature exhibits not simply a higher degree but an altogether different level of complexity." Mandelbrot's *fractal geometry* accounts for this higher degree and different level of complexity. It accounts for the ubiquitous *roughness* in natural objects and natural events. Fractal geometry is the geometry of nature.

Mandelbrot (1982) further said, "Much of fractal geometry could pass as an *implicit* study of texture." By 2004, he had amplified and extended that idea: "I have engaged myself, without realizing it, in undertaking a theory of roughness. Think of color, pitch, heaviness, and hotness. Each is the topic of a branch of physics. Chemistry is filled with acids, sugars, and alcohols; all are concepts derived from sensory perceptions. Roughness is just as important as all those other raw sensations, but was not studied for its own sake." ... "Since roughness is everywhere, fractals—although they do not apply to everything—are present everywhere."[2]

In traditional mathematics, curves/functions typically are smooth, continuous, and differentiable; that is, they have a tangent at (almost) every point. In nature, this is rarely true; natural structure is quite irregular and fragmented. Traditional science dismisses irregularity. This is due in part

[2] Mandelbrot (2004), in an interview with the Edge Foundation.

to an "absence of tools to describe it mathematically" (Mandelbrot, 1982). Fractal geometry provides the tools. We will discuss fractal geometry in the next section.

12.3 FRACTAL GEOMETRY

This section describes the spatial geometry of fractal objects. It is about structure fractals. We describe the concept of fractal dimension, consider both human-made and natural fractal objects, and discuss the fact that fractal objects are self-similar at all spatial scales. Networks are a very important part of the work covered in this book. Relevant aspects of the fractal geometry of networks, therefore, are addressed in some detail.

12.3.1 Fractal Dimension

> *"I argue that roughness happens to be measured consistently by a quantity called fractal dimension, which happens in general to be a fraction, and which one can measure very accurately." ... "The idea is that fractal dimension is a proper measurement for the notion of roughness just as temperature is a proper measure for the notion of hotness."*
>
> ***Mandelbrot (2010b).***

Points, lines, squares, and cubes yield fractal dimensions of 0, 1, 2, and 3, respectively. This is consistent with Euclidian topological dimension measures. Fractal objects *may* have integer fractal dimension, but fractal dimension usually is not an integer. The relationship between fractal dimension and Euclidian topological dimension is

$$D > D_T$$

where D is the fractal dimension, D_T is the topological dimension and

$$\begin{aligned} D_T &= 0 \quad \text{for points} \\ D_T &= 1 \quad \text{for lines and curves} \\ D_T &= 2 \quad \text{for surfaces} \end{aligned}$$

According to Mandelbrot (1982), the types of fractal objects include dusts, fractal curves, and fractal surfaces.

A dust is comprised of a disconnected finite set of points. *Dusts* (usually) have fractal dimensions between 0 and 1:

$$\begin{aligned} D_T &= 0 \\ D &> 0 \end{aligned}$$

Fractal curves are irregular curves with fractal dimension greater than 1:

$$D_T = 1$$
$$D > 1$$

An example of a fractal curve is a coastline. A typical fractal dimension value for a coastline is approximately $D = 3/2$. So-called Koch curves can serve as models of a coastline.

Fractal surfaces have fractal dimension greater than 2:

$$D_T = 2$$
$$D > 2$$

Eddies in turbulent fluid flow can be considered fractal surfaces, with fractal dimension typically in the range 2.5–2.7.

It seems that D_T provides a lower bound on D within a type, but not an upper bound across types. For example, galaxies, with their physical distribution of "point" masses, appear to be "dustlike," yet their fractal dimension D seems to be about 1.23. Furthermore, "a shape with a fractal dimension D between 2 and 3 may be either 'sheetlike,' 'linelike,' or 'dustlike.'"

Mandelbrot (2010b) notes that there are several variants of the fractal dimension definition. He describes the similarity dimension D_s (for line-like self-similar shapes):

$$D_s = \frac{\log N}{\log\left(\frac{1}{P}\right)}$$

where N is the number of replicas of the whole and P is the linear reduction ratio of each replica.

The Koch curve and the Sierpinski gasket are each linearly self-similar with fractal dimension specified by D_s. Those two human-generated fractal objects will be discussed further momentarily. According to Csermely (2006), the fractal dimension of such a fractal object can be written as follows (with some slight notational modifications):

$$g = r^d$$

where g is the number of smaller self-similar structural elements in the next larger self-similar element, r is the ratio of the characteristic measure of the larger to smaller element, and d is the fractal dimension.

The Mandelbrot equation and the Csermely equation are equivalent. This can be shown by noting that $d = D_s$, $g = N$, $r = 1/P$—and by taking the log of both sides of the Csermely equation.

12.3.2 Human-Generated and Natural Fractal Objects

In this section, we will look at some human-made and natural fractal objects in more detail. You will notice two things. First, fractals are generated by simple rules repeated over and over. You will see that explicitly for human-generated fractals. Second, fractals look the same at different scales. You will see that human-generated fractals can appear exactly the same at different scales. Fractals in nature, on the other hand, are usually so only in the statistical sense; that is, their features are statistically, rather than exactly, the same when observed and measured at different scales. Fractal objects exhibit structural self-similarity at all scales; they are scale-invariant.

12.3.2.1 Human-Generated Fractal Objects

Two examples of human-generated fractals are provided here. The generation mechanism takes the form of a recursive process in which self-similar objects of different size are repeated. Figure 12.1 illustrates the development of a Koch curve. (In the early 1900s, Helge von Koch developed a family of curves. The example here is the triadic Koch curve.) Figure 12.2 illustrates the development of a so-called Sierpinski gasket. In each figure, parameter n is the iteration number. The image with the highest iteration number (in each of the two figures) is, of course, the more fully developed view of the fractal object. Clearly, these fractals are generated by simple rules repeated over and over.

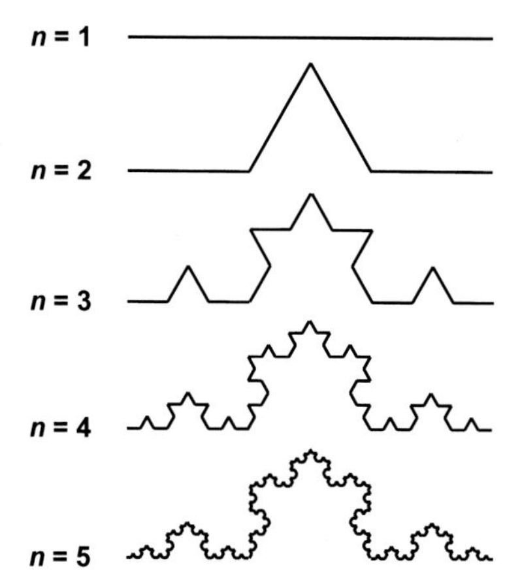

Figure 12.1 Human-generated fractal object—Koch curve.

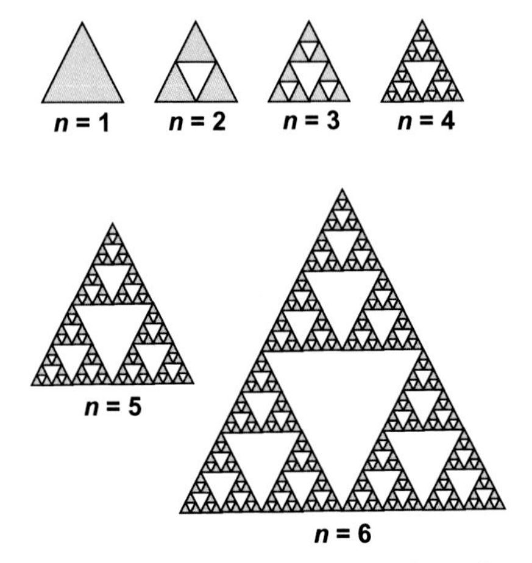

Figure 12.2 Human-generated fractal object—Sierpinski gasket.

Using the Mandelbrot equation given earlier, the fractal dimensions of the two objects can be calculated:

$$D = \frac{\log N}{\log\left(\frac{1}{P}\right)}$$

Koch curve:

$$N = 4 \quad P = 1/3;$$
$$D = \frac{\log 4}{\log 3} \cong 1.2618\ldots$$

Sierpinski gasket:

$$N = 3 \quad P = 1/2;$$
$$D = \frac{\log 3}{\log 2} \cong 1.5849\ldots$$

Solé and Bascompte (2006) derive an expression for the length of a Koch curve fractal and show that the length goes to infinity as the number of iterations goes to infinity:

$$L_n = \left(\frac{4}{3}\right)^n \text{ and } L_n \rightarrow \infty \text{ as } n \rightarrow \infty$$

where L_n is the length and n is the number of iterations.

Koch curves can serve as a model of a coastline. Mandelbrot (1982) says, "The typical coastline's length is very large and so ill determined that it is best considered infinite." Length, therefore, is not an adequate measure in such cases.

Because only a small number of iterations are necessary to demonstrate the Koch curve and the Sierpinski gasket fractal objects, I was able to generate Figures 12.1 and 12.2 manually. When a very large number of iterations are required, an automated method is essential. The iterated function system approach is such a method.

12.3.2.2 Iterated Function Systems

Michael F. Barnsley (1993), the source for this subsection, states that "iterated function systems provide a convenient framework for the description, classification, and communication of fractals." Iterated Function Systems (IFS) are based on mathematical *affine* transformations.

A transformation w of the form

$$w(x_1, x_2) = (ax_1 + bx_2 + e, cx_1 + dx_2 + f)$$

where a, b, c, d, e, and f are real numbers, is called a (two-dimensional) affine transformation, which can be written in matrix/vector notation as

$$w(x) = w\begin{bmatrix} x_1 \\ x_2 \end{bmatrix} = \begin{bmatrix} a & b \\ c & d \end{bmatrix}\begin{bmatrix} x_1 \\ x_2 \end{bmatrix} + \begin{bmatrix} e \\ f \end{bmatrix}$$

There are basically two algorithms for computing fractals from IFS:

1. Deterministic algorithm
2. Random iteration algorithm

We will use the random iteration algorithm and consider hyperbolic IFS of the form

$$\{w_i : i = 1,\ 2,\ \ldots,\ n\}$$

where each mapping is an affine transformation. The random iteration algorithm also requires a probability p_i for each affine transformation such that

$$\sum_i p_i = 1 \text{ and } p_i > 0 \text{ for } i = 1, 2, \ldots, n$$

For a given fractal object, the IFS parameters a, b, c, d, e, f, and p can be provided in an IFS code table in which the columns are the parameters and each row represents one transformation. We will look at two IFS-generated fractal objects in the next subsection.

12.3.2.3 Natural Fractal Objects

The two fractal illustrations provided here are actually not natural fractal objects, but they are *naturalistic*. They are human-generated fractals that follow fractal object construction principles, and they closely mimic natural

Table 12.1 IFS Code for Two "Natural" Fractal Objects

a	*b*	*c*	*d*	*e*	*f*	*p*
Black spleenwort fern						
0.00	0.00	0.00	0.16	0.00	0.00	0.01
0.85	0.04	−0.04	0.85	0.00	1.60	0.85
0.20	−0.26	0.23	0.22	0.00	1.60	0.07
−0.15	0.28	0.26	0.24	0.00	0.44	0.07
Oak tree						
0.050	0.000	0.000	0.600	0.000	0.000	0.167
0.050	0.000	0.000	−0.500	0.000	1.000	0.167
0.460	−0.321	0.386	0.383	0.000	0.600	0.167
0.470	−0.154	0.171	0.423	0.000	1.100	0.167
0.433	0.275	−0.250	0.476	0.000	1.000	0.167
0.421	0.257	−0.354	0.306	0.000	0.700	0.165

fractal objects. It seems to me that this is a very effective way of demonstrating the fractal behavior of natural objects. We already know what a real natural fern or tree looks like. We can compare those natural objects with their naturalistic counterparts that we definitely know to be fractal. We see the obvious similarity/sameness. We see, therefore, that the natural objects are clearly fractal.

I use the approach of Barnsley (1993), as previously discussed, and provide a fern illustration and a tree illustration. The respective IFS code table information is given in Table 12.1.

To generate the fractal images, I have used a Macintosh application developed by J.D. Meiss (2012) for drawing IFS fractals. Meiss says, "this is a rudimentary version of a MacOS application, and as such is to be thought of as a rough outline of a complete program." That is true; however, the program does an effective job of implementing the IFS algorithms and displaying the resulting images.

The fern and tree fractal illustrations are displayed in Figures 12.3 and 12.4, respectively. Consider the Black Spleenwort fern depicted in Figure 12.3. Note that each fern major "branch" (*frond*), each of the leafy "blades" (*lamina*) of each frond, each of the leaflets (*pinnae*) that comprise each of the lamina—as well as the entire fern (the entire set of fronds)—are self-similar. Ferns exhibit self-similarity at all scales; they are fractal. Next consider the oak tree depicted in Figure 12.4. The tree's

Figure 12.3 "Natural" fractal object—fern.

Figure 12.4 "Natural" fractal object—tree.

major branches, minor branches, twigs, and so on, as well as the entire tree, are self-similar. Tree structure is self-similar at all scales (scale-invariant) and is fractal.

Natural fractals are everywhere—in ecological systems and other natural complex systems. Some natural fractal objects have the characteristic of maximum surface area for a given volume. This characteristic is very beneficial for filtering and purification processes. You will not be surprised, therefore, to learn that human lungs and natural wetlands are two more examples of natural structures that are rich in fractals.

12.3.2.4 Fractal Objects and Power Laws

In Chapter 8, we noted that node degree in complex networks is often scale-invariant with a degree distribution that follows a power law. In Chapter 9, we explained that other important network parameters—many in the dynamics realm—can also have scale-invariant/power-law distributions. The term *fractal* is essentially defined as *self-similar at all scales*. The common feature of fractal behavior is the presence of power laws. The reason why these laws are characteristic of fractals is that they are, in fact, the only functions that display invariance under scale change.

Here, in this chapter, we have seen that fractal objects exhibit self-similarity at all scales (scale-invariance). Fractal objects look the same at different scales. You can clearly see that in Figures 12.1–12.4. The self-similar components that comprise a fractal object have a power-law size distribution characterized by the object's fractal dimension. "The power law is a fundamental indicator of fractal-ness."[3] If s = self-similar component size, D = object fractal dimension, and C = a constant, then $N(s)$, the distribution of the self-similar component sizes that comprise the overall fractal object, is given by

$$N(s) = C\, s^{-D}$$

A plot of $N(s)$ versus s yields the familiar long-tailed power-law curve. Figure 12.5 displays power-law plots in both normal coordinates and log-log coordinates.

If you were to view a collection of power-law plots with abscissa values that differ by some scaling factor, you would see that all of the plots (all normal plots or all log-log plots) have the same shape. A change of scale does not modify the distribution's basic behavior.

[3] Flake and Pennock (2010).

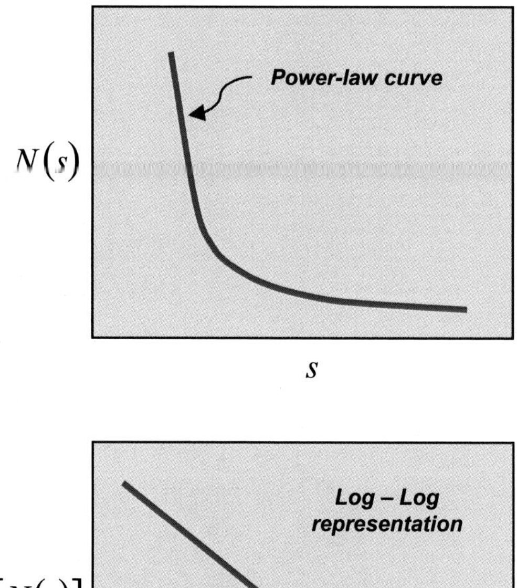

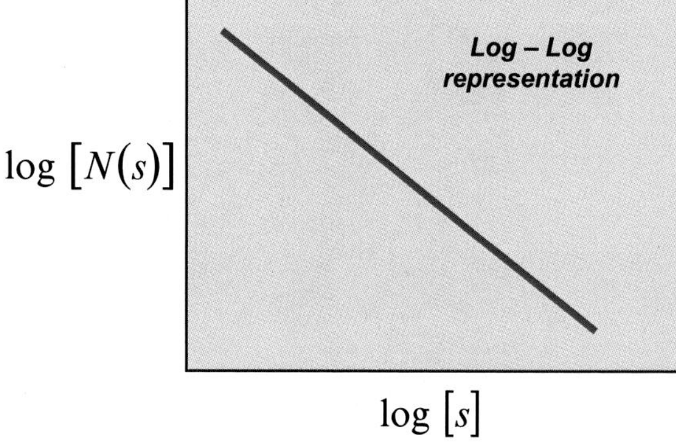

Figure 12.5 Power-law plots.

12.3.3 Fractal Geometry of Networks

Networks are a very important part of my work. The fractal behavior of networks likewise is important. The extant network fractal literature focuses on geometry (i.e., on structure). Networks are viewed as static structures and the fractals considered are network structure fractals. Fractal geometry and network structure fractals are discussed in this section. Later in the book, we will focus on dynamics and network process fractals.

Back in Chapter 8, we showed that many real-world networks (including ecological networks) are scale-invariant with respect to node degree and have power-law degree distributions. This indicates that, for such networks, node degree is fractal. How about network path length? In the following paragraphs, the fractal geometry of network path length is addressed.

First note that, for some network representations, we can visually observe the physical spatial layout of the network. The structural

Figure 12.6 Structure fractal behavior in a percolating lattice network. *(Adapted, with permission, from Albert, R. and Barabási, A-L., Reviews of Modern Physics, Volume 74, January 2002, pp. 47–97. Copyright (2002) by the American Physical Society (http://link.aps.org/)).*

Figure 12.7 Structural depiction of a complex network. *(Adapted with permission from Macmillan Publishers Ltd: NATURE (http://www.nature.com/), Strogatz, S.H., Exploring Complex Networks, Volume 410, Issue 6825, pp. 268–276, March 8, 2001, copyright (2001)).*

self-similarity across scales and, therefore, the structure fractal behavior can be seen explicitly. Such is the case for the simple two-dimensional lattice network at percolation shown in Figure 12.6.[4]

When considering representations of more complex network configurations and considering network path length, it is not so easy to visually observe fractal behavior. One reason is that complex networks are often depicted as shown in Figure 12.7. In typical fashion, this network diagram has been redrawn by rearranging the locations of the nodes in order to emphasize node degree behavior and to reduce or eliminate messy network link crossings. The actual layout of the network is lost. When dealing with network path length, another visual complication is the fact that network

[4] Network percolation is described in Chapter 9.

path length itself is not a physical spatial measure (except in certain regular lattice networks). In general,

- *Path length* is not a physical length; rather, it is the number of links traversed from node i to node j.
- *Shortest path length* from node i to node j is the path length with the smallest number of links.
- *Characteristic path length* is the average of all shortest paths in the network.
- *Network diameter* is the longest of all the shortest paths in the network.

Several methods, however, have been developed to determine the power-law distribution and the fractal dimension of complex networks that exhibit fractal behavior with respect to path length. One approach is the "box covering method."

12.3.3.1 *Description of the* Box Covering Method[5,6]

For a given network and a box size ℓ_B, a box is defined as a set of nodes for which the shortest path length ℓ_{ij} between any two nodes i and j in the box is smaller than ℓ_B. The entire network is covered with such boxes. The minimum number of boxes required to cover the network is denoted by N_B. Note that

When $\ell_B = 1$, $N_B = N$ where N = the number of nodes in the network
When $\ell_B \geq \ell_B^{max}$, $N_B = 1$ where ℓ_B^{max} = the network diameter + 1

Figure 12.8 shows an example of the box covering procedure for a very simple network. For each value of the box size ℓ_B, the number of boxes needed to tile the entire network is determined such that each box contains nodes separated by a distance $\ell < \ell_B$.

If the network is fractal with respect to path length, it can be shown that

$$N_B(\ell_B) \propto \ell_B^{-d_B}$$

where d_B is the network path length fractal dimension. $N_B(\ell_B)$ is a measure of the distribution of path lengths in the network. (More precisely, it is a measure of the distribution of path lengths most likely to be used, that is, the distribution of shortest path lengths in the network.)

Plots of path length distributions using data from several types of real-world fractal complex networks are provided in Figure 12.9. Values of the fractal dimension, d_B, are calculated.

5 Song et al. (2007).
6 Song et al. (2005).

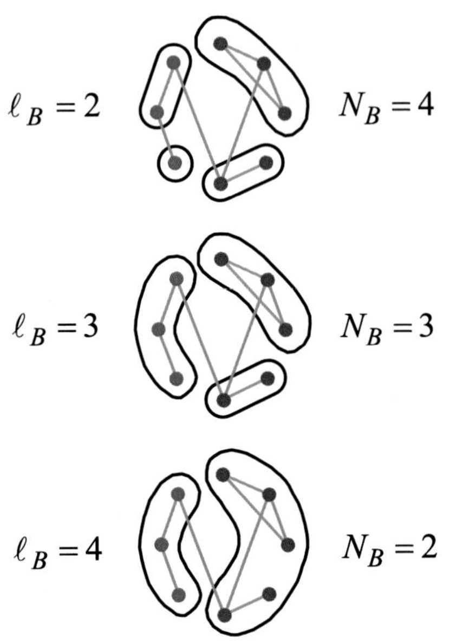

Figure 12.8 Example of box covering procedure. *(Adapted with permission from Macmillan Publishers Ltd: NATURE (http://www.nature.com/), Song, C., Havlin, S., and Makse, H. A., Self-Similarity of Complex Networks, Volume 433, Issue 7024, pp. 392–395, January 27, 2005, copyright (2005)).*

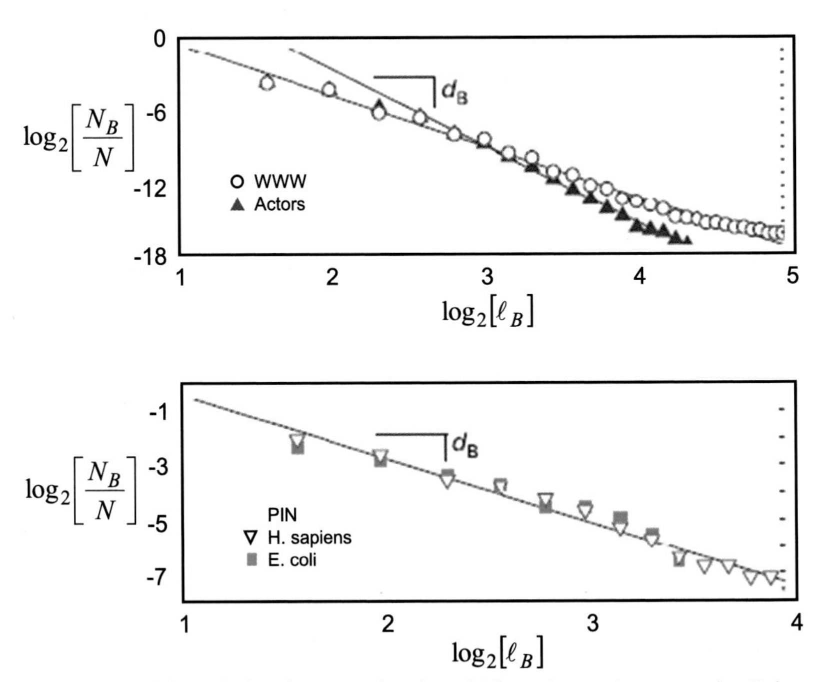

Figure 12.9 Path length distributions of real-world fractal complex networks. *(Adapted with permission from Macmillan Publishers Ltd: NATURE (http://www.nature.com/), Song, C., Havlin, S., and Makse, H.A., Self-Similarity of Complex Networks, Volume 433, Issue 7024, pp. 392–395, January 27, 2005, copyright (2005)).*

The distributions in Figure 12.9 plot N_B/N, where N= the number of nodes in the network, versus ℓ_B. Because these are log-log plots, the straight lines fit to the data represent power laws that indicate network path length fractal behavior:

$$\frac{N_B(\ell_B)}{N} \propto \ell_B^{-d_B}$$

In the upper diagram of Figure 12.9, the World Wide Web (WWW) data is for a portion of the web network composed of 325,729 web pages that are connected if there is a URL link from one page to another. The actors data is for a social network where the nodes are 392,340 actors who are linked if they were cast together in at least one film. Values of the fractal dimensions, d_B, are calculated as

$$\text{WWW}: \; d_B = 4.1$$
$$\text{Actors}: \; d_B = 6.3$$

In the lower diagram of Figure 12.9, the protein interaction network (PIN) data is for the biological networks of protein-protein interactions found in *Homo sapiens* (946 proteins) and *Escherichia coli* (429 proteins). Proteins are linked if there is a physical binding between them. Values of the fractal dimensions, d_B, are calculated as

$$\textit{Homo sapiens}: \; d_B = 2.3$$
$$\textit{Escherichia coli}: \; d_B = 2.3$$

Note that each network path length distribution is driven by the propagation processes occurring on that network over time. The previous analysis treats each network as a resulting static structure and addresses only spatial aspects of the path length distribution. Path length behavior, however, is dynamic and changes with time. To get the full picture of path length fractal behavior, we need to include the temporal aspects. We will do that in Part V of the book, with respect to our model of the propagation process in ecological networks.

12.4 THE MANDELBROT SET

"A priori, one would expect the construction of complex shapes to necessitate complex rules, but surprisingly, it is not so."

Benoit Mandelbrot (2010b)

Benoit Mandelbrot became very interested in the mathematical sets known as Julia Sets (named for Gaston Julia, Mandelbrot's uncle's teacher).

Mandelbrot wanted to know "the overall pattern behind the whole family of Julia Sets. He decided to make a map of them, which map is now called the Mandelbrot Set."[7] The underlying, recursive equation of interest here is

$$z_{n+1} = z_n^2 + c \quad n = 0, 1, 2, \ldots$$

where z and c are complex variables (have real and imaginary parts). This is the simple rule that generates Julia Sets and the Mandelbrot Set. "We apply this rule over and over again, taking the outcome of one transformation as the input for the next. . . . For each point the rule gives us the location of the next point in the sequence."

If we iterate $z = z^2 + c$ for an array of different z_0 values, keeping c fixed, we get a Julia Set for that value of c. Different values of c yield different sets. Julia Sets, therefore, are a family of sets.

If we iterate $z = z^2 + c$ for one particular value of z_0 (that is, $z_0 = 0$, the so-called critical point) and an array of different c values, we get the Mandelbrot Set—one single, unique set.

Figure 12.10 provides a complex plane depiction of the Mandelbrot Set. The abscissa specifies the real part of complex variable c and the ordinate specifies the imaginary part of complex variable c.

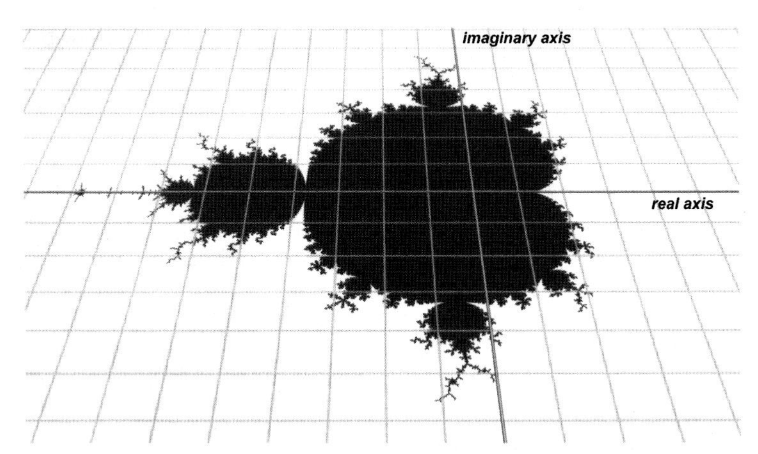

Figure 12.10 The Mandelbrot set. *(Reprinted, with permission, from Rood, W., Fractal Limits, Chapter 5 in the book The Colours of Infinity, N. Lesmoir-Gordon (ed.), Springer-Verlag London Limited, p. 77, 2010. Copyright © Springer-Verlag London Limited 2010).*

[7] This first part of the Mandelbrot Set discussion (and the quotes) is from Will Rood (2010).

For a given value of c (real, imaginary):

If the rule (i.e., $z = z^2 + c$ with $z_0 = 0$) goes to infinity, that c-point is outside the Mandelbrot Set—in the white space of the complex plane in the figure.

If the rule remains bounded, then that c-point is a member of the Mandelbrot Set—represented by the black area in the figure.

Will Rood (2010) describes the Mandelbrot Set this way: "The main body of the Mandelbrot Set consists of a cardioid, or heart-shaped core, surrounded by infinitely many circular buds. Each bud is surrounded by a further infinity of smaller buds, and, at the end of each of these chains of buds, a spiral frond, sometimes lacy and floral, sometimes straight and spiky. The fronds, which comprise the boundary of the Mandelbrot Set, actually consist of infinitely many miniature copies of the whole shape, joined together by bifurcating threads of ever-smaller miniatures."

Benoit Mandelbrot (2010b) provides further explanation of his Mandelbrot Set (M): For a given value of c, the resulting complex-plane "orbit" can either go to infinity or not. If the orbit goes to infinity, then the point corresponding to the given value of c is outside M. If the orbit does not go to infinity, then the point is contained within the set M. M, therefore, is the set of all points c such that the sequence z_n does not escape to infinity. For points inside M, the orbit is orderly (asymptotically periodic). For points outside M, the orbit, although described by a deterministic recursive equation, is chaotic. The boundary of the Mandelbrot Set is the boundary between order and chaos.

The Mandelbrot Set is fractal. There is self-similarity across scales, but in an approximate sense rather than an exact sense. At smaller and smaller scales, we see slightly different miniatures of the whole connected together in various ways. Mitsuhiro Shishikura (1998) has calculated a value of 2 for the (Hausdorff) fractal dimension of the boundary of the Mandelbrot Set.

Recall our analysis of the logistic map equation in Chapter 10. That recursive equation is quite similar to the Mandelbrot Set generating equation. It can be shown that there is a direct correspondence between the Mandelbrot Set and the logistic map bifurcation diagram that we illustrated and discussed in Chapter 10. Both have a defined demarcation between order and chaos. The region outside the Mandelbrot Set boundary is characterized by chaotic behavior and a strange (chaotic) attractor. In Stephen Wolfram's terms, that is class 3 system behavior. Benoit Mandelbrot (1982) notes that this behavior is fractal, and further claims that "every known 'strange' attractor is a fractal." (Recall from Chapter 11 that class 3 and class 4 systems can exhibit fractal behavior.) Mandelbrot says the fractal dimension has been evaluated for many such attractors. "In all cases, $D > D_T$." (D is fractal dimension and D_T is topological dimension.)

The Mandelbrot Set shows us that the criterion separating order and chaos is clear, but the boundary is extremely "messy." It exhibits both self-similarity and change. It is astonishingly complex, yet is generated by a very simple rule repeated over and over. I would like to emphasize further the important point that extraordinary complexity occurs at the boundary between order and chaos. This is where extremely complex behavior lives. We will discuss that further in Chapter 14.

In addition to all of its important scientific implications, the Mandelbrot Set has much esthetic and artistic appeal as well. Will Rood (2010) notes that the behavior of points outside the Mandelbrot Set can be indicated in various arbitrary ways with arbitrary coloring. For example, "colouring these points, according to whether or not they go to infinity, will produce pictures of enormous complexity." Or contours can be generated that "illustrate the level sets of points that escape [to infinity] in equal time." Adding variations in escape time and escape angle behavior yields even more fascinating pictures. All of these can be quite beautiful.

The Mandelbrot Set demonstrates the phenomenon of extreme complexity generated from a simple rule, and characterized by fractal behavior.

12.5 MY PERSPECTIVES

12.5.1 Simple Rules, Complexity, and Fractals

A pillar of Stephen Wolfram's (2002) A *New Kind of Science* is the premise that simple programs with simple rules, repeated over and over, can yield extremely complex system behavior. Benoit Mandelbrot[8] has concluded that "bottomless wonders spring from simple rules ... repeated without end." The bottomless wonders are fractals. Will Rood (2010) concurs and adds that the fractal Mandelbrot Set has "been called the most complex shape known to man" ... "and yet it is generated by a formula of surprising simplicity" applied "over and over again."

The extreme complexity observed by Wolfram in his cellular automata explicit experimentation and the bottomless wonders observed by Mandelbrot in his fractal investigations have the same generating mechanism! The same fundamental simple rules mechanism produces complexity and

[8] Benoit Mandelbrot concluded his talk on *Fractals and the Art of Roughness* at the TED2010 conference in February 2010 with this statement.

fractals. Complexity is widespread in nature. Fractals are widespread in nature. Why? It seems that it is because they have a widespread common source. It should be no surprise, then, that extremely complex natural systems are characterized by fractal behavior.

In my view, the simple rules mechanism and the fractal behavior characterization may be the key parts of the emerging principles of complex ecological system dynamics. I comprehensively address a behavior characterization hypothesis in Parts IV and V of the book. Then, in Part VI, I suggest an integrated mechanism/characteristics hypothesis for going-forward consideration.

12.5.2 On to Ecological Network Dynamics

At the beginning of this chapter, I said that, in my view, there are two fundamental types of fractals: structure fractals and process fractals. Structure fractals are spatial and are usually treated as static. Process fractals are dynamic and have both spatial and temporal aspects. This chapter has focused on static *structure fractals*.

One of our references here has been Benoit Mandelbrot's Chapter 3 (Mandelbrot, 2010b) in the book *The Colours of Infinity*.[9] Mandelbrot had used the material in that chapter previously for a presentation at Gustavus Adolphus College in 1990.[10] Questions and comments from the audience of the presentation have been included at the end of Chapter 3 in *The Colours of Infinity*. One audience member commented, "Somehow the geometry is very static, and to me the static should best be seen with deeper understanding as flowing from the dynamics." More of a dynamical perspective may be needed. Another in the audience agreed. I agree as well.

In his book *The Fractal Geometry of Nature*, Mandelbrot (1982) says that his "work is concerned primarily with shapes in the real space one can see" while dynamics work "is ultimately concerned with the temporal evolution in time . . .." My work is concerned primarily with the latter.

I am concerned primarily with the dynamics of complex ecosystems. I am concerned with the spatial and temporal aspects of ecological network *process fractals*. We will proceed with this focus in Part IV.

[9] Nigel Lesmoir-Gordon (2010).

[10] Benoit Mandelbrot, *A Lecture on Fractals*, presented at the 1990 Nobel Conference at Gustavus Adolphus College, St. Peter, Minnesota.

PART IV

A View of the Characteristics of Ecological Network Dynamics

Based on knowledge of the systems approach (Part I), the function-structure-process dynamics framework (Part II), and applicable complex systems theory (Part III), we synthesize, in this part, a comprehensive view of the characteristics of the propagation dynamics of ecological system networks. This view, in turn, is the basis for the central hypothesis of the book: ecological networks are ever-changing, "flickering" networks with propagation dynamics that are punctuated, fractal, local-to-global, and enabled by indirect effects.

CHAPTER 13

Issues of Human Perception

13.1 INTRODUCTION

In this book, we are attempting to understand the dynamics of complex systems (and, in particular, ecological systems). So far, we have addressed mostly the related "scientific" issues. Before proceeding with that, I think it is important to discuss the human perceptual context in which we are working, especially the human tendency to see smoothness, stability, and continuity in the natural world—even when they are absent.

Here's a list of the topics covered in the chapter:

- Is Reality Knowable?
- Limitations of Human Perception
- Roughness versus Smoothness
  - The Typical Human Perceptual Domain
  - Analysis of the Relationships Between Scale and Perception
- Modeling and Human Perception
  - Models are Simplifications
  - Differential Equation Models and Steady State Behavior

This chapter begins with a question (is reality knowable?) and an attempt to answer it. What is typically called "truth" means agreement with human perception of reality. But that is not absolute; it changes with time, with people, and with paradigms. Reality and absolute truth, therefore, may be unknowable to humans. A lot depends on perceptions. Human perception capabilities, however, have significant limitations. Some specific limitations are discussed and described. Although discontinuity and roughness are pervasive in nature, humans often do not see it. The human perceptual preference for smoothness—while ignoring roughness—is explored and examined in some detail in the chapter. My view of the typical human perceptual domain is discussed. Human "smoothing effects" can be observed in both spatial scales and temporal scales. Two supporting analyses that investigate the relationships between scale and perception are provided.

Understanding Complex Ecosystem Dynamics
http://dx.doi.org/10.1016/B978-0-12-802031-9.00013-9

The preference for smoothness extends into our modeling activities. Most of our system models "smooth" reality in some way. This can be useful to achieve insights. The insights, of course, are valid only when interpreted in the context of the smoothing assumptions.

13.2 IS REALITY KNOWABLE?

David Bohm (1983) posits that there exists a fundamental underlying reality that may be unknowable to humans. We can call it *implicate order*. There exist many different human interpretations or perceptions of this one reality. We can call those *explicate order*. Bohm hypothesizes that implicate order is primary and explicate order is derived. He says, "What we are proposing here is that the implicate order . . . be taken as fundamental." Explicate order "is secondary, derivative, and appropriate only in certain limited contexts." "Explicate order can be regarded as a particular or distinguished case of a more general set of implicate orders from which latter it can be derived." The explicate order is what we are capable of sensing with our human sensory equipment and our human epistemic abilities. There is one universal implicate order, but there are multiple explicate orders. Explicate orders are abstractions that furnish approximations, within limits, in particular domains (sets of conditions). "All our different ways of thinking are to be considered as different ways of looking at the one reality, each with some domain in which it is clear and adequate."

Conflicts in the explicate world can arise and, at least in some cases, can be explained. Bohm (1983) discusses the case of three major theories in physics: classical Newtonian physics, relativity theory, and quantum mechanics. Relativity theory apparently conflicts with quantum theory, and both differ from classical physics. Quantum theory and relativity theory "differ radically in their detailed notions of order." In quantum mechanics, movement "is discontinuous, not causally determinant and not well defined." In relativity, "movement is continuous, causally determinant and well defined." Some view classical physics as a simplified subset of relativistic physics. In Bohm's view, all of these explicate theories are derived from a consistent underlying implicate order and each has validity, but just in its particular limited domain of application. In the explicate world, perhaps "it has no meaning to talk of a fundamental theory, on which all of physics could find a permanent basis, or to which all the phenomena of physics could ultimately be reduced. Rather, each theory will abstract a certain aspect that is relevant only in some limited context . . . ." It seems that "all theories are insights, which are neither true nor false but, rather, clear in certain domains, and unclear when extended beyond these domains."

Are there absolute truths? Philosopher Karl Popper (1999) sheds some light on this topic in the context of the following science development scenario. In the 1500s, the prevailing belief (due to Aristotle, who lived 384–322 BC) was that the earth was the center of the universe. Copernicus (who lived 1473–1543) said no; the earth revolves around the sun. Kepler went further; he determined that the planetary orbits were elliptical and described properties of the orbits. Galileo significantly improved the telescope, observed the motion of the planets, and deduced the principles of gravity. Newton explicitly described and calculated the gravitational forces at work in planetary motion. Einstein showed that Newton's laws were approximations – and reached bold, revolutionary conclusions about force, space, and time. But Einstein felt that his theories were approximations also, and only a piece of a larger, unified field theory. This sequence of *approximations* might be moving in the direction of truth, but is it?

David Bohm (1983) states, "There is evidently no reason to suppose that there is or will be a final form of insight (corresponding to absolute truth) or even a steady series of approximations to this." This is consistent with Thomas Kuhn's views. Kuhn (1996) says that the science developmental process is "a process of evolution *from* primitive beginnings—a process whose successive stages are characterized by an increasingly detailed and refined understanding of nature. But nothing that has been or will be said makes it a process of evolution *toward* anything." Kuhn explains that our discoveries (and errors) depend upon the paradigms we adopt for viewing the world. Different paradigms produce different views of the world and different "truths." William James (who lived 1842–1910), in *Pragmatism*, says, "Truth is a property of certain of our ideas. It means their agreement, as falsity means their disagreement, with reality."[1] What we call "truth" means agreement with human perception of reality. But that is not absolute; it changes with time, with people, and with paradigms.

Here's some food for thought. Is the *process* of perceived truth fractal? As we have seen, perceived truth is not static or persistent—it is dynamic, and changes with time and circumstance. Most changes are small, but a few (e.g., paradigm shifts) are extremely large. This collection of change events may follow a power law. Perhaps perceived truth behavior can be characterized as a process fractal.

Reality and absolute truth apparently are unknowable to humans. We conclude this section with a pertinent quote: "It must be remembered," said Vaihinger, "that the object of the world of ideas as a whole is not the

[1] This William James quote is from *The Practical Cogitator*, Houghton Mifflin Company, 1962.

portrayal of reality—this would be an utterly impossible task—but rather to provide us with an instrument for finding our way about in this world more easily."[2]

13.3 LIMITATIONS OF HUMAN PERCEPTION

We see reality through "many-colored lenses which paint the world their own hue, and each shows only what lies in its focus."[3]

Ralph Waldo Emerson

Henry David Thoreau "became increasingly convinced that we see only what we are prepared to see, that we find, not the world as it is, but the world we look for."[4]

Robert D. Richardson Jr.

We see not reality, but our perceptions of reality. Human perception capabilities, however, have significant limitations. Francis Bacon, in his *Novum Organum* (first published in 1620), addresses these limitations in the *Idols of the Mind*. Stephen J. Gould and Carolyn Bloomer add further thoughts on the subject. Stephen Wolfram expresses his views on the processes of human perception and analysis. These contributions are expanded upon in the next several paragraphs.[5]

A very good analysis of mental biases in scientific work was done long ago by Francis Bacon.[6] This is somewhat ironic because Bacon is very often identified with "objectivity." His reputation for objectivity came, in part, from "the tables for inductive inference" in his *Novum Organum*. Bacon provided tabular procedures "for stating and classifying observations, and for drawing inductive inferences therefrom." This seemed to be a method of pure observation and reason, with no messy human components. In the same work, however, Bacon talked of "mental and social impediments" that prevent pure objectivism—called the *idols*. Bacon was actually a spokesperson for a "concept of science as a quintessential human activity." In *Novum Organum*, Bacon presented his *Idols of the Mind*. In that analysis, he explained that we have two major sensory deficiencies: *destitution*, which involves limits on the physical ranges of human perception, and *deception*, which

[2] This quote is also from *The Practical Cogitator*, 1962.

[3] Ralph Waldo Emerson (1983).

[4] Robert D. Richardson Jr. (1986).

[5] Some of these contributions are also relevant, and mentioned, elsewhere in the book: Chapter 2 (Gould's views) and Chapter 11 (Wolfram's views).

[6] Gould (2003) is the source of this description of Bacon's work.

involves internal biases that we impose on external nature. These deficiencies yield *idols*, and Bacon identified four idols in two categories. The first category is *attracted idols*, which deals with social and ideological biases imposed from without, and includes

- *Idols of the theater*—"imposed by old and unfruitful theories that persist as constraining myths"
- *Idols of the marketplace*—"limitations arising from false modes of reasoning"

The second category is *innate idols* that "inhere in the nature of the intellect." This category includes

- *Idols of the cave*—each individual's biases
- *Idols of the tribe*—biases inherent in the evolved human mind (e.g., linearity biases and a propensity for dichotomy)

Bacon warned, "Beware the idols of the mind." We must understand these idols, so that we can account for them in our thinking.

Stephen J. Gould (2003) adds his further thoughts to Bacon's analysis. Gould agrees that science has a subjective component. There is a powerful myth about scientific procedure, which says that science is an objective activity, with no mental biases as in the humanities. Not true—science does have subjective aspects. Gould says, "The peculiar notion that science utilizes pure and unbiased observations as the only and ultimate method for discovering nature's truth, operates as the foundational (and ... pernicious) myth of my profession." In fact, we look for patterns consistent with our view of some hypothesis (for or against). Bias enters when we are unwilling to abandon preferences that observation refutes. "Universal cognitive biases affect the work of scientists as strongly as they impact any other human activity ... ." Humanists know this, but scientists mostly can't see it.

Gould (2003) also argues that science is a social activity. Social aspects are embedded in the practice of science. Scientists must recognize that science is "a quintessentially human enterprise." Social biases and social judgments show up all the time. Gould provides examples of such biases and discusses, in particular, dichotomy bias (classifying things into two opposing categories). There appear to be "strong human inclinations toward sequential dichotomy, or successive divisions into pairs, as a preferred mental device for classifying complex systems." The dichotomies are very often false.

Carolyn Bloomer (1976) notes that often "our minds are made up before the fact." We "see things only in relation to categories already established in [our] mind." Our conclusions do not "represent objective knowledge about a stimulus but rather the confirmation of a preexistent idea." We "encounter reality with an enormous number of preconceived notions." Beware the power of conception over perception. We can see what we expect to see rather than what is there.

Stephen Wolfram (2002) has expressed his views on the processes of human perception and analysis. "For inevitably our experience of the natural world is based in the end not directly on behavior that occurs in nature, but rather on the results of our perception and analysis of this behavior." And human perception and analysis capabilities are limited. In every kind of human perception, the apparent scheme is to "respond to specific fixed features in the data, and then to ignore all other features." As discussed in Chapter 11, Wolfram has developed ideas about upper bounds and equivalences of complex system capabilities. In his Principle of Computational Equivalence, Wolfram hypothesizes that all very complex systems have essentially equivalent computational capabilities. Because system processes, including perception and analysis, can be represented as computations, it follows (according to Wolfram) that all very complex systems, including humans, have essentially equivalent perception and analysis capabilities. Humans, therefore, cannot understand other complex systems any better than they understand themselves. And humans do not understand themselves very well.

Recall for a moment our (perhaps whimsical) discussion in Chapter 11 of extra-complex, extra-intelligent systems. (See Wolfram, 2002.) Now, because apparently very complex behavior can be generated by very simple rules, that behavior may contain regularities at some level. If such regularities should exist, they have not been detected by human perception and analysis capabilities. What if there were extra-intelligent systems in the universe with higher forms of perception and analysis? Would they be capable of detecting regularities in very complex behavior? Would such extra-intelligent systems then understand everything? Would they have perfect perception? While these conjectures may be intriguing, they do not represent human capabilities. Perhaps our evolved human perception and analysis capabilities just "are what they are." They do not appear to be optimum, but rather "good enough." They "satisfice" (Herbert Simon's terminology), consistent with the "roughness of nature" (Benoit Mandelbrot's terminology).

By the way, it is interesting to note that our descriptions in this section of the limitations of human perception are subject to those very same limitations of human perception.

13.4 ROUGHNESS VERSUS SMOOTHNESS

The structure and dynamics of natural systems are very often characterized by fractal behavior (i.e., by roughness). Discontinuity and roughness are pervasive in nature, yet we typically assume and "see" continuity and smoothness. This section explores the human perceptual preference for smoothness, while ignoring roughness. We begin with some introductory comments from Mandelbrot, Wolfram, Bacon, and Drury. Then we discuss what I claim is the typical human perceptual domain (both the spatial scale and temporal scale). The section concludes with two supporting analyses that investigate the relationships between scale and perception.

Benoit Mandelbrot (1982) says that "a continuous process *cannot* account for a phenomenon characterized by *very sharp* discontinuities." ... "The only reason for assuming continuity is that many sciences tend, knowingly or not, to copy the procedures that prove successful in Newtonian physics." With respect to statistics, Gaussian distributions of events are often assumed. The "rejection of statistical outliers" is the usual way of dismissing discontinuous extreme values. Nature, however, is in fact characterized by (sometimes intense) discontinuous fluctuations. This is either ignored or underestimated in the sciences. The same is true in traditional economics, regarding market fluctuations. I would add that the same is true in ecology, regarding ecosystem fluctuations.

Stephen Wolfram and Francis Bacon have commented on the human propensity for seeing order where there is none. Wolfram (2002) says that human visual perception capabilities seem to recognize patterns in random images where actually none exist. He provides simple examples: the "big dipper" in a star constellation, the "man in the moon," and figures in cloud formations. Bacon (in *Novum Organum*, 1620) states, "The human understanding, owing to its own peculiar nature, easily supposes more order and regularity in things than it finds."

Naturalist William Drury (1998) says, "An important portion of the order ascribed to landscapes is supplied by the perceptions of the human observer." We see order and stability, but it is not there. "Evidence collected by climatologists, oceanographers, and geomorphologists supports the belief that most natural processes are constantly in flux." Drury describes the rocky shores of northern New England, in particular Maine: "The landscape appears to be constant because over the short period of our own experience we are not aware of changes in the seaweed-covered rocks. Our impression is one of consistency and continuity, as long as we don't look too closely." This observation brings us to our next topic.

13.4.1 The Typical Human Perceptual Domain

How do humans perceive reality? *Scale* is an important factor. Human "smoothing effects" can be observed spatially and temporally. *Spatially*: at larger and larger spatial scales, both system structure and behavior appear to be more and more smoothed. Specific details fade into a smoothed whole. *Temporally*: extreme changes in long time intervals (large time scales) can be perceived as small, gradual changes when viewed in short time intervals (small time scales). In sufficiently large spatial scales and sufficiently small time scales, therefore, natural systems appear to be smooth and continuous. I suggest that these large spatial scales and small temporal scales are the scales favored in much of normal human perception.

Consider standing on a mountain top observing the forest below for several minutes. Our large-scale spatial view of the forest system does not reveal the rough, fractal behavior of the forest structure and processes that exists at small spatial scales. Furthermore, the time scale of overall forest system dynamics is much larger than the time scale of normal human perception. Extreme changes at very large time scales are perceived as gradual changes (or no changes at all) by humans operating at small time scales. We see balance and smoothed "steady state" behavior in the forest below. That is not what is there.

Consider the historical side of this story. Geologist/anthropologist Charles Lyell (who lived 1797-1875) formulated the philosophy of uniformitarianism or gradualism, which claimed that smooth gradual processes were at work in natural systems. All things could be explained by linear extrapolation. Darwin seemed to accept this view—he stated that evolution is smooth and gradual. At one point, he even denied the existence of evolutionary mass extinctions. Gould and Eldredge disagreed with this view, and instead offered a "punctuated equilibrium" explanation (e.g., in their classic 1977 paper[7]). "Punctuated equilibrium is the idea that evolution occurs in spurts instead of following the slow, but steady path that Darwin suggested. Long periods of stasis with little activity in terms of extinctions or emergence of new species are interrupted by intermittent bursts of activity."[8] In my view, punctuated dynamics clearly are at work in ecological systems and other natural systems. I know that linear explanations are pervasive in our thinking. Acceptance of nonlinear, punctuated dynamics can be a challenge for some.

[7] Gould and Eldredge (1977).

[8] Quote is from Bak (1996), while discussing the work of Gould and Eldredge.

The normal human perceptual domain (large spatial scales and small time scales) may have evolved to enhance human survival. This operating domain provides us with frequent "snapshots" of our larger surroundings, so that we can assess conditions around us and make timely decisions. We can, for example, detect danger and avoid it in a timely fashion. Perhaps all species of life have developed perceptual devices that enhance their survival.

13.4.2 Analysis of the Relationships Between Scale and Perception

This section concludes with two supporting analyses that investigate the relationships between scale and perception. We look first at Herbert Simon's analysis of the perceived spatial and temporal relationships in natural system hierarchical architecture. Then we consider an analysis that examines the relationship between spatial scale and perceived system complexity, for three important categories of systems.

13.4.2.1 Perceptual Effects in Natural System Hierarchical Architecture

Herbert Simon (1962) has explored the spatial/temporal aspects of natural system hierarchical architecture. Figure 13.1 characterizes system component interactions versus hierarchy level for a simple particle physics system.

Herbert Simon has argued that stronger interactions and "faster" dynamics (higher frequency of occurrence) are observed at smaller spatial scales (lower levels of hierarchy), and weaker interactions and "slower" dynamics (lower frequency of occurrence) are observed at larger spatial scales (higher

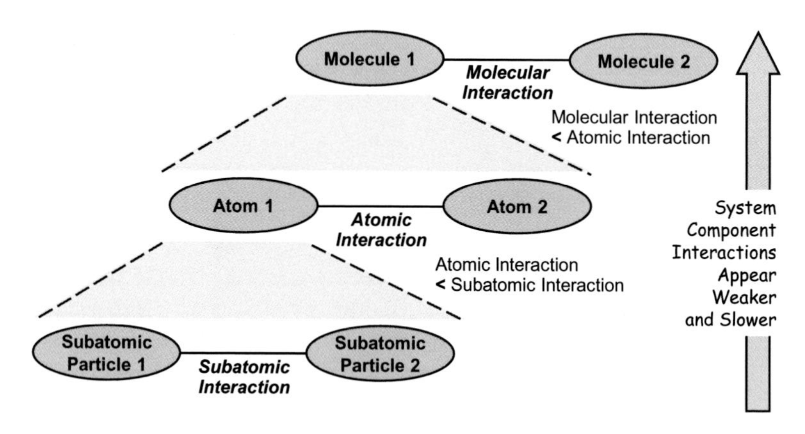

Figure 13.1 System component interactions and hierarchy.

levels of hierarchy). There can be order-of-magnitude differences from hierarchy level to hierarchy level. As we move up the system hierarchy, we eventually enter a spatial/temporal domain in which we perceive what Simon calls a "nearly decomposable" system with nearly independent system components. Component interactions appear to be minimal and gradual. We perceive reduced system complexity. We perceive *smooth* behavior at large spatial scales.

13.4.2.2 Perceived Complexity versus Spatial Scale for Three Important System Categories

The three important system categories I am talking about correspond to Warren Weaver's (1948) ranges of complexity. See Figure 13.2.

Organized simplicity systems are highly ordered, deterministic systems (e.g., Newtonian systems). They are low complexity systems at any scale, and can be described simply and have "laws" at any scale. *Disorganized complexity systems* are systems in disorder with apparent random behavior. At molecular scales these systems are complex, unpredictable, and chaotic, but at larger scales statistical averages apply and simple descriptions and laws (e.g., the ideal gas law) are possible. (The ideal gas law does not address the small-scale details of molecular movement, yet it is a very useful law that describes the fundamentally relevant relationships among pressure, volume, and temperature of gases.) *Organized complexity systems* have the highest complexity overall. This category corresponds to Wolfram's class 4 system category. Many ecological systems are organized complexity/class 4 systems.

Perception of complexity (and simplicity) depends upon the scale of observation. There exists a spatial scale range for which systems in each category—even the highest complexity category—can have relatively simple descriptions. I have adapted and expanded some insights from Bar-Yam (2004) to illustrate this. Consider the system complexity profile chart (plots of perceived complexity versus spatial scale) in Figure 13.3.

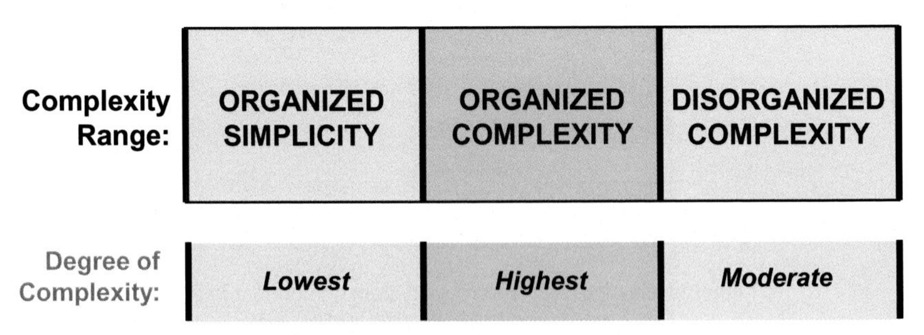

Figure 13.2 Weaver's ranges of complexity.

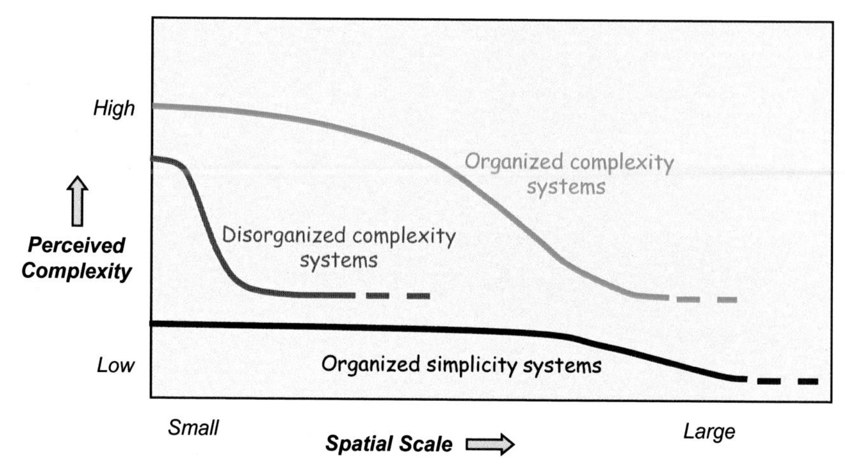

Figure 13.3 System complexity profiles. *(Adapted with permission from Bar-Yam, Y., Making Things Work, NECSI Knowledge Press, Cambridge, MA, p. 55, 2004; http://necsi.edu/publications/mtw/).*

The Figure 13.3 chart indicates that complexity is perceived to decrease at larger spatial scales. The most significant feature of the chart is that it suggests there is a spatial scale range for which perceived complexity of even the most complex class 4 systems (e.g., ecological systems) is reduced.

Human perception is a smoothing process. The resulting perceived continuous/smooth behavior is evident in many of the system models we generate. That is discussed next.

13.5 MODELING AND HUMAN PERCEPTION

The system models we generate begin with our perceptions of reality—our mental models. Mental models "constitute the collected evolutionary perception of the world of the individual, encompassing both knowledge and experience. . . . Other models, developed by individuals or groups, evolve from these mental models."[9] Mental models represent our beliefs. Our beliefs are those things we hold to be true, despite evidence to the contrary.[10] Humans believe that continuity trumps discontinuity in the natural world. Psychologist Walter Mischel says the human mind has a kind of "reducing valve" that "creates and maintains the perception of continuity even in the face of perpetual observed changes in actual behavior."[11] It is not

[9] J.N. Warfield (1976).
[10] Paraphrased from Joseph O'Connor and Ian McDermott (1997).
[11] This quote appears in Malcolm Gladwell (2000).

at all surprising, then, that our system modeling efforts typically are consistent with our perceptual preference for continuous/smooth behavior.

13.5.1 Models are Simplifications

"All theoretical models are wrong, but some are useful."[12]

George E. P. Box

"Such models need not be correct; in fact, one does not so much ask whether or not the model is correct as whether it adequately or inadequately fulfills its purpose. . . . We simply say that all models are 'right' when applied to a purpose for which they are useful."[13]

Coats and Parkin

Models often are simplifications that are consistent with our perceptual preference for continuous/smooth behavior. Hopefully they are useful simplifications. According to Webster's Third International Dictionary, a model is "a description ... to help us visualize, often in a simplified way, something that cannot readily be observed." Models are derived abstractions of reality for the purpose of understanding it. All system models incorporate simplifying assumptions. This is necessary to reduce a complex real-world situation to something that is tractable and amenable to, for example, mathematical or numerical analysis. As we will see in this subsection and the next, most models "smooth" reality in some way, based on human perception of the reality. This can be useful to achieve insights. The insights, of course, are valid only when interpreted in the context of the smoothing assumptions.

Here are some of the more famous examples of models as simplifications. Neils Bohr's simple model[14] of the atom consisted of a nucleus and a series of circles and ellipses that represent in a pictorial way the structure of the atom. Bohr knew the model was incomplete and provisional, but he also knew it was an excellent first step in the iterative process of understanding and defining the atom. Quantum physics, of course, later "upgraded" the model. As noted earlier, Einstein showed that Newton's laws were, in a sense, simplified models. Einstein developed new and revolutionary theories (models), but felt that his work involved simplifications also, and was only a piece of a larger unified field theory. Lotka-Volterra species population (predator/prey) models are oversimplifications. They include only limited

[12] George E.P. Box and Norman R. Draper (1987).

[13] This 1977 quote from Coats and Parkin appears in Malcolm Gladwell (2000).

[14] This account comes from J. Robert Oppenheimer, *A Science in Change*, in the book Galileo's Commandment (1999), edited by Edmund Blair Bolles.

species interactions and no explicit interactions with the environment. These models can, however, provide useful insights.

The general points made by these examples also apply, of course, to traditional ecological network models. Ecological systems are open, nonequilibrium, nonsteady state, nonlinear systems. But equilibrium, steady state, linear ecosystem models are known to provide important insights, even though these models represent substantial simplifications. In my view, however, such models by themselves are not necessarily useful for understanding complex ecosystem real-time dynamics.

Stephen Wolfram (2002) discusses the way in which the process of simplification in modeling involves aspects of human perception and analysis.[15] The role of models "in effect is to take large volumes of raw data and extract from it summaries that we can use. ... There are in general two ways in which data can be reduced by perception and analysis. First, those aspects of data that are not relevant for whatever purpose one has can simply be ignored. And second, one can avoid explicitly having to specify every element in the data by making use of regularities that one sees." (Note that the ignored data that results from both of these "two ways" often includes important extreme values.) The basic goal is "to reduce raw data to a useful summary form. ... In general perception and analysis can be viewed as equivalent to finding models that reproduce whatever aspects of data one considers relevant." Statistical analysis is often employed. "In traditional science statistical analysis has been the most common way of trying to find summaries of data." A usual procedure is to compute simple statistical measures (e.g., mean, standard deviation) from raw data and then apply these measures to known statistical distributions in order to obtain summaries of the data/system behavior. A significant problem with this approach is that the "known distributions" are often Gaussian or other single-peak distributions of similar form, which are not representative of the reality of natural system behavior.

Wolfram (2002) provides further general comments on mathematical models. He says that traditional mathematics and mathematical formulas have been "the primary method of analysis used throughout the theoretical sciences." The usual goal is to try "to find a mathematical formula that summarizes the behavior of a system." Mathematics can work well for less-complex systems, but cannot in general be expected to effectively summarize very complex system behavior. "The evidence is that in the

[15] Wolfram (2002), Chapter 10, Processes of Perception and Analysis.

end there is no way to set up any simple formula that will describe the outcome of evolution for a system like rule 30 [the rule 30 elementary cellular automata system]." As I noted in Chapter 1, however, mathematics can be very useful for certain limited aspects of highly complex systems.

We have to be careful with our modeling simplifications. We are prone to making simplifying assumptions that lead us to mathematical formulations that we know how to solve or at least know how to deal with. Then, with respect to the resulting models, "one typically has rather little idea which aspects of what one sees are actually genuine features of the system, and which are just artifacts of the particular methods and approximations that one is using to study it."[16] This brings us to our next topic.

13.5.2 Differential Equation Models and Steady State Behavior

Benoit Mandelbrot (2010b) says, "Science is expected to be cumulative ... new wisdoms must not deny the old wisdom that the world is made of smooth shapes and involves smooth variation and differential equations." Mandelbrot reinforces the notion that we look for smooth behavior, and he suggests that differential equations can help us find what we seek. My claim is that differential equation models should be expected to yield smooth steady state behavior for adequately ordered (nonchaotic) systems. That steady state outcome, however, is an artifact of the model and not a genuine feature of complex natural systems.

Differential equations are often used to model natural system dynamics. How does one solve these differential equations? A solution requires mathematical integration over time. This mathematical integration is essentially a time-averaging, smoothing operation. It calculates, sums, and essentially averages "dt contributions" over time. It then assigns a resulting cumulative single number (the area under the curve) to each "zero-to-t" time interval of system behavior. As time proceeds and more and more "dt contributions" accumulate, the averaged result becomes less and less sensitive to individual contributions. It becomes less and less sensitive to any extreme values that may be present.

When differential equations cannot be solved in mathematical closed form—which is the case for many complex natural systems of interest—difference equations and computer-based numerical methods are used. As time proceeds and discrete time step "samples" accumulate, the averaged result becomes less sensitive to individual samples. The impact of any specific

[16] Wolfram (2002), Chapter 4.

sample (say a discontinuous extreme value) on the result becomes less and less. The graphical depiction of the difference equation solution (vs. time) typically has three regions. It begins at time zero with system *initial conditions*, then proceeds through a brief *transient response* region where the relatively few individual samples do have an impact, and then settles down to a smoothed long-term response region where any discontinuous extreme values have been averaged out and fade away. That last region is called *steady state*.[17]

Steady state is a *model*. It is a simplified view of reality—it is not reality. It represents a time-averaged and smoothed history of the system. In differential equation solutions, extreme value contributions can have larger effects early in the history (in the transient response region), but increasingly smaller effects as time proceeds and history accumulates (in the steady state region).

Figure 13.4 can help illustrate what I am talking about. The results depicted in the figure are from my ecological network dynamics model (covered in Part V of the book). The plots shown represent system stock dynamics.[18] The upper plot, with ordinate on the left, represents the smoothed steady state dynamics. System initial conditions are depicted at time = 0, the transient response region occurs between 0 and time step 200 or so, and the steady state region is beyond time step 200. The lower plot, with ordinate on the right, represents the punctuated instantaneous dynamics that occur at each point in time. The instantaneous "dt samples" that occur early in time appear to have a significant impact on the smoothed system stock values, while the later instantaneous samples have very little or no impact.

What does all of this say about the many traditional ecological system models that are differential-equation-based steady state models? Ecological systems are not in steady state and their dynamics are not linear, but, as we discussed earlier in the chapter, ecological systems can be perceived that way at sufficiently large spatial scales and sufficiently small time snapshots (the usual human perceptual domain). When we perform traditional ecological system modeling and analysis, we pick "convenient" spatial and temporal domains. We pick a spatial scale large enough to encompass a significant ecosystem. Then, to make things manageable, we define ecological model

[17] According to traditional nonlinear dynamics theory [covered in Chapter 10], for adequately ordered (nonchaotic) systems, steady state can be represented by a single state (a point attractor) or a relatively small set of repeating states (a limit cycle attractor).

[18] Node stock is the biomass/energy contents of a given ecological network node. System stock is the sum of all the node stocks.

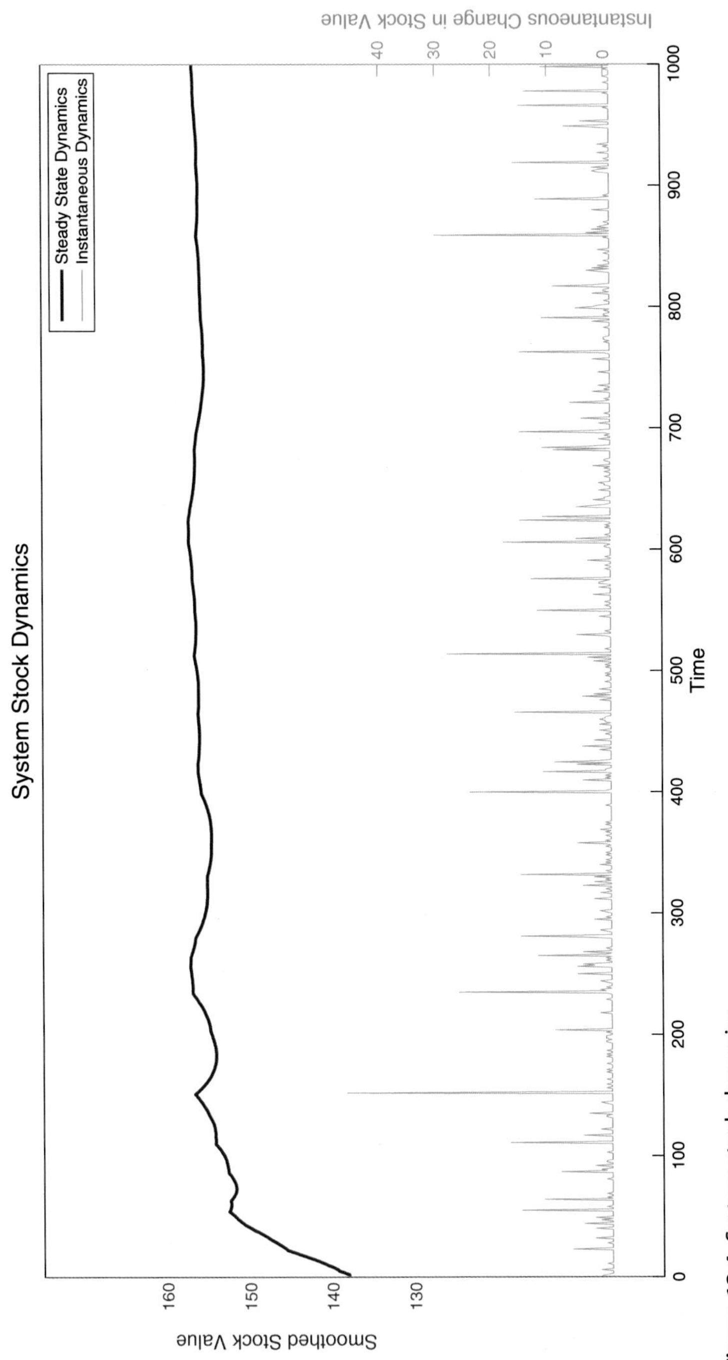

Figure 13.4 System stock dynamics.

compartments and intercompartment flows that are highly aggregated. As a result, the model occupies a relatively high level in the spatial hierarchy. In the temporal domain, we pick a time interval that is short enough so that the system compartments do not change morphologically during the analysis time frame (so that longer term system developmental changes can be considered negligible). The analysis temporal scale, therefore, is often relatively short (e.g., days, weeks, or maybe months). It is just these "convenient" space and time domains, which are consistent with usual human perception, where the steady state and linearity approximations seem, to the observer, to be valid. Are the resulting ecological system models useful, and can they provide important insights? Yes, of course—as long as we recognize that they represent substantial simplifications of reality, and not reality.

Here's the bottom line. Given our perceptual preferences and our associated system models, we must be careful not to draw the flawed conclusion that smooth (steady state) behavior is pervasive in nature. It is not. There is increasing evidence that rough (fractal) behavior may prevail. In my modeling of ecological network dynamics in Part V of the book, I endeavor to avoid the problems of human perception discussed in this chapter. I specifically endeavor to avoid assumptions and preconceptions about ecosystem real-time dynamics outcomes. Rather, I attempt to implement simple rules of interaction, let the dynamics develop, and see what happens.

CHAPTER 14

The Nature of Order and Complexity in Ecological Systems

14.1 INTRODUCTION

We now proceed with our "synthesis of ideas" that yields a view of the characteristics of ecological network dynamics. The focus in this chapter is on the nature of order and complexity in ecological systems—to gain additional systems—to gain additional insights into the behavior of highly complex systems.

Here's a list of the topics covered in the chapter:

Thinking about Order, Complexity, and Systems
Nonlinear Dynamics Attractor Types
Order, Complexity, and Weaver's Ranges
Maximal Complexity and the Edge-of-Order Domain
The Myth of Persistence
Chance and Change: Probabilistic Aspects of Natural Systems
Optimal or Good Enough?

The first part of the chapter focuses on a discussion of topics that are essential to an understanding of system order, complexity, and their relationships. The adequacy of the three traditional nonlinear dynamics attractor types is considered, and an addition to the extant theory to include an attractor that appropriately represents the behaviors of highly complex systems is proposed. I utilize insights from Warren Weaver (1948) and others to specify the proposed attractor, which resides at the edge of order in a region of maximal system complexity. Next, what I call the "myth of persistence" is addressed. The human perceptual preference for steadiness and smoothness does not describe natural system dynamics. To the contrary, these dynamics are ever-changing. Highly complex natural systems can, and do, explore a broad dynamics state space. Such systems can exhibit dynamics throughout the range from very well-ordered dynamics to edge-of-order dynamics. Probabilistic aspects of natural systems are considered next. We find that chance is ubiquitous in natural system dynamics. Finally,

Understanding Complex Ecosystem Dynamics
http://dx.doi.org/10.1016/B978-0-12-802031-9.00014-0

optimality in the structure and dynamics of complex natural systems is addressed. We find not optimal, but rather "good enough" outcomes.

14.2 THINKING ABOUT ORDER, COMPLEXITY, AND SYSTEMS

14.2.1 Nonlinear Dynamics Attractor Types

Conventional nonlinear dynamics theory defines three types of *attractors* for the dynamics of nonlinear systems: point attractors, limit cycle or periodic attractors, and chaotic (strange) attractors. As I discussed in Chapter 11, this classification seems incomplete. It does not include very complex systems such as ecological systems. Stephen Wolfram's[1] cellular automata classes 1, 2, and 3 have dynamics that correspond, respectively, to the three conventional classes of attractors. Class 4 cellular automata, on the other hand, represent much more complex systems. Class 4 system dynamics are not well understood, and they do not correspond to a conventional attractor type. I have proposed a fourth type of nonlinear dynamics attractor and have coined a term for it: *edge-of-order* attractor. I hypothesize that there is a one-to-one correspondence between cellular automata classes and nonlinear dynamics attractors as follows:

Class 1—Point attractor
Class 2—Periodic attractor
Class 3—Chaotic attractor
Class 4—Edge-of-order attractor

I use the term *edge-of-order attractor* because it is associated with what I call the *edge-of-order domain*. I describe this domain of the dynamics of highly complex systems in the following subsections.

14.2.2 Order, Complexity, and Weaver's Ranges

Warren Weaver's[2] ranges of complexity have been influential over the years and are a useful way of thinking about order and complexity. Figure 14.1 illustrates the ranges.

Let's make some additions to Figure 14.1 to further illustrate degree of complexity and degree of order, as a way of introducing the concept of the edge-of-order domain. Figure 14.2 graphically depicts degree of complexity versus Weaver's ranges.

[1] Wolfram (2002).
[2] Weaver (1948).

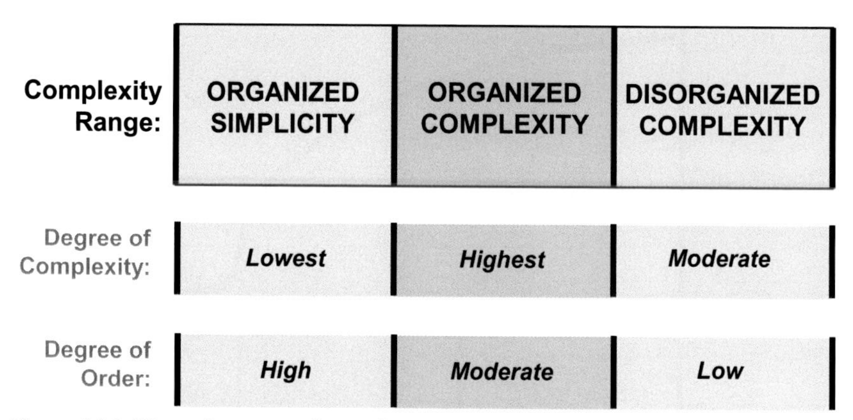

Figure 14.1 Weaver's ranges of complexity.

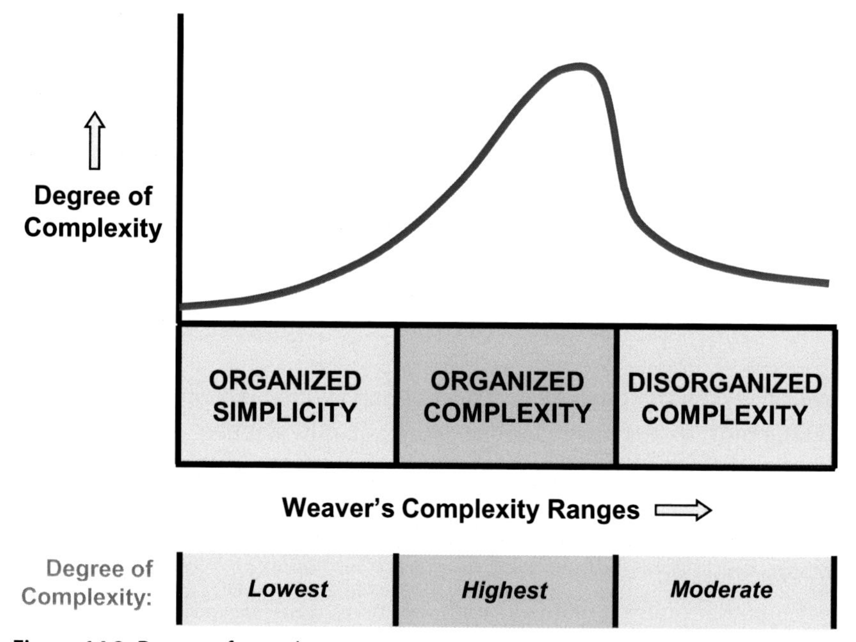

Figure 14.2 Degree of complexity.

Wolfram (2002) and Solé and Goodwin (2000) offer support for this depiction. Stephen Wolfram has noted that "both very ordered and very disordered systems normally seem to be of low complexity, and . . . systems on the border between these extremes—particularly class 4 [systems]—seem to have higher complexity." Solé and Goodwin say, "A properly defined complexity measure . . . should reach its maximum at some intermediate

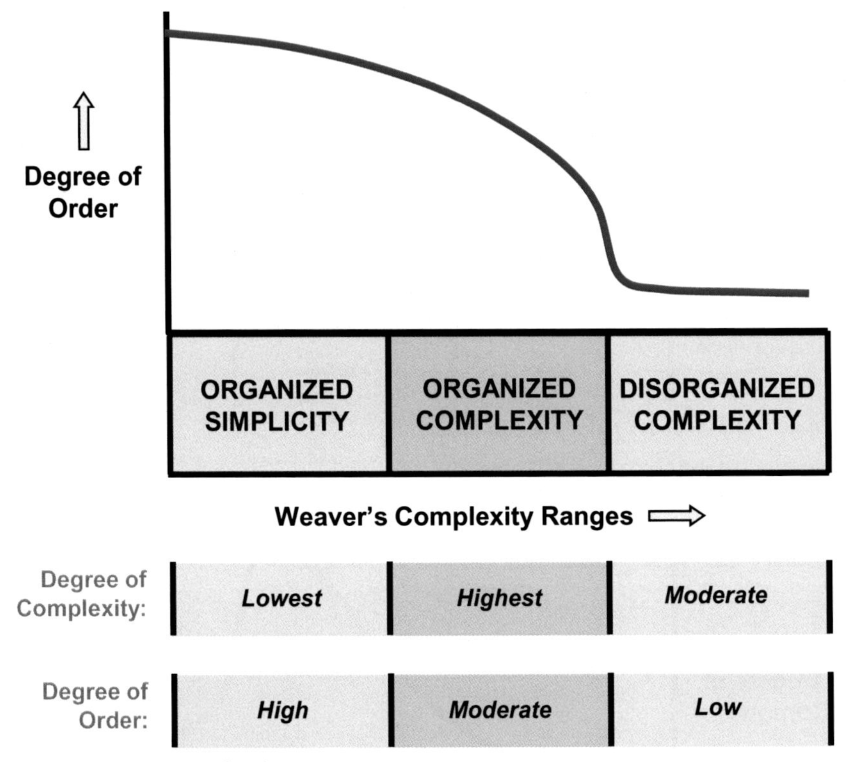

Figure 14.3 Degree of order.

levels between the order of a perfect crystal and the disorder of a gas." ... "The point of maximal complexity is sharply defined. This is where complexity lives."

Consistent with Weaver's original work and the Wolfram/Solé and Goodwin comments, Figure 14.3 graphically depicts degree of order versus Weaver's ranges.

Highly complex organized complexity systems are poised in an agile, responsive domain of "flexible order"—between "too ordered" and "too disordered" behavior. (Note that the right-hand side of the Figure 14.3 curve does not go all the way to zero. In Chapter 13, we saw that, at macro scales, disorganized complexity systems seem to possess some order. In Chapter 10, we saw that, even at micro scales, state space (strange attractor) diagrams and the Feigenbaum bifurcation distance ratio both provide at least suggestions of some order in chaotic disorganized complexity systems.)

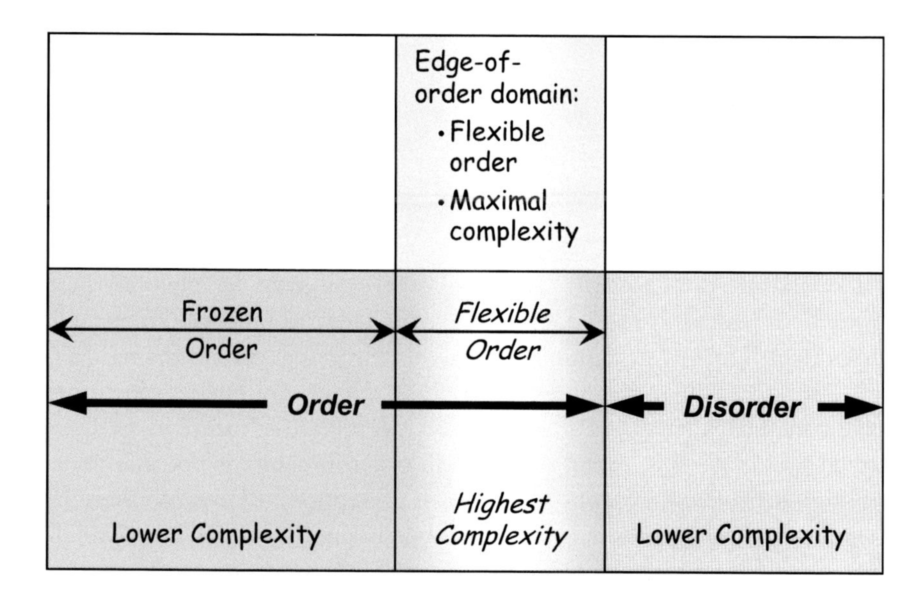

Figure 14.4 The edge-of-order domain.

14.2.3 Maximal Complexity and the Edge-of-Order Domain

Let's now combine these notions of maximal complexity (Figure 14.2) and "just right" flexible order (Figure 14.3), and define what I call the *edge-of-order domain*. See Figure 14.4. The edge-of-order domain is characterized by *maximum complexity* and *flexible order*. This domain is where the edge-of-order attractor for the dynamics of highly complex systems (class 4/organized complexity systems) resides.

Stuart Kauffman offers several thoughts on the edge-of-order location. The edge-of-order is a "good place to be ... where, on average, we all do best."[3] ... "Life evolves toward a regime that is poised between order and chaos." "It is a lovely hypothesis, with considerable supporting data, that genomic systems lie in the ordered regime" near chaos. These systems are not too rigidly ordered, and not unordered, but just right. "How do ... networks achieve both stability and flexibility? The new and very interesting hypothesis is that networks may accomplish this by achieving a kind of poised state balanced on the edge of chaos." Just between order and chaos "the most complex behaviors can occur—orderly enough to ensure stability, yet full of flexibility and surprise. Indeed, this is what we mean by

[3] This first Kauffman quote is from Bak (1996). The remaining quotes in the paragraph are from Kauffman (1995).

complexity." Kauffman posits further that perhaps this edge-of-order location, "ordered and stable, but still flexible, will emerge as a kind of universal feature of complex adaptive systems in biology and beyond."

David Bohm and F. David Peat[4] comment on the "in-between" region of order. They say that in physical systems, there is a whole spectrum of order "with orders of low degree at one end and chaos and randomness at the other. In between are further kinds of order of great subtlety that are neither of low degree nor chaotic." "Science, however, has not yet explored these intermediate orders to any significant extent. They may turn out to be quite important in many areas and indeed life itself may depend on them."

Per Bak (1996) makes observations about the relationship between order and complexity and about chaos. He says, "A picture of nature in 'balance' often prevails." This notion of excessive order in nature is not true (even though it may be consistent with human perception). "How can there be evolution if things are in balance? Systems in balance or equilibrium, by definition, do not go anywhere." Bak observes further that chaotic systems are not highly complex. Not much useful happens in the disordered regime. Here, system behavior exhibits randomness; system fluctuations have a white noise spectrum; there is no memory, no contingency, and no correlation with past events. Bak says, "Chaos signals [vs. time] have a white noise spectrum [vs. frequency], not 1/f." (Note that we'll cover ecological fluctuation signals and their so-called *one-over-f* frequency spectrums in Chapter 15.)

Per Bak is perhaps best known for his work on *self-organized criticality*, a network phase-transition/critical-threshold phenomenon. This phenomenon occurs in what I am calling the edge-of-order domain. Bak's self-organized criticality is prominent in the next section.

14.3 THE MYTH OF PERSISTENCE

We talked quite a bit, in Chapter 13, about the human perceptual preference for order, steadiness, and smoothness. That perception, however, does not describe natural system dynamics. In reality, everything changes all the time. All *is* flux.[5] Persistence should not be expected. As I described in the previous section, highly complex systems can reside in an edge-of-order domain at their edge-of-order attractor. But these systems do not reside

[4] Bohm and Peat (2000, p. 140).
[5] David Bohm (1983).

there persistently. Recall, from Chapter 11, that maximally complex class 4 systems can exhibit not only class 4 behavior, but also, at times, class 3, 2, or 1 behavior. Highly complex systems can, and do, "explore" this broad dynamics state space. Such systems can exhibit dynamics throughout the range from very well-ordered dynamics to edge-of-order dynamics. This range of dynamical behaviors, represented as an event time series, can cover the spectrum from (nearly) null events to modest, gradual, well-ordered events to large, abrupt, extreme events. That spectrum of possible behaviors strongly suggests a *process fractal* description, that is, fractal event dynamics with a punctuated time series and a power-law probability distribution. (See Chapter 15.) Process fractal behavior is the antithesis of persistence.

In light of this nonpersistent dynamics context, how should we think about *self-organized criticality*? When we first discussed self-organized criticality back in Chapter 9 (Section 9.3.3), we noted that Bak et al. (1987) were the first to use the term self-organized criticality and the first to describe the phenomenon. They said that dynamical systems "naturally evolve into a self-organized critical point" and that the critical point "is an attractor reached by starting far from equilibrium." So far, this sounds correct. However, Bak et al. further suggested that self-organized criticality may be the underlying system phenomenon that yields complexity and *generates* spatial and temporal fractals. They hypothesized that self-organized criticality may be the *source* of fractals in dynamical systems. In his 1996 book, Per Bak[6] reinforced these notions of "how nature works" and provided a detailed description of his view of self-organized criticality.

My view of self-organized criticality differs significantly from Bak's. I certainly accept the notion of self-organization and the notion of critical phenomena. I also accept the fact that some form of fractal behavior may be observed at critical points. Bak's claim, however, that self-organized criticality is the fundamental underlying phenomenon that yields complexity and generates fractals is, in my view, inaccurate. It seems to me that Bak has confused cause and effect. Self-organized criticality is not a persistent causal mechanism of system complexity and system behavior, but rather a nonpersistent outcome of complex system behavior. Self-organized criticality comes and goes. A complex system reaches criticality only when required by the real environment and the current status of the system. As I explained in the first paragraph of this section, many complex system dynamical processes exhibit *process fractal* behavior, with the critical point being one of many possible outcomes encountered during the process. Fractal behavior

[6] Per Bak (1996), *How Nature Works - The Science of Self-Organized Criticality.*

is not caused by self-organized criticality. Self-organized criticality is a result (an effect) of fractal behavior.

While self-organized criticality may well be an important and useful concept, it is not (as Bak's book title suggests) *how nature works*. Rather, it is just one of the nonpersistent outcomes of the workings of nature.

14.4 CHANCE AND CHANGE: PROBABILISTIC ASPECTS OF NATURAL SYSTEMS

"A first principle is that chance and change are the rule."
"Chance and change are ubiquitous."

William Holland Drury Jr. [7]

We have already established that change is ubiquitous in natural systems (all *is* flux). Now we will consider the role of chance in natural system dynamics. We will see that chance is also ubiquitous.

William Drury (1998) explains: The role of chance was not a concern when the modern scientific revolution began in the seventeenth century with the Enlightenment in Europe. Three of the major players were Bacon, Descartes, and Newton. Their focus was on what we now recognize as relatively simple physical systems. Accordingly, they created a mechanistic philosophy in which nature "functioned like a great, integrated, well-maintained machine." Consistent with the philosophy, Newtonian physics is well ordered and deterministic. William Drury says, "We can see that [this] philosophy . . . has had a massive impact on the philosophy of science. The orderliness of physical systems, and especially the assuredness and rigor of classical physics, has haunted field biologists for the last two centuries." Because we are now attempting to understand the dynamics of highly complex natural systems, the role of chance is becoming an increasingly important factor. "Nature works on the basis of one-on-one species interactions, variability, and chance." Natural system processes can be stochastic processes. Nature is not deterministic; it is very often probabilistic.

Melanie Mitchell (2009) provides some examples[8] of the probabilistic aspects of natural systems:

- In the human immune system, we see probabilistic behavior of lymphocytes. The human immune system can find, attack, and eradicate specific

[7] Quotes are from Drury (1998).

[8] We previously referred to Mitchell's characterization of the human immune system and ant colonies in Chapter 7.

pathogens that have invaded the body and are causing sickness. In simple terms, immune system lymphocytes have receptors that can recognize (bind with) a particular pathogen, and then produce antibodies to destroy that type of pathogen. Because it is not known *a priori* which type of pathogens will invade the body, the receptor shapes of individual lymphocytes are *randomly* generated. The lymphocytes that successfully recognize the pathogens are reproduced more and more. These successful lymphocytes are *randomly* spatially distributed by the bloodstream for coverage throughout the body, until the invading pathogens are eradicated.

- Ant colonies can very effectively establish food foraging trails. When foraging, ants move randomly in different directions looking for food. Successful ants return to the nest, while depositing a trail of pheromones. Other ants follow the pheromone trails and reinforce them with more pheromones. The process continues until an entire food trail network is established. Initial foraging movement of ants is random; encountering a food source is probabilistic; encountering a pheromone trail is probabilistic.
- Biological metabolism processes have probabilistic aspects. "Cellular metabolism relies on random diffusion of molecules and on probabilistic encounters between molecules, with probabilities changing as relative concentrations change in response to activity in the system."

In each of these examples, and more generally, "it appears that such intrinsic random and probabilistic elements are needed in order for a comparatively small population of simple components . . . to explore an enormously larger space of possibilities . . ." (Mitchell, 2009).

Benoit Mandelbrot (1982) also comments on the probabilistic aspects of nature. He talks about natural fractals and our models of these fractals. Natural fractals are almost always so in the statistical sense. Some human-made fractal models are as well. "The most useful fractals involve *chance* and both their regularities and their irregularities are statistical." The effects of chance can be very substantial. Mandelbrot poses a relevant question and then answers it: "Can chance bring about the strong degree of irregularity encountered, say, in coastlines? Not only does it, but . . . the power of chance is widely underestimated." Our natural system modeling efforts should incorporate probability theory. "The theory of probability is the only mathematical tool available to help map the unknown and the uncontrollable."

Here's an interesting aside in the context of the artistic proclivities of the natural systems called human beings. Denis Dutton[9] discusses the role of chance and contingency in human evolution and human cultural development with respect to the arts. According to Dutton, Charles Darwin agreed that human evolution has been subject to chance influences and, to some degree, is a product of "*prehistoric* contingency." Dutton argues that this prehistoric contingency gave rise to the innate instincts, interests, and preferences of humans, including, notably, the art instinct. Human cultural development has also been subject to chance influences but is not a product of prehistoric contingency. It is a product of "*historic* contingency." Human cultural values "may be products of local cultural contingency—they are 'ours' only through accidents of when and where we were born." Consider the arts of Eurocentric peoples compared to the arts of tribal peoples (e.g., the Carib Indians, the Iroquois, the Maori). All peoples have the art instinct. The sometimes very different expressions of it are cultural. "Art may seem largely cultural, but the art instinct that conditions it is not." Both human evolution and human cultural development exhibit very important probabilistic contributions.

It seems reasonable to conclude that the simple rules of interaction in natural systems, repeated over and over, which can yield extremely complex behavior, may very often include probabilistic aspects. The rules of interaction in my model of complex ecological network dynamics (Part V) include such probabilistic aspects.

14.5 OPTIMAL OR GOOD ENOUGH?

Are the structure and dynamics of complex natural systems optimal or just good enough?

Stephen Wolfram (2002) has some ideas on this subject. Biological system development may not require "maximum fitness" outcomes, but rather "good enough" outcomes. These are outcomes that work but are not optimum solutions. "In the past, the idea of optimization for some sophisticated purpose seemed to be the only conceivable explanation for the level of complexity that is seen in many biological systems." But maybe optimization is not the reason for extreme complexity in living systems. It seems complexity is "easy" to produce (simple programs with simple rules repeated over and over). Yes, the particular simple programs/simple rules that lead to more

[9] Denis Dutton (2009).

successful (but not optimal) organisms will be "selected" and eventually dominate, but perhaps some of the complexity they generate is "free," with perhaps little or no "sophisticated purpose." We may get bonus complexity. Wolfram says, "I believe . . . that the vast majority of the complexity we see in biological systems actually has its origin in the purely abstract fact that among randomly chosen programs many give rise to complex behavior." Note, for example, that "if one looks at mollusks of various types . . . the range of pigmentation patterns on their shells corresponds remarkably closely with the range of patterns that are produced by simple randomly chosen programs based on cellular automata." "If one just chooses [simple] programs at random, then it is very easy to get behavior of great complexity. And it is this that I believe lies at the heart of most of the complexity that we see in nature, both in biological and non-biological systems." Benoit Mandelbrot (1982) has made similar observations. He says that the complications of biological form sometimes seem "to serve no purpose (as is often the case in fairly simple creatures)." . . . "However, the complications in question are often highly repetitive in their structure" and could be due to a systematic and simple generating rule. "The key is that the rule is applied again and again, in successive loops."

But we still have not answered the question of *why* complex natural system structure and processes are not optimum. The likely answer is that finding the optimum requires a comprehensive search through a "design" space (the space of all possible solutions). That search is often not practical and, for high-functioning natural systems, not possible. As suggested by Mitchell (2009) in the previous section, it is not possible for a "comparatively small population of simple components . . . to explore an enormously larger space of possibilities" in any deterministic and comprehensive way. So, perhaps a natural system "tries" (at random) a simple program with simple rules. If it works satisfactorily, it is used. Sometimes we get bonus complexity and bonus features. A "best" simple program cannot, in any practical sense, be selected *a priori*. One has to run it and see what happens. If it "works," then it may be good enough. If not, "selection" will find one that is.[10] Herbert Simon (1996) has noted that "in the face of real-world complexity, [one turns] to procedures that find good enough answers to questions whose best answers are unknowable."[11]

[10] For some, use of the term "selection" implies long-term biological evolution. Natural systems, however, "evolve" and change structurally and dynamically in both long time frames and short time frames. The comments in this section generally apply to both.

[11] We first referred to Simon's thoughts in this area in Chapter 1 in the context of engineering system design, and then again in Chapter 7 in the context of an evolution process model.

Here's an interesting point to ponder. Perhaps less-complex natural system behaviors can be (nearly) optimized. This might be a reason why biological organisms are often modularized into subsystems. In humans, for example, the subsystems would include the major organ subsystems (e.g., cardiovascular, respiratory, renal, etc.). Natural selection (and self-organization) might have a better chance of optimizing these somewhat less-complex modules (less complex when compared with the total organism). That's how we develop engineering systems. We partition, modularize, and attempt to move toward optimality in each module.[12] In some ways, perhaps biological "design" and engineering design are indeed analogous.

Herbert Simon (1996) has coined a term for finding solutions that are *good enough*. The term is *satisficing*. Simon's focus in this work was on the design of artificial (human-made) systems. "Satisficing" techniques compare design alternatives against a whole array of requirements and constraints. The techniques are used when it is not possible to explore the full set of alternatives (the full design space). As we discussed in Chapter 7, here is Simon's suggested approach:

- Define an aspiration level for each dimension of system satisfaction, and use these levels to evaluate system alternatives.
- If an alternative meets or exceeds the aspiration level along each dimension, then it satisfices and is good enough.
- If not, the alternative must be modified or rejected.

Perhaps natural systems use some version of this approach.

Next, in Chapter 15, we want to synthesize a view of the characteristics of ecological network dynamics based on the complex system behaviors discussed in this chapter and in previous chapters.

[12] I acknowledge, of course, that engineering systems are much less complex than biological systems.

CHAPTER 15

Characteristics of Ecological Network Dynamics

15.1 INTRODUCTION

It is now possible to describe a comprehensive view of the behavioral characteristics of ecological network dynamics which is the basis for a characteristics hypothesis—the central hypothesis of the book. We have been building to this point throughout the book. In the context of a systems and engineering approach, two closely related and, I think, very important sets of ideas have been explored. We have learned about a hypothesized mechanism for generating the extreme complexity we see in nature (i.e., simple rules, repeated over and over). We have learned a great deal about the behavioral characteristics of the resulting highly complex natural systems. In this chapter (the last of Part IV) and in all of Part V, the focus is on the behavioral characteristics. We synthesize (this chapter) and analyze (Part V) a characteristics hypothesis in detail. In Part VI, the mechanism aspects are added to create an overall hypothesis.

So far in the book, a whole collection of important insights and ideas relating to the characteristics of complex systems and complex system dynamics have been discussed. For example, we have seen that complex systems in general—and ecological systems in particular—take the form of networks. "Whenever we see life, we see networks."[1] Network thinking is crucial for understanding complex ecosystem dynamics. We have seen that, in nature, everything changes all the time. All *is* flux.[2] Persistence should not be expected. We have noted that ecological system dynamical processes very often seem to exhibit *process fractal* behavior, that is, fractal event dynamics with a punctuated time series and a power-law probability distribution.

In the current chapter, our synthesis of a characteristics hypothesis, based on all of these important dynamics insights and ideas, reaches fruition. It is

[1] Fritjof Capra (2002).
[2] David Bohm (1983).

Understanding Complex Ecosystem Dynamics
http://dx.doi.org/10.1016/B978-0-12-802031-9.00015-2

now time to specify a comprehensive view of the characteristics of ecological network dynamics. The work here is based further on the systems approach (Part I), the function-structure-process dynamics framework (Part II), applicable complex systems theory (Part III), and the insights regarding human perception and the nature of ecosystem order and complexity provided by the previous two chapters (Part IV).

Here are the components of the resulting characteristics view. I claim that the defining characteristics of complex ecological system dynamics are:

- Flickering Networks
- Punctuated Dynamics
- Fractal Behavior
- Local-to-Global Interaction
- Indirect Effects

In Sections **15.2–15.6** of this chapter, respectively, these five characteristics are described and discussed. In Section **15.7**, we pull it all together into a characteristics hypothesis.

15.2 FLICKERING NETWORKS

The traditional view of ecological (and other) networks, usually for the purposes of modeling and analysis, assumes a static network structure, steady state network stocks and flows, and (sometimes) linear system behavior. All of these are simplifying assumptions that are regarded as necessary to reduce a complex real-world situation to something that is tractable and amenable to mathematical or numerical analysis. My view of network dynamics is quite different. I envision an ever-changing network structure, nonsteady state stocks and flows, and nonlinear system behavior. Odum and Barrett (2005) say, "There are no equilibriums [in ecological networks], but there are *pulsing* balances." If we look closely, and override our perceptual preference for smoothness, we can see the pulsing dynamics. We see not steady state behavior, but fluctuating dynamics in space and time.

How do ecological networks operate? Ecological networks "flicker."[3] A description of my hypothesized view follows.

Reality is dynamic. There are families of natural system networks in play at all times. The network nodes are available for interaction, and any given node can participate in multiple networks. In response to some

[3] Here's the definition from Merriam-Webster online (http://www.merriam-webster.com/): *flicker*—**1.** to move irregularly or unsteadily: flutter, **2.** to burn or shine fitfully or with a fluctuating light.

stimulus—say an input of energy or biomass from the environment—connections between nodes become active and begin formation of a network. The ecological network can evolve to become a critically connected network with high complexity and flexible order. The nodes of the network operate and interact locally and/or globally as necessary until the "processing" of the stimulus is completed. If the system input is removed, the network connections eventually become inactive once more. *Network connections are not persistent; they are ever-changing—active when needed and inactive when not needed. The network "flickers."* The above scenario is repeated over and over again, in space and in time, as required by the real environment.

Ecological network operation is illustrated graphically in Figure 15.1. The figure displays two different moderate-sized propagation events that occur at two different instants of time. At time t_1 a unit of input from the environment is applied to a randomly selected input node. The input causes that node to propagate (and turn red or "light up," as depicted in Figure 15.1). Propagation results in flows to nearby nodes. Some of those downstream nodes then propagate to other nodes, and so on. The process continues until no further propagation flow occurs. We see that node interaction begins locally, but can extend globally. As a result of this propagation event, the states of all the involved nodes have changed. At the next instant (time t_2), we have a different set of network initial conditions, potentially a different input node, and therefore a different propagation event. Different nodes will "light up." The network *flickers*. (A good visual metaphor would be blinking lights on a holiday tree.)

15.3 PUNCTUATED DYNAMICS

As noted in Chapter 13, geologist/anthropologist Charles Lyell formulated, in the nineteenth century, the philosophy of uniformitarianism or gradualism, which claimed that smooth gradual processes were at work in natural systems. A small cause yielded a small effect. All things could be explained by linear extrapolation. Darwin apparently accepted this view, at least initially, and stated that evolution is smooth and gradual. Much later, Gould and Eldredge disputed the gradualism view and instead offered a "punctuated equilibrium" explanation (e.g., in their classic 1977 paper[4]). "Punctuated equilibrium is the idea that evolution occurs in spurts instead of following the slow, but steady path that Darwin suggested. Long periods of stasis with

[4] Gould and Eldredge (1977), *Punctuated Equilibria: The Tempo and Mode of Evolution Reconsidered.*

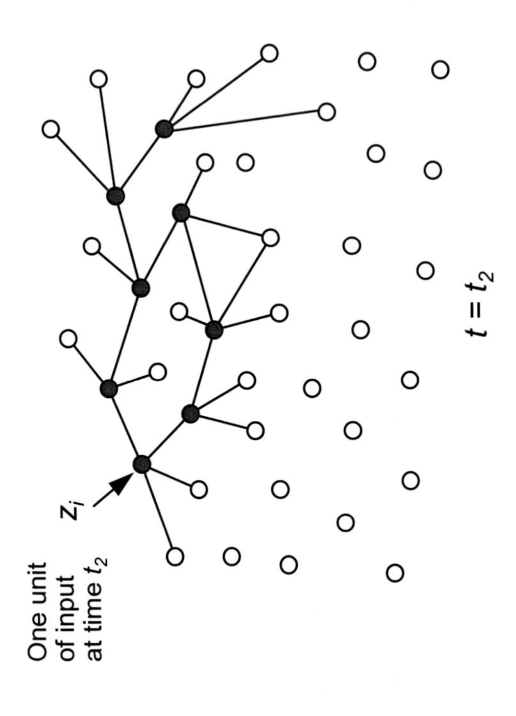

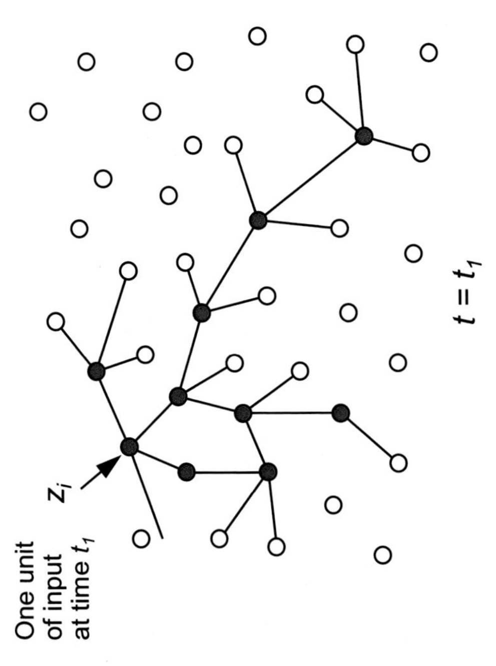

Figure 15.1 Ecological network operation.

little activity in terms of extinctions or emergence of new species are interrupted by intermittent bursts of activity."[5]

My view is that punctuated dynamics, both in long developmental time frames and in short operational time frames, are at work in ecological systems. I claim that the dynamics of complex ecological networks exhibit such behavior. The time series of ecological system events are punctuated and the event probability distribution follows a power law. There is a wide distribution of events, from the ordinary (gradual and expected) to the extreme (abrupt and unexpected).

Figure 15.1, which illustrates ecological network operation, displays two different moderate-sized propagation events that occur at two different instants of time. Events of all sizes, however, can occur. Smaller events involving just a few nodes are most likely, but extreme events involving nearly all nodes in the network can also occur.

15.3.1 Black Swans

"Black swans" are extreme events—unlikely, unexpected, high-impact events. They reside at the "long tails" of power-law/fractal event distributions. The black swan metaphor, perhaps mentioned first in John Stuart Mill, *A System of Logic* (1860), has been used many times since. Currently, it is often associated with Karl Popper and his work on the problem of induction: I see only white swans; therefore all swans are white; except when they are not, etc. (The Problem of Induction was perhaps first addressed by David Hume in the eighteenth century.)

A pictorial illustration of the black swan region of a power-law event distribution is provided in Figure 15.2. The flat portion of the curve on the left (expected events) side of the diagram is in the so-called cutoff region of the distribution, where the power law does not apply.

Marsh and Pfleiderer (2012) discuss the impacts of black swan extreme events on the financial markets and the Standard & Poor's (S&P) 500 Index. They note that many existing financial models use simple, unconditional Gaussian assumptions that ignore or otherwise minimize extreme events. As a result, these models are subject to serious failures. Marsh and Pfleiderer go on to describe modeling methods that may deal more effectively with extreme events. As part of their analysis, they provide a plot of maximum drawdown (over intervals of 24 months) on the S&P 500 Index. Maximum drawdown is a percentage measure of the decline of a variable from a

[5] Quote is from Bak (1996).

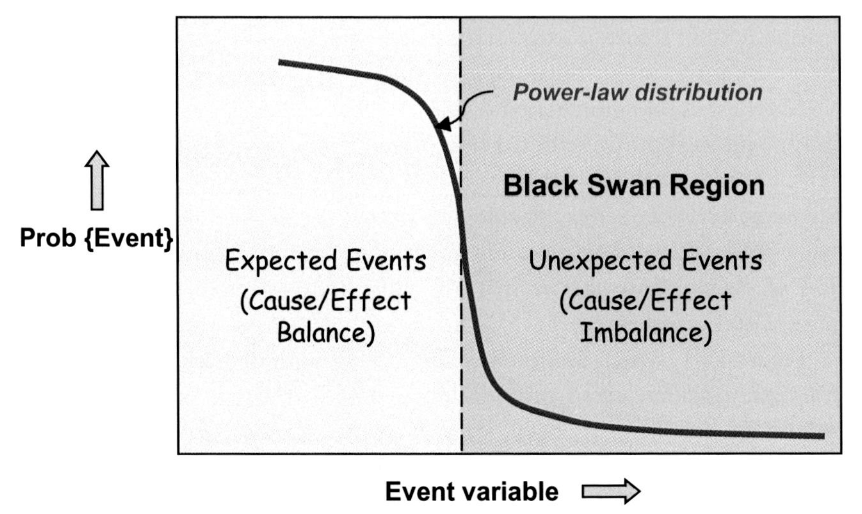

Figure 15.2 Black swan region.

previous peak. Their plot very well illustrates financial market dynamics and the presence of black swan events. An adaptation of the Marsh and Pfleiderer plot is provided here as Figure 15.3.

The figure covers the time period from 1801 to beyond 2008. Although S&P 500 history dates back only to 1923, Marsh and Pfleiderer have extended the chart back to 1801 using data from the Global Financial Database.

The time series plot of Figure 15.3 shows a wide range of fluctuating event sizes. There are a few extremely large events. The reference line on the figure is the drawdown of 50.95% that occurred during the financial crisis of 2008. Events of 2008-magnitude have occurred three times in

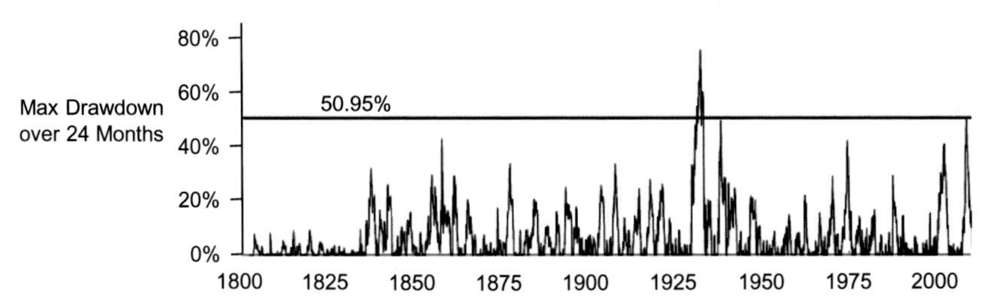

Figure 15.3 Black swans in the stock market. *(Adapted, with permission, from "Black Swans" and the Financial Crisis, T. Marsh and P. Pfleiderer, Review of Pacific Basin Financial Markets and Policies, Volume 15, Issue Number 02, Copyright © 2012 World Scientific Publishing).*

211 years. There is also a considerable number of medium-size to large-size events, but most events are small. The event dynamics are clearly punctuated. This behavior is consistent with the black swan power-law probability distribution of Figure 15.2.

Nassim Taleb (2007) emphasizes that large, extreme, high-impact black swan events are *not* anomalies. They are a very important part of system behavior, and can change the playing field. Taleb states, "By removing the ten biggest one-day moves from the U.S. stock market over the past fifty years, we see a huge difference in returns—and yet conventional finance sees these one-day jumps as mere anomalies. . . . In the last fifty years, the ten most extreme days in the financial markets represent half the returns. Ten days in fifty years."

15.4 FRACTAL BEHAVIOR

Fractal behavior is perhaps the dominant characteristic of complex ecological network dynamics. It is central to our understanding of these dynamics. This section addresses the types of fractals, the mathematical and graphical representation of fractals, and the very widespread occurrence of fractals in nature.

Ian Stewart (2010) seems to view fractals as a *mathematical concept* that can describe real-world physical structures and processes. He says, "What makes them [fractals] so useful in today's scientific research is that they have opened up entirely new ways to model nature. They give scientists a powerful tool with which to understand processes and structures hitherto described merely as 'irregular,' 'intermittent,' 'rough,' or 'complicated.'" While this (mathematician's) view has validity, I think the converse perspective is more accurate. I view fractals as a *real-world concept* that happens to have very useful mathematical descriptions. I think that Benoit Mandelbrot would have agreed. He began with a mathematics perspective, but evolved toward a natural systems perspective. Mandelbrot (2004) says, "I have engaged myself, without realizing it, in undertaking a theory of roughness. Think of color, pitch, heaviness, and hotness. Each is the topic of a branch of physics. Chemistry is filled with acids, sugars, and alcohols; all are concepts derived from sensory perceptions. Roughness is just as important as all those other raw sensations, but was not studied for its own sake. . . . Since roughness is everywhere, fractals—although they do not apply to everything—are present everywhere."

15.4.1 Two Types of Fractals

Recall the systems triad discussions and diagrams from Chapter 3. Figure 15.4 essentially reproduces Chapter 3's Figure 3.6. The figure depicts system function and the implemented form that delivers the functions. System *form* is comprised of physical *structure* and the system *processes* that execute on that structure. Consistent with the systems triad, then, we would expect two fundamental types of fractal behavior: *structure fractals* and *process fractals*. In my view, that is the correct categorization.[6] Structure fractals are spatial and are treated by most investigators as static (at least in the short term). Process fractals are dynamic and have both spatial and temporal aspects.

The work of Mandelbrot (and others) discussed in Chapter 12 focuses on structure fractals. In his book *The Fractal Geometry of Nature*, Mandelbrot (1982) says that his "work is concerned primarily with shapes in the real space one can see" and not on dynamics, which "is ultimately concerned with the temporal evolution in time." (Mandelbrot, however, did perform some dynamics work in his analyses of fractal behavior in the financial markets.)

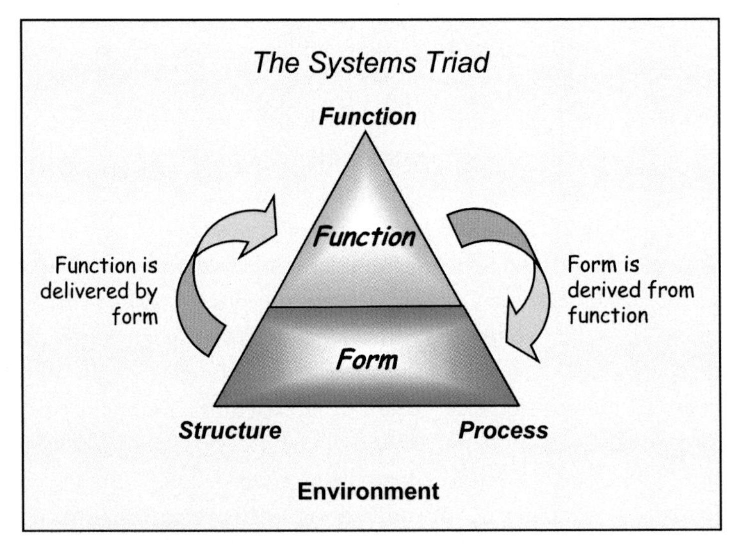

Figure 15.4 Systems triad—function and form.

[6] From a systems and engineering perspective, the conventional *spatial fractal* and *temporal fractal* categorization reflects a poor, overlapping partitioning. Crucially important system dynamics (and associated processes) are both spatial and temporal.

My work focuses on *process fractals*. I am concerned primarily with understanding the dynamics of complex ecosystems. I am concerned with the combined spatial and temporal aspects of ecological system dynamic behavior. Process fractals characterize system dynamics; they characterize system events over time. Recall that system events are outcomes of process over structure and that event sets comprise functions (see the systems triad). Events are spatial (they occur in space) and temporal (they change with time). Process fractals, therefore, result from an integration of system behavior over space and time.

Interestingly, note that *process* fractals can characterize changes in system *structure* over time. Mandelbrot (and others) treat structure fractals as static but, with respect to natural systems, they are not. Structure fractals continue to develop and change, but usually in longer time frames. Process fractals can reflect those changes. Both structure and process can be considered dynamic—structure in longer developmental time frames and process in shorter operational time frames as well as longer developmental time frames (see my function-structure-process framework in Chapter 5). Natural system structure fractals are not static; they just typically change more slowly. Everything is dynamic.

We can now begin discussing the very useful mathematical representations of process fractals.

15.4.2 Mathematical Representations of Process Fractals: Event Time Series and Event Distribution

Figure 15.5 graphically illustrates a typical process fractal time series and distribution associated with complex ecological network process activity. System events are outcomes of process. The event time series (upper diagram in Figure 15.5) plots event quantity versus time. For the purposes of discussion, let's assume that the event quantity of interest is event size. The time series plot shows a wide range of fluctuating event sizes. Most events are small, some are medium sized, and a few are extremely large. The event dynamics are clearly punctuated. One can take this event time history and generate an event distribution (the lower diagrams in Figure 15.5). The event distribution indicates the likelihood of an event having a given event quantity value (in our example case, a given event size). As shown in the left-hand-side normal-coordinates diagram, the distribution

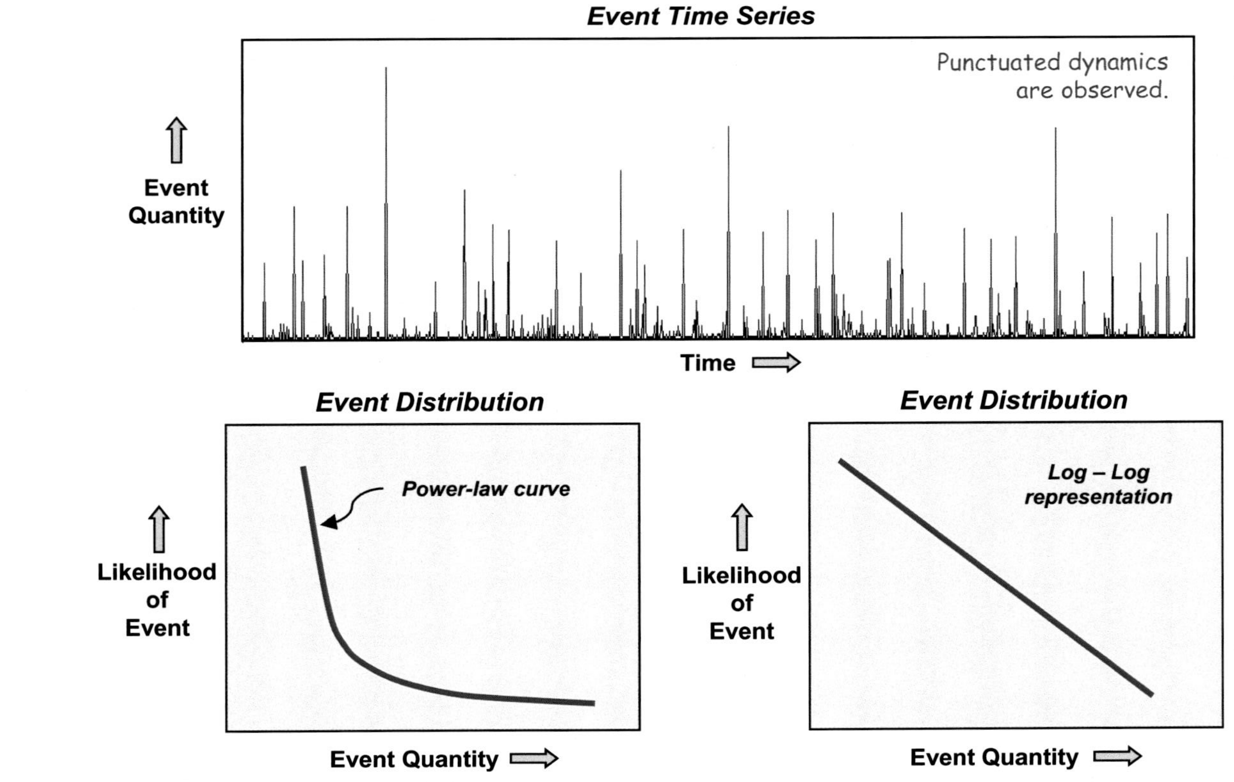

Figure 15.5 Process fractals—time series and distribution.

follows a power-law curve.[7] Consistent with the time series, the distribution shows a wide spectrum of possible event sizes. Small events are more likely, but large events can occur and do occur. Small events are expected; large events should be expected as well. The right-hand-side event distribution diagram in Figure 15.5 is a log-log representation. Power-law curves are straight lines in log-log coordinates. It is easy to show that, as follows: The power-law curve of the normal-coordinates event distribution can be specified by the equation

$$N(s) = Cs^{-\gamma}$$

Here s is event size, C is a constant, and γ is the fractal exponent. Taking the logarithm of both sides of the equation, we obtain

$$\log N(s) = \log C - \gamma \log s$$

This, of course, is the equation of a straight line in log-log space with

$$\text{ordinate intercept} = \log(C)$$
$$\text{slope} = -\gamma$$

Process fractals have the property of scale-invariance, that is, self-similarity across scales. It is straightforward to show that process fractal event distributions are scale-invariant (see Solé et al., 1999). Again, these distributions follow a power law:

$$N(s) = Cs^{-\gamma}$$

If we look at a larger or smaller scale, that is, if we take

$$s' = as$$

where a is a multiplicative factor, it is not difficult to see that

$$N(s') = cN(s) \quad \text{where } c \text{ is another constant}$$

or, in other words, a change of scale in the event quantity s does not modify the basic statistical behavior. We have self-similar behavior at all event quantity scales.

Process fractal time series are also scale-invariant. Self-similarity across time scales can be observed. Later, in Section **15.4.3**, in the context of

[7] Mandelbrot (1982) refers to these, more generally, as hyperbolic probability distributions. A random variable U is called hyperbolic when $P(u) = \Pr(U > u) = ku^{-D}$ where k is a constant and D is the hyperbolic exponent.

process fractal frequency spectrum representations, I will show mathematically why self-similarity in time occurs. In Section **15.4.4**, we will discuss heart rate dynamics and provide illustrations of both process fractal frequency spectrum behavior and self-similarity across time scales.

In my view, self-similarity across scales is a reflection of the universality of fractal behavior. Process fractal behavior is not dependent on event scale. It is not dependent on time scale. One can observe it at all scales (within the fractal validity range of a given process). In Section **15.4.5**, we will even see behavioral similarity with respect to *different* processes that occur at *vastly different* space and time scales. Perhaps we should call that not *self*-similarity, but *cross*-similarity. The point is that fractal behavior is very widespread in nature. It is a universal concept. As Mandelbrot (2004) has said, "fractals . . . are present everywhere."

15.4.3 Mathematical Representations of Process Fractals: Frequency Spectrum

The frequency spectrum representation of process fractals takes the signal processing perspective. My earlier degrees in electrical engineering and my earlier work at Bell Labs involved communication theory and signal processing theory. In signal processing work, one is often interested in the frequency content of a given signal. From the "noise-like" signal of a process fractal time series, we can derive a process fractal frequency content distribution, also known as a frequency spectrum representation. Interestingly, this process fractal frequency spectrum takes the form of a power-law curve. The frequency perspective yields additional insights into process fractal behavior and, therefore, into ecological system dynamical behavior. We'll cover the mathematical derivations and their meanings in the following paragraphs.

At the beginning of Section **15.4.2**, I presented a figure (Figure 15.5) that graphically illustrates a typical process fractal event time series and distribution. The time series diagram in that figure could very well represent ecological system fluctuations (in fact it does). That time series diagram is reproduced here as Figure 15.6. The time series signal is noise-like; its appearance suggests a noise signal. (You'll see why I point that out in a moment.)

From signal processing theory, we know that we can transform a time domain representation (time series) into a frequency domain representation (frequency spectrum) via the Fourier transform, defined by

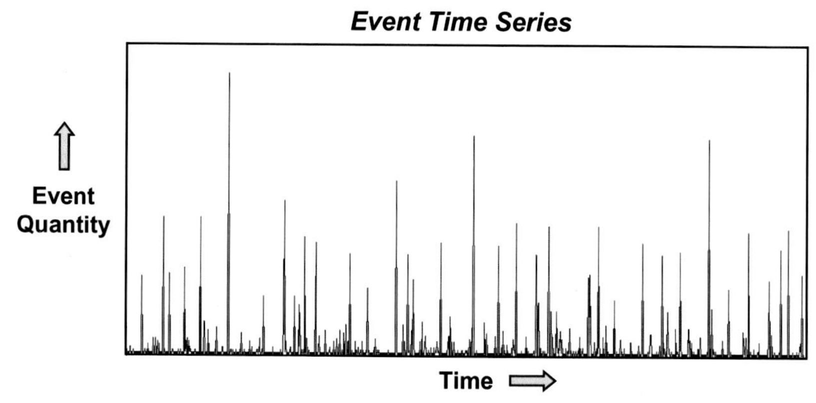

Figure 15.6 Ecological system event fluctuations over time.

$$X(f) = \int_{-\infty}^{+\infty} x(t)e^{-i2\pi ft}dt \quad \text{where}$$

f is frequency
t is time
X is a function of frequency
x is a function of time

The resulting frequency spectrum of a fluctuating noise-like signal has one of a family of forms depicted in the graph of Figure 15.7. The family of frequency spectra shown in the graph can be expressed as

$P(f) = Cf^{-\beta}$ where
$P(f) =$ the amplitude of the frequency contribution
$f =$ frequency
$\beta =$ the spectral exponent
$C =$ constant

The behavior here, therefore, follows a power law.

Because the expression for $P(f)$ can, of course, be written as

$$P(f) \propto \frac{1}{f^{\beta}}$$

these dynamics are often called "$1/f$ noise."

Different values of the spectral exponent correspond to different "colors" of noise:

$\beta = 0 \Rightarrow$ white noise
$0 < \beta < 2 \Rightarrow$ pink noise
$\beta = 2 \Rightarrow$ brown noise
$\beta > 2 \Rightarrow$ black noise

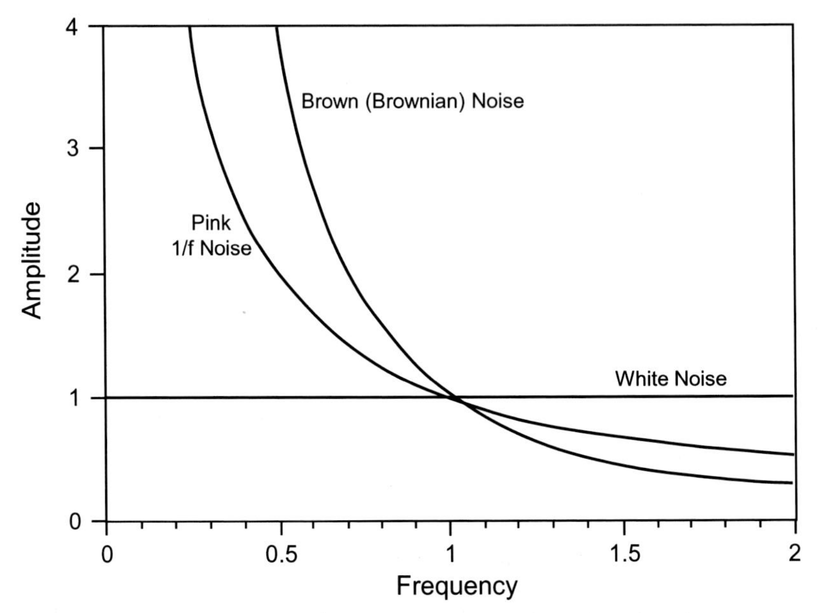

Figure 15.7 Frequency spectra of noise-like time signals. *(Adapted, with permission of Elsevier, from Halley, J. M., Ecology, Evolution and 1/f-noise, Trends in Ecology and Evolution, Volume 11, Number 1, January 1996, pp. 33–37 (http://www.sciencedirect.com/science/journal/TREE)).*

The white and pink "colors" are assigned by analogy with the visible light spectrum. White light contains equal amounts of all visible light frequencies. By analogy, white noise has a flat spectral density, containing equal amounts of all frequencies. As you can observe from the graph of Figure 15.7, the pink noise spectrum emphasizes the lower frequencies. By analogy with the visible spectrum, where the lower frequencies are red, pink noise is "reddened" relative to white noise; hence it is pink. Brown noise (from Brownian motion) has a spectrum that corresponds to that of a signal doing a random walk. Mandelbrot (1982) has said, "Many [such] noises have remarkable implications in their fields, and their ubiquitous nature is a remarkable generic fact."

"Pink $1/f$ noise" is the frequency spectrum representation of process fractals; it is the frequency spectrum representation of the time signals of complex ecological system fluctuations.

The pink $1/f$ noise frequency spectrum of Figure 15.7 yields important insights into ecological system process fractal behavior. From the figure, we can see that events with low frequency content (gradual changes in time) are more likely than events with high frequency content (abrupt changes in time), but both can occur and do occur. In ecosystems, gradual behavior occurs more often and *is expected*, but abrupt, extreme behavior also occurs and *should be expected*. (Recall the black swans.) It is well known from signal processing theory that any fluctuating signal in time can be represented by a sum of sine waves. The frequency of each of these component sine waves is equal to the inverse of its period, that is, the time it takes for one complete oscillation:

$$f = \frac{1}{T} \quad \text{where}$$
$$f = \text{frequency}$$
$$T = \text{time period}$$

In a fluctuating signal, low-frequency content means larger time periods and more slowly changing dynamics. High-frequency content means smaller time periods and more rapidly changing dynamics.

Other investigators have commented on pink $1/f$ noise behavior. Csermely (2006) says, "Pink noise is encountered in a wide variety of systems . . . and is suggested to be a characteristic feature of system complexity." Halley (1996) says, "There are good reasons to believe that the structure of environmental fluctuation is well described by a phenomenon called $1/f$-noise. . . . Recent analyses of data, results of models, and examination of basic $1/f$-noise properties suggest that pink $1/f$-noise, which lies midway between white noise and the random walk, might be the best null model of environmental variation [fluctuations in time]." Environmental fluctuation "processes behave in an essentially fractal way, having statistical self-similarity on all scales." Halley comments further on the wide range of possible events associated with ecosystem process fractal behavior. He says, "Ecologists expect both rare and common events to be important. The diversity of a desert ecosystem, for example, will be influenced by numerous small changes each day. Some rare events, such as desert storms, will have longer-lasting influence. [Pink] $1/f$-noise is a way of describing these kinds of events."

We can now show analytically why process fractal self-similarity across time scales occurs. We have just established that process fractal frequency behavior follows a power law and is therefore scale-invariant. Because $f = 1/T$, time behavior also follows a power law and is scale-invariant:

$$P(f) = Cf^{-\beta}$$
$$f = \frac{1}{T}$$
$$P(T) = C\left(\tfrac{1}{T}\right)^{-\beta}$$
$$P(T) = CT^{\beta}$$

P(T) follows a power-law curve that is flipped horizontally relative to the *P(f)* power-law curve. If we now look at a larger or smaller time scale, that is, if we take

$$T' = bT$$

where b is a multiplicative factor, it is not difficult to see that

$$P(T') = c\,P(T) \quad \text{where } c \text{ is another constant}$$

or, in other words, a change of scale does not modify the basic statistical behavior. We have fractal self-similar behavior across time scales.

15.4.4 Heart Rate Dynamics

Ary L. Goldberger, MD, is the director of the Rey Laboratory and a professor of medicine at Harvard Medical School. He is also a core faculty member of the Wyss Institute for Biologically Inspired Engineering at Harvard University. Dr. Goldberger and his colleagues have helped pioneer the application of fractals to physiology and medicine over the past two decades. They published the first papers describing the fractal nature of normal cardiac electrophysiology and fractal mechanisms in a variety of other physiological systems. We will take a look at some of the heart rate dynamics study results from Goldberger and colleagues in this subsection. They have used the tools that we have been discussing here in Section **15.4**. Their heart rate results provide illustrations of process fractal time series, frequency-spectrum behavior, and self-similarity across time scales.

Consider heart function. In healthy human beings, heart rate is not steady; it shows significant spontaneous variation. As part of the body's physiological network, a healthy circulatory subsystem is responsive to what is going on in the body's other subsystems and in the body's environment. Healthy heart function has characteristics of agility and responsiveness. Figure 15.8 shows the heart rate dynamics of a healthy young subject in the supine position.[8] The time series in the upper portion of the figure displays instantaneous heart rate over a 5-min interval. The healthy heart exhibits variability and fluctuation (punctuated dynamics).

In the lower portion of Figure 15.8, Fourier analysis has been applied to the time series to yield a frequency spectrum. There is a broad range of

[8] Lipsitz et al. (1990).

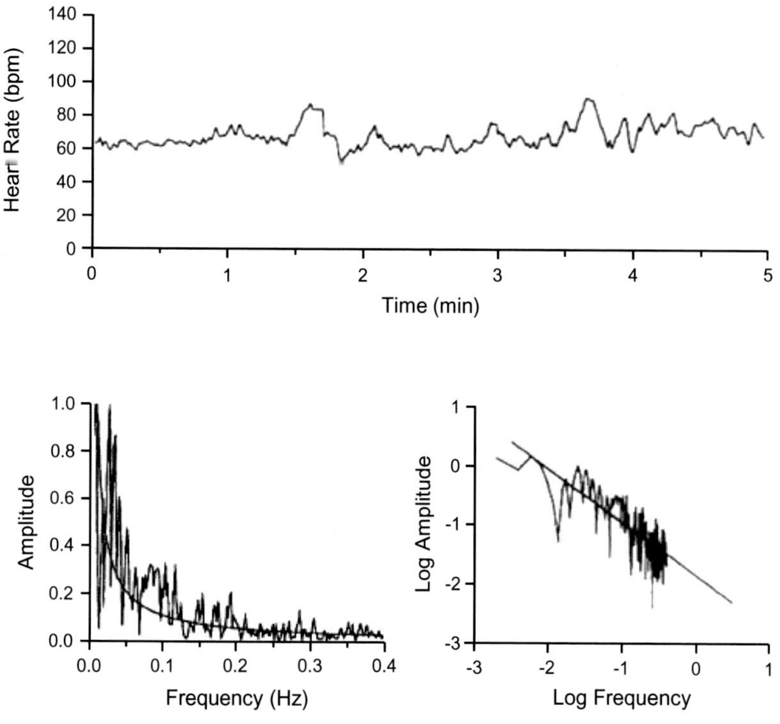

Figure 15.8 Heart rate time series and frequency spectrum. *(Reprinted, with permission, from Lipsitz L. A., Mietus J., Moody, G. B., and Goldberger, A. L., Spectral Characteristics of Heart Rate Variability Before and During Postural Tilt, Circulation, Volume 81, Issue 6, June 1, 1990, pp. 1803–1810. Copyright © 1990, Wolters Kluwer Health).*

frequencies, with emphasis on the lower frequencies. A power-law curve was fit to the frequency spectrum representation on the left. A power-law straight line was fit to the log-log frequency spectrum representation on the right. These results indicate "pink $1/f$ noise" behavior. They indicate process fractal behavior.

To illustrate healthy heart self-similarity in time, Ary Goldberger (1996) examined different time scales, as shown in Figure 15.9. The one-tenth boxed area of an upper time series plot is expanded in the next lower time series plot. The resulting three time series at three different time scales are statistically self-similar. At each scale, we see similar fluctuations.

15.4.5 Cross-similarity

Fractals are everywhere. Ecological system networks can exhibit similar fractal behavior for very different processes that occur in very different spatial

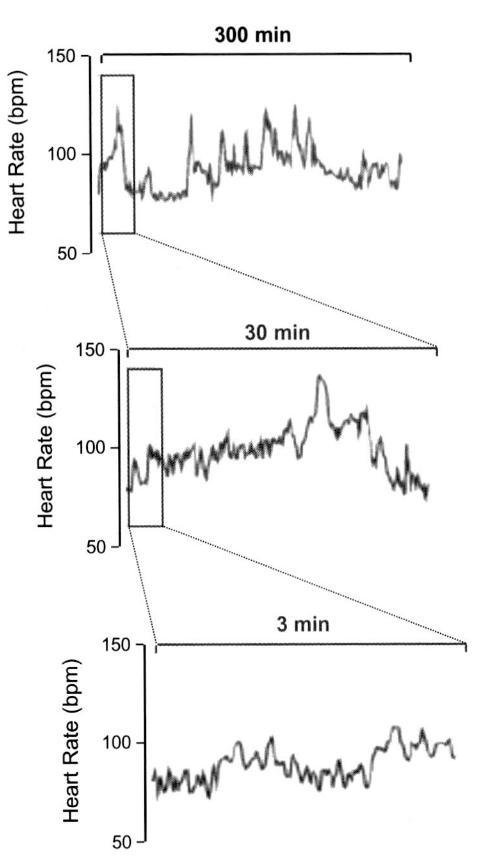

Figure 15.9 Healthy heart self-similarity in time. *(Reprinted from Non-linear dynamics for clinicians: chaos theory, fractals, and complexity at the bedside, Goldberger, A. L., The Lancet, Volume 347, Issue 9011, Pages 1312–1314, Copyright (1996), with permission from Elsevier. (http://www.sciencedirect.com/science/journal/lancet)).*

frames and very different time frames. In this sense, the dynamics of short-term operational processes that occur in smaller spaces can match the dynamics of long-term developmental processes that occur in much larger spaces. We will call that effect not *self*-similarity, but *cross*-similarity.

Consider an ecological network of species and two quite different normal processes: an operational biomass/energy propagation process and a developmental macroevolution process. We know that we can characterize these processes in terms of their outcomes (i.e., the events that occur), and that we can plot the corresponding time series of events. A short-term propagation event time series and a long-term macroevolution event time series are shown in Figure 15.10. I generated the (upper) propagation time series

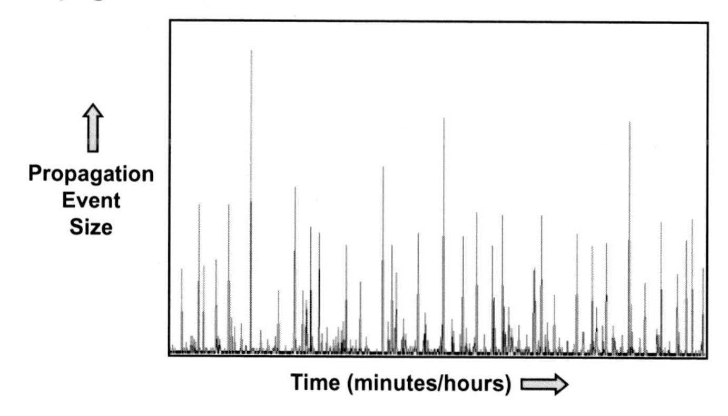

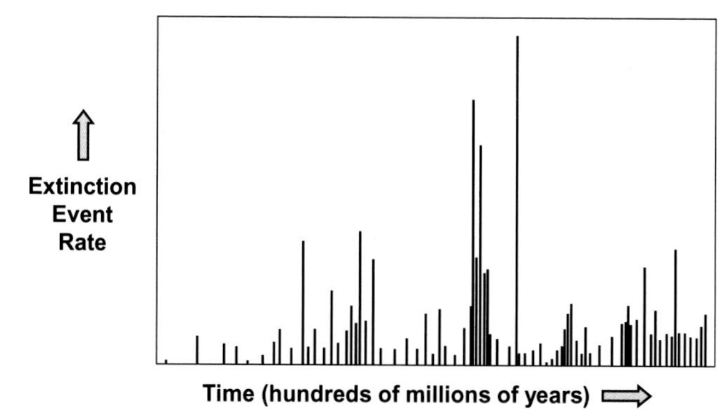

Figure 15.10 Dynamical equivalence across process, space, and time.

from my ecological network dynamics model data. I generated the (lower) macroevolution time series from Phanerozoic eon data in the fossil record, which can be obtained from *The Fossil Record 2* website at http://www.fossilrecord.net/fossilrecord/summaries.html. (See also Benton, 1993, 1995.)

The two time series exhibit similar fractal behavior, irrespective of the vast differences in process, spatial frames, and time frames. The two graphs reflect the same statistical principles. There seems to be equivalence—*universality*—across process, space, and time in ecological network dynamics.

15.4.6 Commentary

Can we define necessary and sufficient conditions for process fractal behavior?

There exist multiple cases for which one can take a collection of physical quantities, plot them as a distribution, and find that the distribution follows a power law. One set of examples[9] involves the so-called allometric scaling laws, which plot various organism physical measures versus organism mass. The plots turn out to be power-law curves. The allometric scaling laws can be written as

$$
\begin{aligned}
P &= cM^{\alpha} \quad \text{where} \\
P &= \text{the physical measure} \\
c &= \text{a constant} \\
M &= \text{mass of the organism or organelle} \\
\alpha &= \text{the scaling exponent}
\end{aligned}
$$

The concept seems very general—for example, for metabolic rate the scaling exponent is $\alpha = 3/4$, for lifespan $\alpha = 1/4$, and for heart rate $\alpha = -1/4$. Another set of examples involves the Pareto law (1897), which takes the form of a power law and is perhaps best known for quantifying the distribution of wealth over populations. It is sometimes called the 80-20 rule; that is, 20% of the people have 80% of the wealth. (Surely, with respect to wealth, those values need to be updated.) The Pareto rule also seems general and applies much more widely than just to wealth. None of these cases, however, involve process fractal behavior. There is no identifiable process over structure that yields function (per the systems triad). Because a system is defined as a group of interacting, interrelated, or interdependent components that together form a complex whole (see Chapter 1), there is, in the aforementioned cases, no identifiable system and no identifiable system network of components. There is just a collection of physical quantities that follow a power law but are not process fractals. We have, of course, already clearly established that process fractals have power-law distributions. One can say, therefore, that a power-law distribution is a necessary condition for process fractal behavior, but not a sufficient condition.

Process fractals characterize the behavior of bona fide dynamic processes operating in bona fide systems. Process fractals involve more than power-law distributions. Process fractals have a dynamic, punctuated, fluctuating time series. From this time series, one can directly derive an event distribution (e.g., number of occurrences vs. event quantity), which follows a power law. One can also take the discrete Fourier transform of the time series

[9] The examples in this paragraph are from Csermely (2006).

discrete time signal and derive a frequency content distribution (frequency spectrum), which also follows a power law.

Although I cannot provide an explicit formal proof at this time, I strongly suspect that, for highly complex natural systems (e.g., ecological systems) and their processes, punctuated time series and the associated power-law distributions comprise *necessary and sufficient* conditions for process fractal behavior.

15.5 LOCAL-TO-GLOBAL INTERACTION

Complex ecological networks are characterized by local-to-global interaction. In this section, a brief operational view of local-to-global dynamics is provided, and then the facilitators of local-to-global interaction are discussed in some detail.

15.5.1 An Operational View

Back in Chapter 6, we referred to ecosystem local-to-global dynamics as a relaxation phenomenon that involves the processing and propagation of inputs (beneficial and otherwise) that impinge upon an ecological network. To survive, every living system must respond effectively to stimuli from its environment. The stimulus can be considered a source of tension, and the response a relaxation. When an ecological network receives a stimulus (e.g., energy, biomass, information) from its environment, the input is often processed locally, yielding local propagation and relaxation in the system. When some stimulus cannot be handled locally, tension persists. Local tensions may develop to a point where global propagation and processing suddenly occur. Over time, therefore, both local and global network propagation events are required to respond effectively to the ongoing environmental stimuli. Any propagation event can have small or large spatial extent, as well as anything in between. Peter Csermely (2006) says that living system functions "cannot be performed in the absence of widespread network communication." They cannot be performed without local-to-global dynamics.

Stephen Wolfram (2002) relates local-to-global communication to the system classes discussed in Chapters 11 and 14. In class 1 systems, information is not communicated across system components. In class 2 systems, some information is communicated, but only locally and not globally. Class 3 systems exhibit random global communication across all parts of the system. In class 4 systems, coherent global communication occurs, but not always. These "differences in the handling of information are in some respects particularly fundamental." Recall that, in my view, complex ecological systems are class 4 systems. These systems can move to the edge-of-order domain (per Chapter 14) and exhibit extremely complex behavior, including local-to-global dynamics.

I'll conclude this operational view with two observations. First, an ecosystem propagation event time series (say, like the one in Figure 15.10) reflects local-to-global behavior. Observe that many propagation events are small and mostly local, but some are large and global. The extremely large events can span (nearly) the entire ecological network. Second, as previously discussed, network local dynamics can spawn global dynamics. The global propagation and processing potentially changes the state of each involved network node, which can affect subsequent local processing. Global dynamics, therefore, impact local dynamics. We actually have local-to-global-to-local behavior.

15.5.2 Facilitators of Local-to-Global Interaction

Two important facilitators oflocal-to-global interaction are (1) the small-world property of many networks and (2) network critical connectivity. The two facilitators are discussed in the following two subsections, respectively.

15.5.2.1 The Small-World Network Property

Once again, the "small world" property of networks is pertinent. Many real-world complex network systems, including ecological networks, have the small-world property. As discussed in Chapter 8, small-world networks exhibit *high clustering* and *short characteristic path lengths*. Recall that, with respect to clustering, the relevant network parameter is *clustering coefficient*. The clustering coefficient of a network node can be defined simply as the number of links that actually exist among the node and its neighbors divided by the total number of links that could possibly exist among the node and its neighbors. The network clustering coefficient can be regarded as an average of node clustering coefficients. Path length is defined as the number of links traversed along a path between two network nodes. The shortest path length is the shortest route (smallest number of links) between two nodes. The characteristic path length is the average of all the shortest path lengths in the network.

The structural, geometry-related small-world network attributes of high clustering coefficient and short characteristic path length have important implications for complex network dynamics. In ecological systems and other complex networks, high clustering (which can be viewed as modularization) facilitates local propagation and processing. Short path lengths can provide efficient propagation channels between distant parts of the network, thereby facilitating any dynamical process taking place on the network that requires global propagation and processing.

Back in Chapter 8, we looked at empirical data from real networks. A table (Table 8.1) of some basic statistics for several published real networks in various disciplines was provided. That table is reproduced here as Table 15.1. For each of the networks, the table indicates network size (number of nodes), characteristic path length ℓ, and clustering coefficient C. The characteristic path length ℓ_{random} and the clustering coefficient C_{random} of a comparable random network (i.e., a random network having the same number of nodes and links as the real network) are also included for comparison purposes.

Focus on the two ecological food web entries in Table 15.1 (the shaded entries). In Chapter 8 we explained that, for small-world networks, the short characteristic path length would be close to that of a comparable random network and the high clustering coefficient would be much higher than in a comparable random network. These effects can be clearly seen for the food webs in Table 15.1. There is empirical evidence that real-world ecological networks (in this case ecological food webs) exhibit the small-world property.

Other investigators offer their perspectives on the small-world property and local-to-global interaction. Watts and Strogatz (1998) and Strogatz (2001) suggest that small-world dynamical systems would display enhanced signal propagation speed as compared with regular lattices of the same size. The reasoning is that the short paths could provide high-speed communication channels between distant parts of the system, thereby facilitating any dynamical process that requires global coordination and information flow. Newman (2003) agrees and says that the small-world property has obvious

Table 15.1 Empirical Data for Several Real Networks

Network	Size (#Nodes)	ℓ	ℓ_{random}	C	C_{random}
World Wide Web	153,127	3.1	3.35	0.1078	0.00023
Los Alamos Lab co-authorship	52,909	5.9	4.79	0.43	1.8×10^{-4}
Medical co-authorship	1,520,251	4.6	4.91	0.066	1.1×10^{-5}
Physics co-authorship	56,627	4.0	2.12	0.726	0.003
Neuroscience co-authorship	209,293	6	5.01	0.76	5.5×10^{-5}
E. coli substrate graph	282	2.9	3.04	0.32	0.026
E. coli reaction graph	315	2.62	1.98	0.59	0.09
Ythan estuary food web	134	2.43	2.26	0.22	0.06
Silwood Park food web	154	3.40	3.23	0.15	0.03
C. elegans	282	2.65	2.25	0.28	0.05
Electric power grid	4941	18.7	12.4	0.08	0.005

Adapted, with permission, from Albert, R. and Barabási, A-L., Reviews of Modern Physics, Volume 74, January 2002, pp. 47–97. Copyright (2002) by the American Physical Society (http://link.aps.org/).

implications for the dynamics of processes taking place on networks. For example, if one considers the spread of information, or indeed anything else, across a network, the implication is that the spread will be fast on small-world networks.

So, complex ecological networks appear to possess the necessary (small-world) infrastructure to support local-to-global interaction. But what *dynamical property* of these networks actually facilitates the abrupt change from mostly local flow to widespread global flow? The answer has to do with network phase transitions and the achievement of network critical connectivity in the edge-of-order domain.

15.5.2.2 The Network Critical Connectivity Property

Network critical connectivity is achieved at a network phase transition that takes place in the nonlinear dynamics edge-of-order domain. Critical connectivity is the dynamical property that facilitates the abrupt change from local flow to global flow that can occur in complex networks. It facilitates complex ecological network local-to-global dynamics. To provide an overview of this important network property, we'll summarize salient points from our Chapter 9 discussions of critical connectivity phase transition investigations—specifically random network unconnected-to-connected transitions and lattice network percolation—and then we'll look at some Boolean network investigations by Stuart Kauffman. (These are the investigations covered in the extant literature.) I expect that the general concepts and principles developed in these investigations will be useful for networks of all sorts, including ecological networks.

15.5.2.2.1 Random Network Unconnected-to-Connected Transition

As the number of links in a random network increases, the network rather abruptly transitions from fragmented to connected. When the number of links, m, is small, the network is likely to be fragmented into many small clusters of nodes. As m increases, the clusters grow, at first by linking to isolated nodes and later by coalescing with other clusters. A phase transition occurs at

$$m = \frac{n}{2} \text{ where}$$
$$m = \text{the number of links in the network}$$
$$n = \text{the number of nodes in the network}$$

when many clusters cross-link spontaneously to form a single giant cluster. Furthermore, all nodes in the giant cluster are connected to each other by short paths. (Note that as n increases, the network characteristic path length grows slowly, like log n.)

The network transition can also be viewed from a connection probability perspective. Consider a network of n nodes in which every pair of nodes is connected with probability p. As the probability p grows and reaches the critical probability $p_c(n)$, the partially connected network abruptly transitions to a connected network. There is a phase transition from a low-density, low-p state in which there are few links and all clusters are small to a high-density, high-p state in which an extensive fraction of all nodes are joined together in a single giant cluster. At that time, the expected value of the number of network links and the expected value of node degree also reach their critical points.

Here's a summary of the particulars of the phase transition to a connected network. This applies to *undirected* random networks with large n (say $n \geq 100$).

$$E\{m\} = p\left[\frac{n(n-1)}{2}\right] \quad \text{and}$$

$$E\{k\} = \langle k \rangle = p(n-1)$$

where

$n =$ the number of network nodes

$p =$ the connection probability of any pair of nodes

$m =$ the number of network links

$k =$ node degree $=$ the number of links connected to a node

$E\{m\} =$ the expected value of random variable m

$E\{k\} = \langle k \rangle$ is the expected value of random variable k (the mean degree)

Critical connectivity occurs, for large n, when the expected number of links equals $n/2$. Therefore,

$$E_c\{m\} = \frac{n}{2}$$

$$E_c\{m\} = p_c\left[\frac{n(n-1)}{2}\right] = \frac{n}{2}$$

and therefore

$$p_c(n-1) = 1$$

It follows that

$$E_c\{k\} = \langle k \rangle_c = p_c(n-1) = 1$$

and

$$\langle k \rangle_c = 1$$

$$p_c = \frac{1}{(n-1)}$$

where the subscript c indicates the critical connectivity values

15.5.2.2.2 Lattice Network Percolation

The abrupt phase transition from an unconnected network to a connected network is central to the percolation phenomenon. The critical probability previously derived applies to undirected random graphs. Percolation, however, was originally studied in physics and statistical mechanics rather than in graph theory. Traditional percolation theory is usually based on regular lattice structures and not on random graphs. For general lattices, the transition point (the critical probability p_c) is much more difficult to determine. In a few cases, the critical probability can be calculated explicitly. For example, for a square lattice in two dimensions, $p_c = 0.5$.

Figure 15.11 shows two snapshots of a regular square two-dimensional lattice whose edges (links) are present with probability p and absent with probability $1 - p$. Percolation theory studies the emergence of paths that percolate through the lattice (starting at one side and ending at the opposite side).[10] For small p, only a few edges are present and only small clusters of nodes connected by edges can form. At larger values of p—specifically at critical probability p_c, called the *percolation threshold*—a percolating cluster of nodes connected by edges appears. Global flow occurs.

For the illustration of percolation in Figure 15.11, nodes are placed on a 25×25 square lattice, and two nodes are connected by an edge with probability p. In the snapshot on the left, which is below the percolation threshold of $p_c = 0.5$, the connected nodes form isolated clusters. In the snapshot

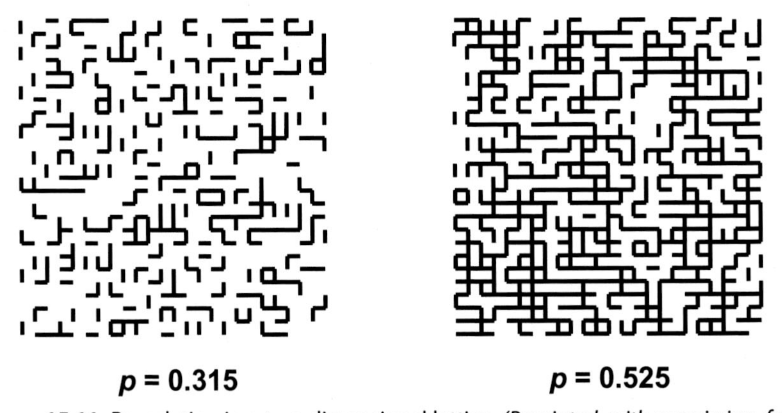

Figure 15.11 Percolation in a two-dimensional lattice. *(Reprinted, with permission, from Albert, R. and Barabási, A-L., Statistical Mechanics of Complex Networks, Reviews of Modern Physics, Volume 74, January 2002, pp. 47–97. Copyright (2002) by the American Physical Society (http://link.aps.org/)).*

[10] Benoit Mandelbrot (1982) notes that *percolate* is from the Latin, where *per* means *through* and *colare* means *to flow*, that is, *flow through*.

on the right, which is slightly above the percolation threshold (slightly above critical connectivity), the network percolates.

Percolation is sudden—a threshold phenomenon—separating two well-defined phases. In the subcritical phase ($p < p_c$), there are short-range (local) interactions. In the supercritical phase ($p > p_c$), there are long-range (global) interactions. At criticality, we get processes with global correlations arising from the local interactions. Benoit Mandelbrot (1982) says, "A fascinating discovery of lattice physics, and one that deserves to be known widely, is that under certain conditions it happens that purely local interactions snowball into global effects."

15.5.2.2.3 Boolean Network Investigations

Stuart Kauffman (1995) has modeled and analyzed large, complex, randomly constructed Boolean (2 states per node) *directed* networks.[11] His "NK model" analysis considers networks with n nodes and k connections per node (here, k is the node *mean* degree). Sample networks with given n and k values are assembled, and the behavior of those networks is studied. For large n, the results indicate:[12]

$k=1$	Nothing interesting happens—"too orderly" (in *system class* terms, this is class 1 or 2 behavior)
$k=n$	Massively chaotic—small perturbations cause massive behavioral changes (class 3 behavior)
$k=4$ or 5	Chaotic (class 3 behavior)
$k=2$	Flexible "order arises, sudden and stunning" (class 4 behavior)

In Kauffman's words, with large n and $k=2$, the network *looks* like a "scrambled jumble" but it *behaves* in an orderly fashion. "These networks are not *too* orderly. Unlike the $k=1$ network, they are not frozen like a rock, but are capable of complex behaviors." Kauffman says that his basic behavioral results "apply to networks of all sorts." This "is almost certainly merely the harbinger of similar emergent order in whole varieties of complex systems."

Kauffman (1995) says further that "astonishingly simple rules, or constraints, suffice to ensure that unexpected and profound dynamical order

[11] Kauffman has used random Boolean networks to model genetic regulatory networks, where genes are the nodes and a node A to node B link means gene A regulates gene B. He has used Boolean networks to represent macroevolution processes as well. Also see Solé and Goodwin (2000) for descriptions of some of Kauffman's work.

[12] You will probably notice a factor of 2 discrepancy between these directed network results and the previous undirected network results. I'll explain momentarily.

emerges spontaneously. . . . If the network is 'sparsely connected' [i.e., $k=2$], then the system exhibits stunning order. . . . Our intuitions about the requirements for order have, I contend, been wrong for millennia. We do not need careful construction; we do not require crafting. We require only that extremely complex webs of interacting elements are sparsely coupled."

Note that, for Kauffman's Boolean networks, a node mean degree value of 2 corresponds to edge-of-order critical connectivity. In our earlier discussion of unconnected-to-connected network phase transitions, we said that the critical value of node mean degree equals 1. In both cases, we are talking about random networks. The factor-of-two difference is explained by the fact that the earlier discussion applies to *undirected* random networks and the Boolean discussion here applies to *directed* random networks. Each adjacent connection in an undirected network equals two adjacent connections in a directed network.

In our ecological network modeling and analysis work in Part V of the book, we will look for and evaluate the local-to-global interactions and network critical connectivity discussed here in Section **15.5**. We will also check out the claim that network critical connectivity is sparse connectivity. Perhaps ecological networks are sparsely connected networks, with just enough connections to provide full or nearly full node-to-node access via mostly indirect paths (indirect effects). We will see if that is so, and we will discuss *indirect effects* next.

15.6 INDIRECT EFFECTS

Indirect effects are an important characteristic of complex ecological network behavior. They play a prominent role in the node-to-node interactions of network dynamics.

Indirect effects, and direct effects, are defined simply in terms of the connection path length associated with a node-to-node interaction. Direct effects have a connection path length of 1 (a direct path). Indirect effects have a connection path length greater than 1 (an indirect path). A network propagation event, for example, can involve many node-to-node interactions, many of which involve indirect paths. So, indirect effects dynamics must be related to path length dynamics as well as overall propagation event dynamics. We have seen (in Chapters 4 and 12) that ecological network path length can be fractal with a power-law distribution. We have suggested (in this chapter) that, over time, ecological network propagation event dynamics are punctuated, fractal, and local-to-global. Do indirect effects dynamics possess these same features? I hypothesize that indirect effects dynamics are indeed punctuated and fractal, and support local-to-global interactions. We will have to wait to see the modeling and analysis results of Part V to determine whether or not this is so.

Recognition of the importance of indirect effects in ecological networks is due, in large part, to the work of Dr. Bernard C. Patten and colleagues at the University of Georgia and elsewhere. The Patten-inspired group of researchers uses a linear steady state modeling and analysis approach. The insights obtained by the Patten group have proven very useful. Many of the group's results indicate a dominance of indirect effects over direct effects. In complex ecological networks, indirect effects are always prominent. Are they also dominant? Can we define conditions for prominence vs. dominance? Let's use some mathematics in an attempt to gain some insights into these questions.

15.6.1 Indirect Effects: Prominence or Dominance?

As we know, ecological systems take the form of networks. Network node-to-node connections and their associated connection path lengths can be characterized as random variables. Any two nodes in the network (any given pair) are either connected or not connected. If a node pair is connected, the connection can be either *direct* or *indirect*. The path length is 1 for a direct connection and greater than 1 for an indirect connection.

Let's represent the connection path length by a discrete random variable X with probability distribution function $P\{X=x_i\}$:

$$
\begin{aligned}
&P\{X=x_i\} = P\{x_i\} \quad \text{where} \\
&\qquad x_i \in \{1,2,3,\cdots,N-1\} \\
&\qquad \text{and} \\
&\qquad N = \text{the number of nodes in the network}
\end{aligned}
$$

I claim that a power-law distribution is an appropriate and suitable choice for the probability distribution function $P\{x_i\}$ (see Chapters 4 and 12 for the rationale and real-world network supporting evidence). $P\{x_i\}$, therefore, is given by:

$$
\begin{aligned}
&P\{x_i\} = Cx_i^{-\gamma} \quad x_i \in \{1,2,3,\cdots,N-1\} \\
&\qquad \text{where} \\
&\qquad C \text{ is a constant} \\
&\qquad \text{and} \\
&\qquad \gamma = \text{the network path length scaling exponent}
\end{aligned}
$$

Using data from complex biological networks, Song et al. (2005), Song et al. (2007), and Vishwanathan et al. (1999) have calculated values of γ that are in the neighborhood of 2, that is, $(1+) < \gamma < (2+)$.

Because $P\{x_i\}$ is a probability function, its values must sum to 1:

$$\sum_{x_i} P(x_i) = \sum_{x_i} C x_i^{-\gamma} = 1$$
$$C \sum_{x_i} x_i^{-\gamma} = 1$$

This sum has the same form as the so-called *p-series*:

$$\text{For any positive real number } p:$$
$$p - \text{series} = \sum_{n=1}^{\infty} n^{-p}$$
$$\text{For } p > 1:$$
$$\sum_{n=1}^{\infty} n^{-p} = \zeta(p)$$

The *p-series* always converges if $p > 1$ (in which case it is called the *over-harmonic series*). When $p > 1$, the sum of this series turns out to be equal to the *Riemann zeta function* evaluated at p, that is, $\zeta(p)$.

Note that, while the *p-series* is infinite, our path length probability series is finite. However, even for small to moderate size networks, the value of the path length series is close to the convergent value of the *p-series*. For example, consider an exponent of 1.7. The value of the *p-series* is $\zeta(p) = \zeta(1.7) = 2.054$. For a 25-node network ($N - 1 = 24$), the value of the path length series $= 1.902$. For a 100-node network ($N - 1 = 99$), the value of the path length series $= 1.997$. We will use "approximately equal" symbols when appropriate as we proceed.

Continuing with the development, the *p-series* and the Riemann zeta function can be used to calculate constant C:

$$C \sum_{x_i} x_i^{-\gamma} = 1$$

$$C \zeta(\gamma) \cong 1$$

$$C \cong \frac{1}{\zeta(\gamma)}$$

The path length probability sum relationship can be rewritten in the following way:

$$\sum_{x_i} P(X=x_i) = P(X=1) + \sum_{x_i>1} P(X=x_i) = 1$$
$$P(X=1) = C(1)^{-\gamma} = C \cong \frac{1}{\zeta(\gamma)}$$
$$P(X=1) \cong \frac{1}{\zeta(\gamma)}$$
$$\sum_{x_i>1} P(X=x_i) = 1 - P(X=1) \cong 1 - \frac{1}{\zeta(\gamma)}$$
$$\sum_{x_i>1} P(X=x_i) \cong 1 - \frac{1}{\zeta(\gamma)}$$

Relating this to direct and indirect connections, we have

$$P(X=1) \cong \frac{1}{\zeta(\gamma)}$$
$$\sum_{x_i>1} P(X=x_i) \cong 1 - \frac{1}{\zeta(\gamma)}$$
$$P(X=1) = \text{Prob}(\textit{direct connection})$$
$$\sum_{x_i>1} P(X=x_i) = \text{Prob}(\textit{indirect connection})$$
$$\textit{When } \zeta(\gamma) \cong 2:$$
$$\text{Prob}(\textit{direct connection}) = \text{Prob}(\textit{indirect connection})$$
$$\textit{When } \zeta(\gamma) > 2:$$
$$\text{Prob}(\textit{indirect connection}) > \text{Prob}(\textit{direct connection})$$

Indirect paths dominate

$$\textit{When } \zeta(\gamma) < 2:$$
$$\text{Prob}(\textit{direct connection}) > \text{Prob}(\textit{indirect connection})$$

Direct paths dominate

Figure 15.12 is a graph of the Riemann zeta function for real arguments >1. The ordinate is $\zeta(\gamma)$. The abscissa is γ—that is, values of the path length distribution scaling exponent. Typical path length scaling exponents are in the range

$$(1+) < \gamma < (2+)$$

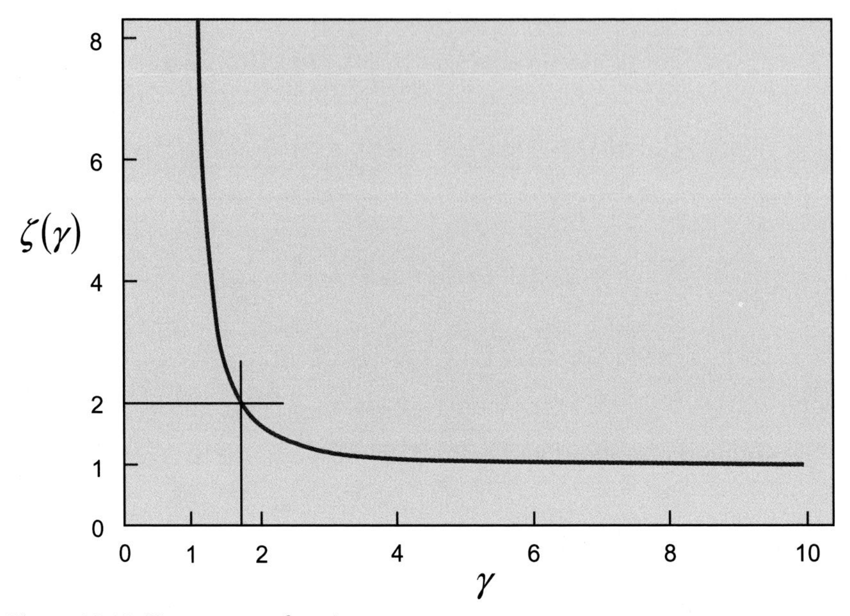

Figure 15.12 Riemann zeta function.

We observe that

When $1 < \gamma < 1.7$
$\zeta(\gamma) > 2$
Indirect effects dominate

When $\gamma > 1.7$
$\zeta(\gamma) < 2$
Indirect effects do not dominate
Indirect effects are still prominent

In our ecological network modeling and analysis work in Part V of the book, we will test these results.

15.7 THE CHARACTERISTICS HYPOTHESIS

We now pull everything together and state the characteristics hypothesis. I hypothesize that the defining characteristics of complex ecological network dynamics are

- Flickering Operation
- Punctuated Dynamics
- Fractal Behavior

- Local-to-Global Interaction
- Indirect Effects

I envision flickering networks that continually change with time. The network dynamics are punctuated and they are fractal. Fractal behavior is perhaps the dominant characteristic of complex ecological network dynamics. It is central to our understanding of these dynamics. Specifically, I claim that ecological network dynamics exhibit *process fractal* behavior with punctuated event time series, power-law event probability distributions, and power-law event frequency spectra. Node-to-node interactions in ecological networks are local when that is adequate, but can abruptly blossom into global interactions when that is required. Finally I claim that, in ecological networks, indirect effects are prominent and, in many cases, dominant.

We must now test and corroborate (or falsify) this hypothesis. To do so, we will model the operation of a complex ecological network. We will proceed to collect operational data and then perform analysis and reach conclusions in five areas, as outlined here:

- *Network operational propagation flow*
  - Construct node-and-link flow diagrams at selected model time steps.
  - Compile histories of network inputs and outputs, node stock (per-node biomass/energy contents), and node-to-node biomass/energy flow.
- *Propagation events*
  - Determine propagation event size for each model time step.
  - Construct propagation event time series.
  - Construct propagation event distributions.
- *Path length*
  - Calculate path lengths.
  - Construct path length time series.
  - Construct path length distributions.
- *Indirect effects*
  - Use path length data to perform indirect effects analysis.
  - Demonstrate prominence/dominance of indirect effects.
- *Network connectivity*
  - Construct node degree distributions.
  - Examine network critical connectivity and connection densities.

The network node-and-link flow diagrams and the input, output, stock, and flow value histories will provide direct visual evidence of network operation. We will see if the network exhibits flickering operation. The propagation event and path length results will test for the presence of punctuated, fractal, and local-to-global dynamics. In addition, the propagation event data will be used to test frequency spectrum behavior and the path length data will be used to perform indirect effects analysis. The indirect effects results will test for indirect effects dominance and will indicate whether indirect effects dynamics are consistent with propagation event and path length dynamics, that is, whether indirect effects are punctuated (rather than continuous), fractal, and enablers of local-to-global propagation. Network connectivity results will test for node degree fractal behavior and will facilitate examination of network critical connectivity. We will also be able to check for complex ecological network sparse connectivity.

Our ecological network dynamics characteristics hypothesis, therefore, will be fully and comprehensively tested. That is accomplished in Part V of the book.

PART V

Modeling Ecological Network Dynamics and the Generation and Analysis of Results

In this part, the requirements, design, and development of an innovative ecological network dynamics model are described. We run the model, generate results, and analyze the results in order to test the ecosystem dynamics characteristics hypothesis. The specific dynamics results categories are: operational propagation flow, network propagation events, propagation path length, indirect effects, and network connectivity. The characteristics hypothesis is fully tested—and corroborated.

CHAPTER 16

A New Approach to Modeling Ecological Network Dynamics

16.1 INTRODUCTION

The objective of a "good model" is to provide "an abstract representation of effects that are important in determining the behavior of a system."[1]

To test the dynamics characteristics hypothesis of Chapter 15, we model the operation of complex ecological networks, collect operational data from the model, and analyze the data to determine ecosystem dynamical behavior. The requirements for the ecological network dynamics modeling and analysis are covered here in Chapter 16.

The chapter provides overview discussions and then detailed descriptions of the model features. A fundamentally important feature is that the model is cellular automata based. Cellular automata can provide very effective representations of complex networks. For propagation of biomass and/or energy through an ecological network, the model must implement realistic node-to-node propagation neighborhoods and propagation preferences within the neighborhoods. These neighborhoods and preferences are described. A key outcome of this modeling effort is an analysis of ecosystem dynamical behavior. Accordingly, a summary of model analysis requirements is provided.

The approach here is to model complex ecological networks as discrete dynamic systems that emulate relevant features of real-world ecological networks. Traditionally, much network modeling and analysis *assumes* a static network structure, steady state network flows, and (sometimes) linear system behavior. I do not make such *a priori* assumptions about network dynamics. The objective here is to model and analyze ecological networks in order to *determine* the nature of their dynamics. I let the dynamics develop as the model simulation runs.

[1] Stephen Wolfram (2002), *A New Kind of Science*.

Understanding Complex Ecosystem Dynamics
http://dx.doi.org/10.1016/B978-0-12-802031-9.00016-4

16.2 OVERVIEW OF MODEL FEATURES

We want to model complex real-world ecological networks as discrete dynamic systems in order to observe and scrutinize their behavior at each model simulation time step, as well as their composite behavior over all model simulation time steps. An overview of the model features is provided in this section.

Model feature summary list:

- Cellular automata based
- Both aggregate and individual modeling views
- Realistic node neighborhoods for propagation
- Realistic propagation preference capabilities
- Probabilistic aspects consistent with real-world ecological networks
- Underlying ecological network compartment model

Discussion of features:

The model (actually a family of models) shall be cellular automata based. Cellular automata provide an excellent metaphor for complex ecological networks. The cells represent the network nodes and the cell interactions represent the network links. Much traditional network modeling (ecological and otherwise) does not include physical space and distance considerations. For example, the path length between any two nodes is not a physical distance, but rather is defined as the *number of links* traversed. Because our cellular automata based model will be defined on a physical spatial grid, both link-quantity and physical space/length aspects of ecological networks will be explicitly represented.

Both aggregate and individual modeling views shall be represented. For ecological networks, I consider *compartment* (e.g., groups of species) models to be aggregate-based views and *node* (e.g., species) models to be individual-based views. Traditional ecological network models (stock and flow models) are typically aggregate compartment models. My primary focus is ecological networks of nodes; however, I shall represent both aggregate compartments and the individual nodes within the compartments, and gather data and observe behavior at both levels.

Regarding propagation of biomass or energy through an ecological network, the model shall implement realistic node-to-node propagation neighborhoods and realistic compartment-to-compartment and node-to-node propagation preference capabilities. For node interaction, local von

Neumann neighborhoods and extended local von Neumann neighborhoods as well as a nonlocal neighborhood concept will be used. Compartment and node preferential attachment will be implemented in all neighborhoods.

The model shall include real-world probabilistic aspects of ecological networks. Ecological network behaviors are not, in general, deterministic. The spatial distribution of network nodes (and, therefore, compartments), the initial conditions of the nodes and compartments, the compartment-to-compartment and node-to-node interactions, as well as the environmental inputs to the network all involve probabilities. Our model shall reflect these probabilistic aspects.

The ecological network dynamics model shall be linked to an underlying ecological network compartment model. Many preexisting compartment models are candidates for this. The compartment model diagram, adjacency matrix, input vector, and output vector are particularly relevant. For our model development effort, I have selected a specific compartment model, and I will describe it shortly.

Additional detail on the model features is provided in Sections 16.3–16.5.

16.3 CELLULAR AUTOMATA BASED MODEL

My ecological network dynamics model is cellular automata based. As previously stated, cellular automata provide a very effective metaphor for complex networks. The cells represent the network nodes and the interactions among the cells represent the network links. A cellular automaton can be described as an array of interacting *finite state machines*. In the network context, each finite-state-machine node is a discrete dynamic system. The total network (array), therefore, can be a very complex discrete dynamic system. Note that I do not adhere strictly to all definitional restrictions of cellular automata. For example, I do not require that all cells be identical and that they all be updated simultaneously at each model time step.

An important aspect of cellular automata models is that they do not have the limitations (discussed in Chapter 13) of mathematical differential equation models. Furthermore, cellular automata can be used to model ecological system dynamics when it is difficult or infeasible to solve or even formulate the system differential equations.

Here are some perspectives from other investigators on cellular automata modeling. Xin-She Yang (2006) says, "A cellular automaton is a rule-based computing machine, which was first proposed by von Neumann in the early 1950s and the systematic studies were pioneered by Wolfram from the 1980s.

Since a cellular automaton consists of space and time, it is essentially equivalent to a dynamical system that is discrete in both space and time. The evolution of such a discrete system is governed by certain updating rules rather than differential equations. Although the updating rules can take many different forms, most common cellular automata use relatively simple rules." Phillip Bonacich (2003) comments on cellular automata usage in network research: Cellular automata can be used in simulations of network processes and network evolution by identifying vertices in a network with cells in a cellular automaton. Stephen Wolfram (2002) reinforces these views. He says, "A network system is fundamentally just a collection of nodes with various connections between these nodes, and rules that specify how these connections should change from one step to the next." (A cellular automaton has fundamentally the same description.) Traditional science models typically are "collections of mathematical equations," which certainly are complicated, but do not adequately represent system complexity. In many cases, mathematical descriptions are not even possible. It is better to use cellular automata, that is, "models that are based on programs which can effectively involve rules of any kind." System behavior can be observed directly just by running the model.

A simple example can help us understand the methods of cellular automata modeling. I choose a well-known example—the so-called sandpile cellular automaton. The sandpile model is often used in an attempt to demonstrate persistent "self-organized criticality" behavior. I don't think it does that. (See my comments in Chapter 14, Section 14.3.) The sandpile model, however, is a good starting point for a model of propagation in dynamic networks (the sandpile cells are the network nodes, the cell interactions are the network links, and the grains of sand are the network flow currency). Figure 16.1 illustrates propagation in the sandpile cellular automaton model. Here's the setup. The sandpile resides on a cell grid. Inputs are grains of sand from the external environment. A single grain of sand is input to a randomly selected cell at each model time step. When a cell critical "sand threshold" is exceeded, grains of sand propagate.

Here's a mathematical description of the model (from Bak, 1996). The sandpile resides on a two-dimensional grid ($N \times N$), with cells having coordinates (x,y). The state $Z(x,y)$ of any cell is a number from 0 to 4:

$$Z(x, y) \in \{0,1,2,3,4\}$$

To apply input to the system, choose a cell randomly and increase Z by 1:

$$Z(x, y) \rightarrow Z(x, y) + 1$$

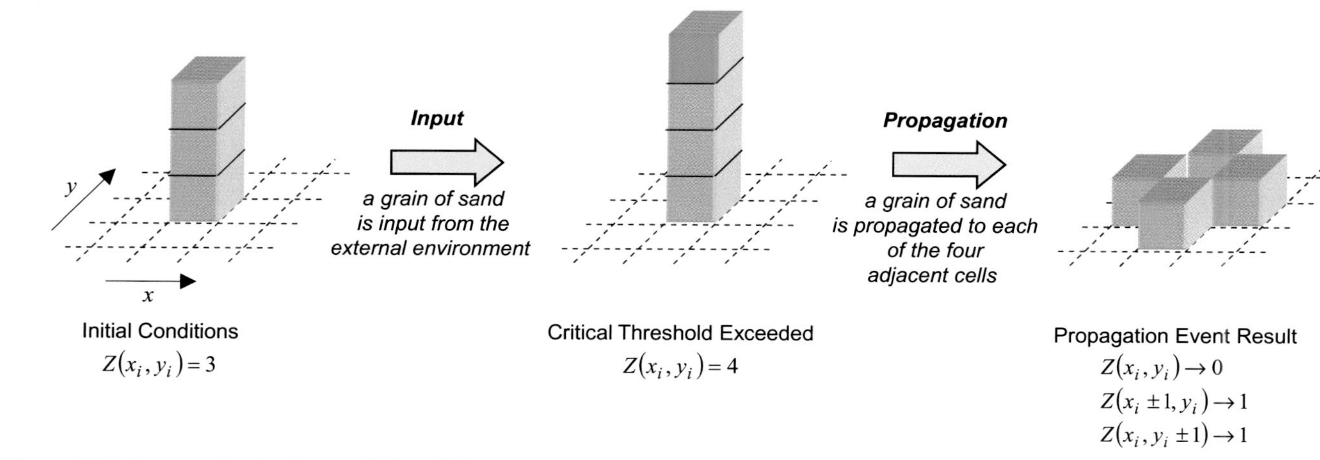

Figure 16.1 Propagation in the sandpile cellular automaton.

Repeat the input process at each model time step (iteration step). Grains of sand are propagated through the network via a propagation rule. Here is our example rule:

If $Z(x, y) > Z_{crit}$ (say $Z_{crit} = 3$)
Then propagate 1 unit to each of four neighbors:
$Z(x, y) \rightarrow Z(x, y) - 4$
$Z(x \pm 1, y) \rightarrow Z(x \pm 1, y) + 1$
$Z(x, y \pm 1) \rightarrow Z(x, y \pm 1) + 1$
If $x = 1$ *or* N and/or $y = 1$ or N (at system boundary)
Then units leave the system (output)

The propagation events are called *avalanches*. (We usually think of avalanches as large extreme events, but these avalanches can be of any size.) Here, the defined size of an avalanche equals the number of cascading propagation events that occur at a given time step.

Bak (1996), among others, has shown that the sandpile cellular automaton avalanche events (propagation events) exhibit power-law/fractal behavior. The avalanche event time series reflects punctuated dynamics. The avalanche event distribution follows a power law. The behavior of the simple model is extremely complex.

16.4 PROPAGATION NEIGHBORHOODS AND PREFERENCES

For propagation of biomass and/or energy through an ecological network, the model must implement realistic node-to-node propagation neighborhoods. The model's cellular automata structure can most straightforwardly model lattice networks. Such networks have a high *clustering coefficient* (i.e., a node's neighbors are likely to also be neighbors of each other) and high *characteristic path length* (defined as the average shortest path length taken over all pairs of nodes in a network). Most real-world ecological networks, however, are not lattice networks; they are small-world networks with high clustering and *low* characteristic path length. The high clustering is consistent with the modularity found in many real networks. The low characteristic path length facilitates the efficient global (network-wide) interaction required by real networks. To achieve these real-world structure features in our network model, the usual cellular automaton neighborhoods must be augmented, following the lead of Watts and Strogatz (1998) and Newman and Watts (1999). For node interaction, local von Neumann neighborhoods, extended local von Neumann neighborhoods, and a

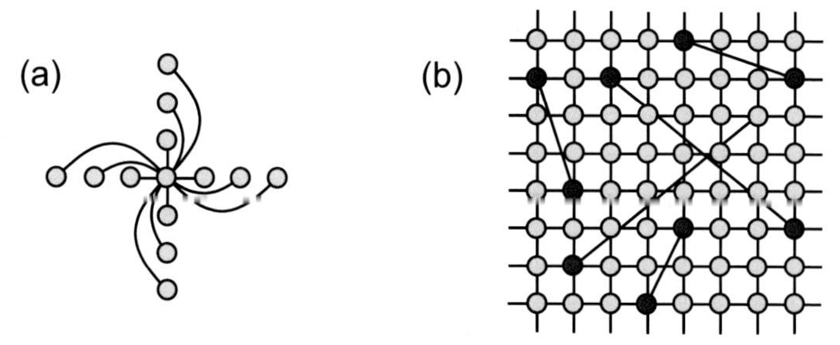

Figure 16.2 Neighborhoods for node interaction. *(Adapted, with permission, from Newman, M. E. J. and Watts, D. J., Physical Review E, Volume 60, Issue 6, December 1999, pp. 7332–7342. Copyright (1999) by the American Physical Society (http://link.aps.org/)).*

nonlocal neighborhood concept will be used. Figure 16.2 illustrates these neighborhoods for node interaction.

Part (a) of Figure 16.2 illustrates three types of von Neumann neighborhoods, with grid spacing ranges from the center node of 1, 2, and 3, respectively. We will refer to the range = 1 nearest-neighbor case as a *local neighborhood* and the range = 2 next-nearest-neighbor case as an *extended local neighborhood*. We will not use the range = 3 case. Part (b) of Figure 16.2 illustrates some node-to-node *nonlocal neighborhood* connections.

Within the context of these neighborhoods, the model must implement realistic compartment-to-compartment and node-to-node propagation preference capabilities. Albert and Barabási (2002) were among the first to identify "preferential attachment" in networks. Their objective was to identify a mechanism that explained why many real networks—human-made and natural—are "scale free" (i.e., scale-invariant with respect to node degree). My objective is somewhat different. I am interested in the natural compartment-to-compartment and node-to-node propagation preferences exhibited by many ecological networks. For example, in a trophic (nutrient) context, any given compartment (e.g., species group) will typically interact only with certain other compartments and will display preferences among those compartments (first choice, second choice . . .). Furthermore, nodes (e.g., species) within a compartment will exhibit node interaction preferences among the nodes within the target preferred compartment. There are, as a result, both compartment-to-compartment and node-to-node propagation preferences.

In our ecological network dynamics model, these propagation neighborhoods and preferences are implemented. Figure 16.3 illustrates the implementation.

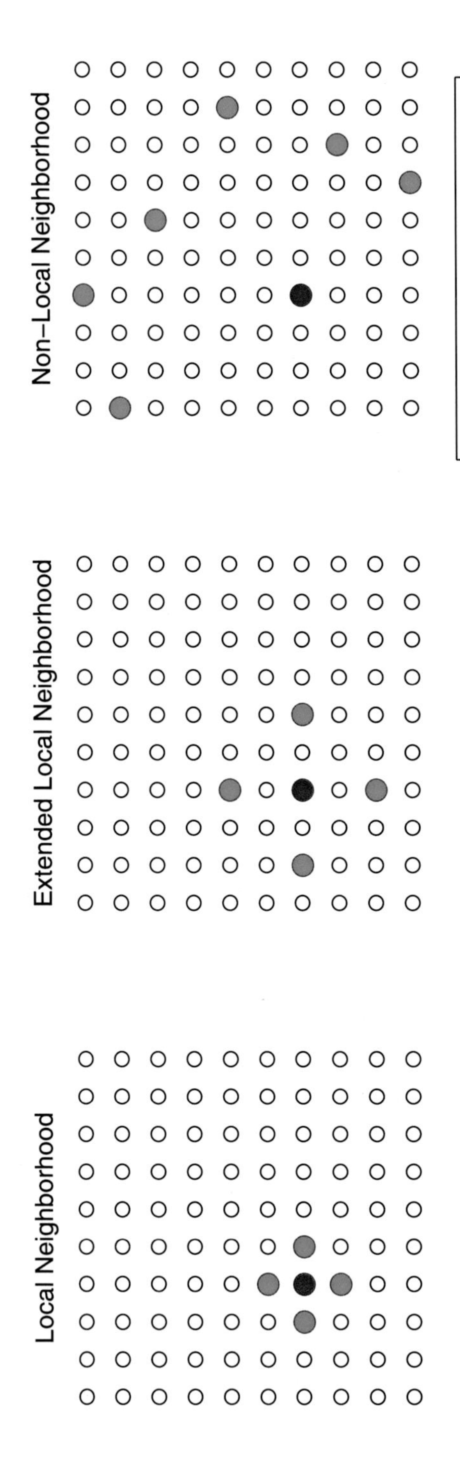

Figure 16.3 Propagation neighborhoods and preferences.

Propagation proceeds by neighborhood, specified compartment preference, and specified node preference, in that order. We begin with propagation to the local neighborhood and, if necessary, proceed to the extended local neighborhood and then to the nonlocal neighborhood until the propagation "currency" (stock) is exhausted. (Why do we need to proceed from neighborhood to neighborhood? As one example, suppose the local neighborhood contains no nodes in preferred to-compartments. We would then have to proceed to the wider neighborhoods looking for preferred to-nodes in preferred to-compartments.) Figure 16.3 depicts a propagation from-node (in red) and its potential propagation to-nodes (in blue) for the three neighborhood levels. For propagation, we must know the assigned node identifier and compartment identifier of the from-node and each of the potential propagation to-nodes, and we must know the assigned compartment preference value and node preference value of each of the potential propagation to-nodes.

Various node stock and flow rules can be implemented. Our models are discrete dynamic models. The node stock and flow rules are therefore discrete. A node will accumulate stock in unit increments until a threshold (*thd*) is reached. At that point, the node will propagate a unit flow to each of a number of node destinations based on neighborhood and preference criteria defined and specified in the model (as previously discussed), until the total propagation flow quantity ($npfq$, where $npfq \leq thd$) is exhausted. The values of *thd* and *npfq* can be varied. Note that stock and flow rules and values are kept quite simple in our model. Stock values can be 0, 1, 2, ..., *thd*. All individual flows are unit (value = 1) flows.

16.5 UNDERLYING ECOLOGICAL NETWORK COMPARTMENT MODEL

The ecological network dynamics model is linked to an underlying ecological network compartment model. For our model development effort, a preexisting estuary aquatic compartment model[2] is used. The compartment model diagram, adjacency matrix, input vector, and output vector are particularly relevant and are provided here. The compartment model diagram is shown in Figure 16.4. The adjacency matrix, input vector, and output vector follow.

[2] This compartment model was developed by Chip Small, Nicole Gottdenker, and Bill Yackinous in 2005 as a systems ecology course project at the Odum School of Ecology at the University of Georgia.

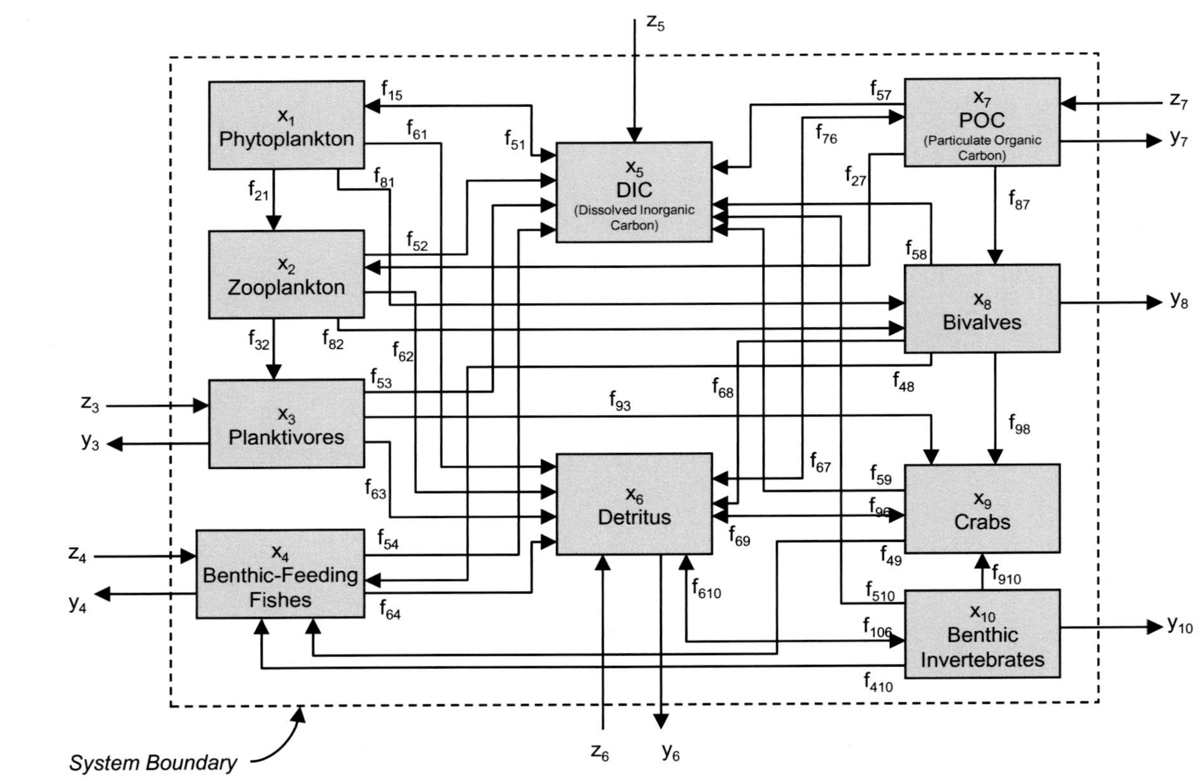

Figure 16.4 Estuary aquatic compartment model.

Adjacency Matrix:

$$\mathbf{A} = (a_{ij}) = \begin{bmatrix} 0 & 0 & 0 & 0 & 1 & 0 & 0 & 0 & 0 & 0 \\ 1 & 0 & 0 & 0 & 0 & 0 & 1 & 0 & 0 & 0 \\ 0 & 1 & 0 & 0 & 0 & 0 & 0 & 0 & 0 & 0 \\ 0 & 0 & 0 & 0 & 0 & 0 & 0 & 1 & 1 & 1 \\ 1 & 1 & 1 & 1 & 0 & 0 & 1 & 1 & 1 & 1 \\ 1 & 1 & 1 & 1 & 0 & 0 & 1 & 1 & 1 & 1 \\ 0 & 0 & 0 & 0 & 0 & 1 & 0 & 0 & 0 & 0 \\ 1 & 1 & 0 & 0 & 0 & 0 & 1 & 0 & 0 & 0 \\ 0 & 0 & 1 & 0 & 0 & 1 & 0 & 1 & 0 & 1 \\ 0 & 0 & 0 & 0 & 0 & 1 & 0 & 0 & 0 & 0 \end{bmatrix}$$

Input Vector:

$$\mathbf{Z} = (z_j) = \begin{bmatrix} 0 \\ 0 \\ z_3 \\ z_4 \\ z_5 \\ z_6 \\ z_7 \\ 0 \\ 0 \\ 0 \end{bmatrix}$$

Output Vector:

$$\mathbf{Y} = (y_i) = [0 \;\; 0 \;\; y_3 \;\; y_4 \;\; 0 \;\; y_6 \;\; y_7 \;\; y_8 \;\; 0 \;\; y_{10}]$$

16.5.1 Compartment-to-Node Transition

We will model the dynamics of individual nodes within the compartments, and we will maintain a view of the dynamics of the aggregate compartments as well. The node network contains 100 nodes and there is an average of 10 nodes per compartment. The transition from network compartments to network nodes includes some important connection-based considerations. Network connection density (also called connectance) is given by

$$\text{Network density} = \frac{\langle k \rangle}{n-1} \cong \frac{\langle k \rangle}{n}$$

where

$\langle k \rangle$ = mean vertex (compartment or node) degree

n = number of vertices in the network

In our case, the number of vertices is being increased by a factor of 10. To keep the network density constant, the mean vertex (node) degree must also *increase* by a factor of 10. That would occur in the following scenario: if compartment C1 connects to compartment C2, then every node in C1 connects to every node in C2. On the other hand, to keep mean vertex degree constant, network density would *decrease* by a factor of 10. The most realistic situation may lie somewhere between these two extremes. In a real-world ecological network, we would not expect every individual in C1 to interact with every individual in C2. Initially at least, we will hold the adjacency matrix network density constant, while keeping in mind that this assumption may yield higher values of mean node degree. (We will not know for sure until we run the model simulation. See the next subsection.)

16.5.2 Candidate and Operational Adjacency Matrices

There are also considerations regarding the interpretation of adjacency matrices in the transition from conventional compartment modeling to the network dynamics modeling presented here. In conventional compartment modeling, direct propagation links are determined *a priori* and are represented in the adjacency matrix. My network dynamics modeling approach differs. The operational propagation links among nodes are not determined *a priori*; they are part of what is *being modeled* and are known only *a posteriori*. In this approach, the usual adjacency matrix represents *candidate* propagation links. In addition to candidate adjacency matrices, later we introduce the concept of *operational* adjacency matrices, which represent realized propagation links.

The values of network density and mean node degree associated with a candidate adjacency matrix will, in general, differ from the values associated with operational adjacency matrices. Candidate adjacency matrix values serve as an upper bound for operational adjacency matrix values.

16.6 SIMPLE RULES

Our ecological network dynamics model is cellular automata based. As noted in Section 16.3, "a cellular automaton is a rule-based computing machine" that uses "relatively simple rules" (Xin-She Yang, 2006). We have seen, particularly in Chapter 11, that Stephen Wolfram (2002) has a lot to say about simple programs with simple rules. "It takes only very simple rules to produce highly complex behavior" in cellular automata. Wolfram then makes the connection with natural systems. "To what extent [is this] similar to behavior we see in nature? . . . I suspect . . . that it reflects a deep correspondence between simple programs and systems in nature. . . . This suggests that a kind of universality exists in the types of behavior that can occur, independent of the details of underlying rules" and "quite independent of the details of particular systems. . . . And if this is the case, then it means that one can indeed expect to get insight into the behavior of natural systems by studying the behavior of simple programs. For it suggests that the basic mechanisms responsible for phenomena that we see in nature are somehow the same as those responsible for phenomena that we see in simple programs." These ideas certainly have helped motivate my modeling approach.

Wolfram goes on to talk about a simple rules threshold phenomenon. "Using more complicated rules may be convenient if one wants, say, to reproduce the details of particular natural systems, but it does not add fundamentally new features. . . . Once the threshold for complex behavior has been reached, what one usually finds is that adding complexity to the underlying rules does not lead to any perceptible increase at all in the overall complexity of the behavior that is produced." I may have exceeded this threshold somewhat in an attempt to add realism to my model of ecological network dynamics.

The rules embodied in my ecological network dynamics model, however, are in fact simple, but not trivial. For propagation, there is a simple node stock threshold and simple node-to-node flow patterns. The propagating from-node "looks" closer for available to-nodes before looking farther. When there are multiple available to-node candidates within a given neighborhood, the from-node looks for the more preferred to-node. Simply stated, for propagation, closer is better and more preferred is better.

16.7 SUMMARY OF ANALYSIS REQUIREMENTS

A key objective of this modeling effort is to analyze model behavior at each model simulation time step as well as composite behavior over all simulation time steps. A brief summary of model analysis activities is provided in this section. Further analysis details are provided in the discussions of Chapters 17 and 18.

Dynamics model analysis activities:

- Generate network node-and-link propagation flow diagrams. Generate diagrams that apply to individual time steps and display them at selected time steps. Does the network "flicker"? Also generate diagrams that are cumulative over time steps and display them at various time steps.
- Determine network cumulative flow values. The flows shall be graphically depicted and displayed at selected time steps.
- Track and record network input, output, and stock values at each time step, and plot and display cumulative values over time.
- Calculate and plot network propagation event data. Network propagation event size is equal to the number of nodes involved in propagation, at a given time step. Develop and plot the event time series (event size vs. time). Check for punctuated dynamics. Develop and plot the event distribution (number of events vs. event size) using both normal coordinates and log-log coordinates. Check for power-law/fractal behavior. Take the discrete Fourier transform of the network propagation event time series to obtain its frequency spectrum. Check for "pink *1/f noise*" behavior. ("Pink 1/f noise" is the frequency spectrum representation of process fractals.)

- Calculate network propagation path lengths. Plot the path length time series (path length vs. time). Plot path length distributions. Does propagation path length exhibit punctuated dynamics and power-law/fractal behavior?
- Use path length data to calculate network indirect effects and direct effects. Plot the results. Check for punctuated, fractal behavior and dominance of indirect effects.
- Analyze network connectivity. Develop and plot network node degree distributions. Do they exhibit degree scale-invariance (fractal behavior)? Calculate network node mean degree and network connection density for each time step. At any given time step, does the network achieve critical connectivity and percolate? At what value of node mean degree? (Recall that the theoretical value for a directed random network is node mean degree = 2.) What is the size of the resulting giant cluster, that is, the proportion of network nodes linked together?

16.8 COMMENT ON INTENTION

From the model requirements discussed in this chapter, it should be clear that I endeavor to rigorously define, develop, and analyze a realistic model of complex ecological network dynamics. I am not, however, attempting to describe detailed behavior of a particular real-world ecological network. For example, our relatively simple input, stock, and flow rules and values are not intended to match any specific ecosystem network. I am developing a new approach here that differs in many respects from conventional ecological network modeling and subsequent analysis.

My intention is to provide a new perspective on the behavioral dynamics of ecological networks. The focus of my modeling effort is to determine general behaviors and universal characteristics of the dynamics across a spectrum of ecological networks.

In my view, the conventional compartment modeling/analysis perspective and my behavioral dynamics modeling/analysis perspective can be considered complementary. Each can contribute meaningful and useful insights. The conventional perspective typically employs steady-state/linear assumptions and effectively models the mean-value behavior of an ecological network over time. My dynamics approach models real-time behavior of ecological networks. It does not utilize steady-state/linear assumptions but rather allows the real-time pulsing, punctuated dynamics often observed in ecological networks to develop. The results from each perspective should help in interpreting and understanding the results from the other.

CHAPTER 17

Model Software Design and Development

17.1 INTRODUCTION

The software design and development of the ecological network dynamics model is covered in this chapter. The software implements the ecological network operational model, the required analysis activities, and the needed graphics capabilities. The resulting model software is substantial—it consists of more than 750 lines of programming code plus an even greater number of comment lines that describe and explain the code. The MATLAB R2009a programming environment[1] has been used to develop the dynamics model.

A high-level summary of the model software is provided here in Chapter 17. The complete software programming code is provided on the book's companion website. The dynamics model software has four major subsets:

1. Model network structure, parameters, and relationships
2. Ecological network propagation process
3. Analysis activities
    - Network input, output, stock, and flow analysis
    - Network propagation event analysis
    - Path length analysis
    - Indirect effects analysis
    - Network connectivity analysis
4. Graphics generation
    - Network propagation diagrams
    - Comprehensive set of other analysis graphics

Here's an overview of the full set of dynamics model software materials (parts A through D) that can be found on the book's companion website. Within each of the model's four major subsets, I establish and initialize the associated program variables. Part A of the website materials contains a *glossary*

[1] The MathWorks, Inc., February 12, 2009.

Understanding Complex Ecosystem Dynamics
http://dx.doi.org/10.1016/B978-0-12-802031-9.00017-6

of these variables. The glossary includes a set of naming conventions for the variables and a definition for each of the more than 100 variables used in the model. Part B of the website materials contains the software *master m-file,* which implements major subsets 1 and 2. Part C contains the *analysis m-file* that implements the analysis activities major subset. Part D contains the *graphics m-file* that implements the graphics generation major subset. The development of each of the dynamics model software m-files proceeds in manageable groupings called MATLAB *program cells*. Furthermore, each of the m-files is heavily commented to describe and document the software.

The Appendix section of this book provides some selected excerpts of the MATLAB code that implements the model software. My intention is to offer some examples of the code without requiring the reader to go through the full set of MATLAB m-files at this time. Now and then, as we proceed through this chapter and the next (Chapter 18), I will point to these examples.

The high-level summaries of the four major subsets of the dynamics model software follow in Sections 17.2–17.5, respectively.

17.2 MODEL NETWORK STRUCTURE, PARAMETERS, AND RELATIONSHIPS

The *model network structure, parameters, and relationships* subset describes the model network and sets the stage for propagation processing. In this portion of the software, I specify the underlying ecological network compartment model (i.e., specify its adjacency matrix, input vector, and output vector); build the model network node grid and its relationship to the compartments; identify the network input nodes and output nodes; define variables for the values of model network inputs, outputs, and stocks as they change with time; and create the needed node adjacency matrices.

We are modeling the dynamics of propagation in complex ecological networks. This, of course, requires that we define a network node adjacency matrix that is derived from the underlying compartment adjacency matrix. The node adjacency matrix is a "candidate" adjacency matrix that represents candidate propagation links. (The compartment to node transformation approach that is being used preserves network density from the compartment adjacency matrix to the node candidate adjacency matrix.) Per our model requirements, however, we need more "adjacency" information than that provided by the basic node adjacency matrix. In total, the following is needed:

1. For each from-node, must know the candidate connect-to nodes (the basic node adjacency matrix).

2. For each of the candidate connect-to nodes, must know the adjacency type (local, extended local, or nonlocal) in order to implement the neighborhood concept.
3. For each candidate connect-to node, must know the home compartment number in order to implement the compartment preference concept.
4. For each of the candidate connect-to nodes in each of the candidate connect-to compartments, must know the node attachment preference strength in order to implement the node preferential attachment concept.

Item 1 is the basic node candidate adjacency matrix. Items 2 through 4 represent persistent data that could be stored, for convenience, in adjacency-like matrices, but with differing element definitions. I have, therefore, defined four persistent adjacency/adjacency-like matrices:

- Matrix 1 (named *AA*)—basic node adjacency
- Matrix 2 (named *AAT*)—connect-to node adjacency type
- Matrix 3 (named *ACN*)—connect-to node compartment number
- Matrix 4 (named *AAP*)—connect-to node attachment preference strength

Because the basic node adjacency matrix *AA* is a *candidate* adjacency matrix, I have also defined a set of network node *operational* adjacency (*AAO*) matrices and arrays that represent the actual propagation links formed as the network operates and develops over time. Furthermore, because it is necessary to keep track of the amount of flow between nodes and compartments over time, I have defined flow-value arrays for that purpose.

In this portion of the software, the node stock and propagation rules and the number of model simulation time steps are also specified.

See the Appendix for two excerpts from the MATLAB programming code that apply to the topics discussed in this section. Code example 1 shows the assignment of network nodes to compartments. Code example 2 illustrates the creation of the node adjacency-type "adjacency matrix" (AAT).

17.3 ECOLOGICAL NETWORK PROPAGATION PROCESS

The *propagation process* subset is the core of our software model of propagation dynamics in ecological networks.

17.3.1 Propagation Process Flow

The propagation process proceeds by model simulation time step, by stages within a time step, and by node propagation events within a stage. Figure 17.1 provides a simple depiction of propagation process flow.

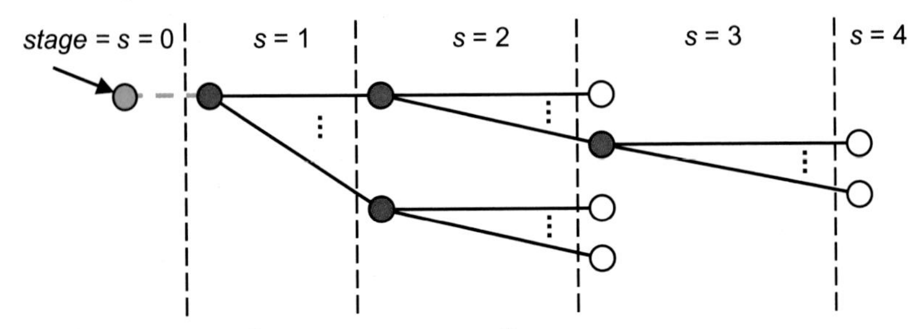

Figure 17.1 Depiction of propagation process flow.

In the figure, the input node is green, propagating nodes are red, and nonpropagating nodes are white. Note that this "tree" approach allows a given node to appear in more than one stage and, therefore, accommodates cycling.

At each model simulation *time step*, a unit input is applied to a randomly selected input node. That input may or may not cause the node to propagate (per the node stock and propagation rules). If the node propagates, we enter *stage* 1 of the time step. We have a *node propagation event* and a unit of stock flows to each of several connect-to nodes.[2] Each unit flow represents a *node propagation instance*.[3] If the flow causes one or more of the connect-to nodes to propagate, we enter *stage* 2, and so on. When we reach a stage where there are no more propagating nodes, the simulation time step concludes. The propagation process, therefore, has three primary nested loops:

1. Time step loop
2. Stage loop
3. Node propagation event loop

Throughout the total propagation process, there are loop control constructs as required.

17.3.2 Implementing the Propagation Process

The propagation process proceeds by model simulation time step, by stages within a time step, and by node propagation events within a stage. For each node propagation event, the from-node propagates by neighborhood and then by compartments within a neighborhood and finally by to-nodes within a compartment, according to compartment and node preferential

[2] A node propagation *event* involves one "from" node and multiple "to" nodes.

[3] A node propagation *instance* is a subset of a node propagation event and involves one "from" node and one "to" node.

attachment probabilities. A brief narrative description, and then a graphical summary, of the propagation process nested loops follow.

17.3.2.1 Time Step Loop

A model simulation run consists of *NumTS* time steps. The time step loop, therefore, has *NumTS* iterations. Each model simulation time step can consist of multiple stages. At stage 0, the input stage, a unit input is applied to a randomly selected input node. Every time step has a stage 0 input. Propagation stages begin with stage 1.

17.3.2.2 Stage Loop

Stage 1 is the first potential propagation stage. Stage 1 can have zero or one node propagation event—zero if the input node does not propagate and one if it does. If the node does propagate, we move to stage 2. Stage 2 (and any subsequent stages) can have zero, one, or more than one node propagation events. Whenever a given stage has one or more propagating nodes, we move to the next stage. (The presence of multiple propagation stages indicates that a "cascade" is in progress.) When we reach a stage where there are no more propagating nodes, the simulation time step concludes. The number of propagation stages for the time step is the last stage number minus 1.

17.3.2.3 Node Propagation Event Loop

Any given propagation stage can have zero, one, or more node propagation events. The extent of a node propagation event is determined by the total flow from the propagating node. This is specified by variable *npfq* (node propagation flow quantity), which is determined from the node stock and propagation rules. Each individual event corresponds to an iteration of the node propagation event loop.

Here's how a node propagation event proceeds. The from-node first attempts to propagate within its local neighborhood, then its extended local neighborhood, and finally its nonlocal neighborhood. This is *neighborhood selection* processing. Within each neighborhood, the from-node attempts to propagate to its preferred compartments in rank order. This is *compartment selection* processing. Within each neighborhood and compartment, the from-node propagates to its preferred to-nodes according to their attachment preference probabilities. This is *node selection and propagation instance* processing. A node propagation event proceeds in this manner until the available flow quantity (per the node stock and propagation rules) is exhausted. The node propagation event loop, therefore, contains three additional interior nested loops:

a. Neighborhood selection loop
b. Compartment selection loop
c. Node selection and propagation instance loop

A graphical summary of all six propagation process nested loops is provided in Figure 17.2.

Here is how overall *propagation process* development proceeds. I begin with a high-level view of the program components and flow logic and progress to the detailed MATLAB code. In this progression, I first define the basic propagation process variables that show up in the high-level view. Next, the subprocesses (and their variables) that are named in the high-level view are defined. There are four such subprocesses. They are input node processing, propagating node processing, output node processing, and node propagation instance processing. As development proceeds, further details for these and other processing activities in the program flow are added as required. Additional variables needed to store data for subsequent analysis and graphics generation activities are defined as well. This is a top-down

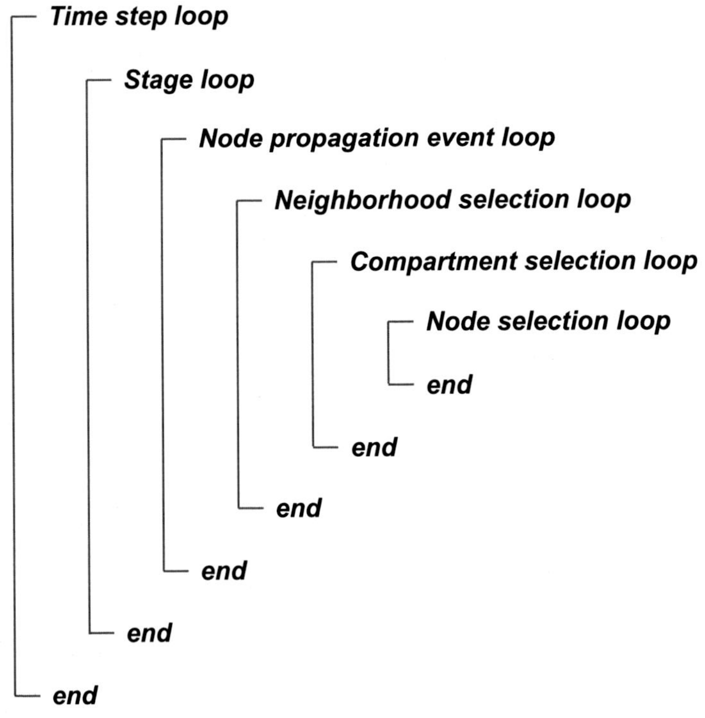

Figure 17.2 Propagation process nested loops.

approach that adds programming code detail as it proceeds. The progressively developed flow logic and code yields the complete MATLAB implementation of the propagation process.

See the Appendix, Code example 3 for a high-level view of the program components and flow logic for the propagation process (early in the development procedure).

See the book's companion website for the MATLAB *master m-file,* which provides the detailed implementation of both the model network (discussed in Section 17.2) and the ecological network propagation process (discussed here in Section 17.3).

17.4 ANALYSIS ACTIVITIES

The *analysis activities* major subset of the model software covers the analysis of the dynamics of propagation in complex ecological networks. The analysis categories are:

- Network operational propagation flow analysis
- Network propagation event analysis
- Path length analysis
- Indirect effects analysis
- Network connectivity analysis

These five categories are summarized in the following five subsections, respectively.

17.4.1 Network Operational Propagation Flow Analysis

An important vehicle for depicting network operational propagation flow dynamics is the network node-and-link propagation flow diagram. The variables and data needed to produce these network diagrams are generated in the first two major subsets of the model software (as implemented in the model master m-file). The actual production of the diagrams is accomplished in the graphics generation major subset of the software (the graphics m-file).

History diagrams are also needed. An effective means of depicting network input, output, stock, and flow value histories includes the use of 3-D grid bar charts displayed at selected time steps. We use an appropriate MATLAB 3-D/discrete-surface plotting capability. For this set of history diagrams, the necessary variables and data are created in the master m-file and the chart production is accomplished in the graphics m-file.

Further description of the development of these propagation flow analysis diagrams is provided in Section 17.5 of this chapter (Graphics

Generation). The actual diagrams are provided, and analyzed, in Chapter 18 (Results). The full set of network operational propagation flow diagrams will either corroborate or refute the continuing ecological network fluctuations and flickering operation that I have hypothesized.

17.4.2 Network Propagation Event Analysis

For the entire simulation run (after all time steps), network propagation event analysis (over time) is performed and the results are plotted and displayed. Network propagation event size per time step (*NetPESize_TS*) is calculated in the master m-file. The network propagation event size at a time step is equal to the total number of node instances involved in propagation at that time step, that is, the number of node propagation instances (variable *nnpi*) plus one (the input node).[4] Because any given node can participate in a network propagation event multiple times (due to cycling), network propagation event size can be larger than the number of physical nodes involved in the event and even larger than the total number of nodes in the network.

A network propagation event time series (event size vs. time) is generated. Punctuated dynamics may be observed. A network propagation event distribution (number of events vs. size of events) is then developed as follows:

- Sort the elements of *NetPESize_TS* from smallest to largest.
- Define a size interval (≥ 1), partition the size domain into intervals, and count the number of events in each interval.
- Generate a distribution with the ordered event size intervals as abscissa and the number of events in each of those size intervals as ordinate.
- Plot and display the event distribution in both normal coordinates and log-log coordinates.

The resulting network propagation event distribution is tested for power-law/fractal behavior.

Discrete Fourier Transform analysis of the network propagation event time series is performed to obtain its frequency spectrum. The spectrum is then tested for "pink $1/f$ noise" behavior and, therefore, process fractal behavior. (Pink $1/f$ noise is the frequency spectrum representation of the time signals of real-world complex ecological system fluctuations.)

[4] To be clear on terminology, a *network* propagation event includes all nodes across the entire network that participate in propagation at a given time step. A *node* propagation event (defined earlier) is a subset of a network propagation event and involves one "from" node and multiple "to" nodes. A node propagation *instance* (also defined earlier) is a subset of a node propagation event and involves the one "from" node and a single "to" node.

All of the network propagation event analysis plots are produced using the graphics m-file.

17.4.3 Path Length Analysis

Network path length analysis is performed for the entire simulation run (after all time steps). Path length time series and path length distributions are developed for that purpose. Path length analysis is also performed at each time step. Individual-time-step path lengths are calculated and the associated distributions are generated. For the entire simulation run and for selected time steps, the results are plotted and displayed. The source data needed to do this has been generated in the master m-file. The plots are produced using the graphics m-file.

17.4.3.1 Path Length Calculations

Figure 17.3 depicts propagation flow and illustrates the path length calculations.

The equations in the lower portion of the figure calculate $N_{\ell,s}$– the number of paths of length ℓ at stage s. (The other equation parameters are defined shortly.) Using such equations, the path length numbers can be calculated at each stage of a time step and for each completed time step. The needed path length time series and distributions can then be generated.

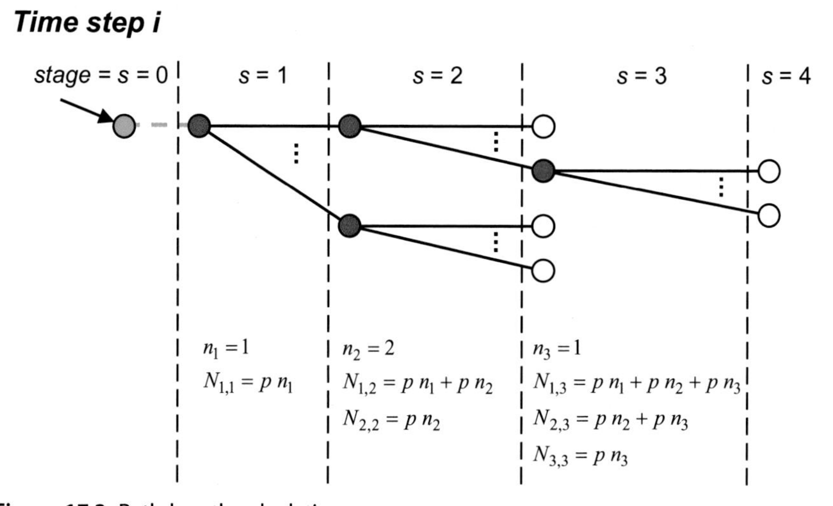

Figure 17.3 Path length calculations.

The equation development follows (with reference to Figure 17.3).

$s =$ simulation time-step stage
$s = 0$ input stage
$s \geq 1$ propagation stages
$n_j =$ number of propagating nodes at stage j where $j \geq 1$
$\ell =$ path length
$N_{\ell,s} =$ number of paths of length ℓ at stage s
$p =$ number of unit flows per propagating node

$$N_{1,s} = \sum_{j=1}^{s} pn_j \quad N_{2,s} = \sum_{j=2}^{s} pn_j \quad N_{3,s} = \sum_{j=3}^{s} pn_j \quad \text{with } s \geq j$$

$$N_{\ell,s} = p\sum_{j=\ell}^{s} n_j \text{ for } 1 \leq \ell \leq s \text{ and } N_{\ell,s} = 0 \text{ for } \ell > s$$

At each completed simulation time step:
$s = \max(s) = S =$ total number of stages for completed time step
$N_\ell = N_{\ell,S} =$ number of paths of length ℓ for completed time step

$$N_\ell = p\sum_{j=\ell}^{S} n_j \text{ for } 1 \leq \ell \leq S \text{ and } N_\ell = 0 \text{ for } \ell > S$$

Note that the number of internal flows from a propagating node is reduced by 1 for output nodes (because one unit of flow is output to the external environment from output nodes). We need to distinguish between output propagating nodes and nonoutput propagating nodes and their respective flow quantities.

$nnpe_j =$ total number of propagating nodes at stage j where $j \geq 1$
$nopn_j =$ number of output propagating nodes at stage j where $j \geq 1$
$nnopn_j =$ number of nonoutput propagating nodes at stage j where $j \geq 1$
$nnopn_j = nnpe_j - nopn_j$
$npfq =$ number of unit flows per nonoutput propagating node
$npfq - 1 =$ number of unit flows per output propagating node

$$N_\ell = (npfq - 1)\sum_{j=\ell}^{S} nopn_j + npfq\sum_{j=\ell}^{S} nopn_j$$

$$N_\ell = (npfq - 1)\sum_{j=\ell}^{S} nopn_j + npfq\sum_{j=\ell}^{S} (nnpe_j - nopn_j)$$

$$N_\ell = (npfq - 1)\sum_{j=\ell}^{S} nopn_j + npfq\sum_{j=\ell}^{S} nnpe_j - npfq\sum_{j=\ell}^{S} nopn_j$$

$$N_\ell = npfq\sum_{j=\ell}^{S} nnpe_j - \sum_{j=\ell}^{S} nopn_j$$

$$N_\ell = npfq\sum_{j=\ell}^{S} nnpe_j - \sum_{j=\ell}^{S} nopn_j \text{ for } 1 \leq \ell \leq S \text{ and } N_\ell = 0 \text{ for } \ell > S$$

These equations define the number of paths of each length ℓ for the time step.

The equations are used to generate per-time-step path length distributions and path length integrated time series and distributions. The distributions are plotted and displayed in both normal coordinates and log-log coordinates. The time series are tested for punctuated dynamics and the distributions are tested for power-law/fractal behavior.

17.4.3.2 Integration Over Space and Time

Much network analysis is performed on a model of a network that is assumed to be at "steady state." The network does not change with time—at least not over the analysis observation interval. The network nodes and their connections are persistent. Analysis, therefore, is somewhat simplified. Path length analysis becomes a *spatial* exercise over static network structure. One essentially identifies the paths of various lengths across the steady-state network and counts them. (Similarly, as we will discuss in Section 17.4.5.1, to perform node degree analysis one would count the number of connections from/to each node in the steady-state network.)

One spatial path length analysis technique is the "box covering method." (It is described in Chapter 12.) For a given network, one generates boxes of size ℓ_B. Each box contains a set of nodes for which the path length ℓ_{ij} between any two nodes *i* and *j* in the box is smaller than ℓ_B. The entire network is then covered with such boxes. The boxes of size ℓ_B required to cover the network are counted and the sum is denoted as N_B. This procedure is performed for $\ell_B = 1, 2, 3, \ldots$ and continues until reaching the point where one box covers the entire network. A plot of the resulting N_B vs. ℓ_B values provides the path length distribution for the steady-state network.

For my analysis work, the previous spatial perspective is not sufficient. An important objective of mine is to model and analyze the changing dynamics of ecological networks. In my view, networks change with time—sometimes dramatically so. Path length behavior is dynamic and changes with time. To get a relevant path length distribution that reflects this behavior, I must include a temporal perspective and integrate over both space and time.

How do I accomplish integration over space and time? I start by performing spatial analysis at each individual point in time (each simulation time step). My initial idea for integrating across time steps was to calculate over-time cumulative path length data and use that. While the cumulative data has value, that approach does not work when attempting to develop an integrated path length distribution. The resulting distribution is essentially a distribution of average numbers of path lengths (vs. path length) over time. A distribution of over-time averages is not particularly meaningful or useful when attempting to capture punctuated dynamical behavior.

A better idea for achieving integration over space and time is to devise a spatial/temporal analog to the conventional spatial box covering approach previously discussed. Instead of generating a collection of boxes that covers a steady-state network and then counting the number of boxes over space, we can generate one spatial ℓ_B box at each point in time (each simulation time step) and then count the number of such spatial boxes of each size that have occurred over time (over the simulation run). The result is a path length distribution that is integrated over space and time. This distribution, and corresponding time series, effectively captures the path length punctuated dynamics. In the model software, I develop this distribution and time series for the entire simulation run.

To test the robustness of this approach, I have performed the integrated path length analysis first using the individual path lengths at each simulation time step, and then using the mean path length at each time step. These mean path lengths are spatial means calculated at a point in time; they are not averages over time. The two analyses yield similar and consistent time series/distribution results. The approach, therefore, appears to be robust.

Table 17.1 provides a point-by-point comparison of a version of the conventional spatial box covering method and my integrated spatial/temporal box covering method.

17.4.4 Indirect Effects Analysis

A comprehensive set of indirect effects analyses are performed. Individual-time-step and over-time-cumulative indirect effects indicators are defined and calculated, and simulation-run time series are generated for each of these indicators. An indirect-path-quantity time series and distribution, a direct-path-quantity time series and distribution, a cumulative indirect-path-quantity time series, and a cumulative direct-path-quantity time series are also developed for the simulation run. All of this can help us understand indirect effects dynamics in ecological networks. The needed source data have been generated in the *path length analysis* program cells of the analysis m-file. The graphics are produced using the graphics m-file. Descriptions of the analyses follow.

17.4.4.1 Indirect Effects and Direct Effects Indicator Analysis

For each time step, individual-time-step and over-time-cumulative indirect effects and direct effects indicators are calculated. The direct effects ratio (DER), indirect effects ratio (IER), and indirect effects index (IEI) are

Table 17.1 Comparison of Box Covering Methods.

Conventional Spatial Method	**Integrated Spatial/Temporal Method**
Static network	Dynamic network
Spatial perspective	Spatial + Temporal perspectives
Define a collection of spatial boxes at one "steady-state" point in time	Define one spatial box at each of a collection of dynamical points in time (the time steps)
Each box contains a set of nodes for which $\ell_{ij} \leq \ell_B$ where ℓ_{ij} is the path length between any two nodes i and j in the box, ℓ_B is the box size, and $\ell_B = 1, 2, \ldots,$ network maximum realized path length	Each box contains a set of nodes for which $\ell_{ij} \leq \ell_B$ where ℓ_{ij} is the path length between any two nodes i and j in the box, ℓ_B is the box size, and $\ell_B =$ time step maximum realized path length
Tile the network *spatially* with boxes of size ℓ_B	Tile the network *spatially* and *temporally* with boxes of size ℓ_B
Count the number of boxes of size ℓ_B and denote the sum as N_B	Count the number of boxes of size ℓ_B and denote the sum as N_B
The path length distribution is given by N_B (ordinate) vs. ℓ_B (abscissa)	The integrated path length distribution is given by N_B (ordinate) vs. ℓ_B (abscissa)
If the steady-state network is fractal with respect to path length, the path length distribution is a power law with a fractal dimension exponent	If the dynamic network is fractal with respect to path length, the integrated path length distribution is a power law with a fractal dimension exponent

calculated using the following expressions. First, here are the individual-time-step expressions.

At each completed simulation time step:
$\ell =$ path length
$N_\ell =$ number of paths of length l for completed time step
$S =$ total number of stages for completed time step
DER = Direct Effects Ratio
DER = number of paths of length 1 divided by total number of paths

$$\mathrm{DER} = \frac{N_1}{\sum_{\ell=1}^{S} N_\ell}$$

IER = Indirect Effects Ratio

IER = number of paths of length > 1 divided by total number of paths

$$\mathrm{IER} = \frac{\sum_{\ell=2}^{S} N_\ell}{\sum_{\ell=1}^{S} N_\ell}$$

IEI = Indirect Effects Index

IEI = number of paths of length > 1

divided by number of paths of length 1

$$\mathrm{IEI} = \frac{\sum_{\ell=2}^{S} N_\ell}{N_1}$$

Next, here are the over-time-cumulative expressions.

Cumulative values over simulation time steps:

$$\mathrm{DER} = \frac{\sum_{\substack{\text{Time}\\\text{steps}}} N_1}{\sum_{\substack{\text{Time}\\\text{steps}}} \sum_{\ell=1}^{S} N_\ell}$$

$$\mathrm{IER} = \frac{\sum_{\substack{\text{Time}\\\text{steps}}} \sum_{\ell=2}^{S} N_\ell}{\sum_{\substack{\text{Time}\\\text{steps}}} \sum_{\ell=1}^{S} N_\ell}$$

$$\mathrm{IEI} = \frac{\sum_{\substack{\text{Time}\\\text{steps}}} \sum_{\ell=2}^{S} N_\ell}{\sum_{\substack{\text{Time}\\\text{steps}}} N_1}$$

Time series of the indirect effect and direct effect indicators—both the individual-time-step values vs. time and the over-time-cumulative values vs. time—are developed in the analysis m-file and are plotted and displayed using the graphics m-file.

17.4.4.2 Additional Analysis

I also develop (and plot) the simulation-run indirect-path-quantity time series (number of indirect paths at each time step vs. time) and distribution as well as the direct-path-quantity time series (number of direct paths at each time step vs. time) and distribution. The results can be used to determine whether indirect effects (and/or direct effects) are punctuated and fractal. A cumulative indirect-path-quantity time series and a cumulative direct-path-quantity time series are also developed for the simulation run. Each provides the cumulative number of paths at each time step vs. time. These cumulative time series can be used to determine whether indirect effects are dominant. Recall that, in Chapter 15, we mathematically derived conditions for indirect effects dominance. That mathematical result can be used to test the indirect effects analysis outcomes.

17.4.5 Network Connectivity Analysis

Both per-time-step and over-time network connectivity analysis is performed. The source data needed to do this is generated in the master m-file and stored as arrays *AAO_t_TS* and *AAO_tc_TS*:

- *AAO_t_TS*=per-time-step operational node adjacency multidimensional array that provides an operational adjacency matrix for each individual time step in a simulation run (100x100x*NumTS*)
- *AAO_tc_TS*=per-time-step operational node adjacency multidimensional array that provides an over-time-cumulative operational adjacency matrix at each time step in a simulation run (100x100x *NumTS*)

From these arrays, node degree, network connection density, and other network connectivity information can be determined. The network connectivity analysis plots are produced using the graphics m-file.

17.4.5.1 Node Degree Analysis

Comprehensive node degree analysis is performed. Note that along with network propagation events, path length, and indirect effects, node degree may exhibit punctuated dynamics. To visually observe the degree dynamics, node degree grids (node degree overlays on the network grid) are developed and plotted at different points in time. To generate the time-varying node degree grids, individual-time-step node degree vector arrays and node degree grid arrays are developed. In addition to the node degree grids, we need to develop node degree time series and distributions that capture the dynamics (whatever they turn out to be) in an integrated fashion over space and time.

17.4.5.1.1 Integration over Space and Time

As discussed in Section 17.4.3.2, much network analysis is performed on a model of a network that is assumed to be at "steady state." Such analysis is spatial in nature. To perform steady-state node degree analysis, one counts the number of connections from/to each node in the static network space. My analysis work does not make the steady-state assumption, and it has both spatial and temporal dimensions. In my analysis, I need to integrate over both space and time.

With respect to node degree distribution analysis, I start by performing spatial analysis at each individual simulation time step. Similar to the path length distribution case, my initial idea for integrating the time steps was to calculate over-time cumulative node degree data and then generate the distributions associated with that data. That approach did not work in the path length distribution case and it does not work for node degree distributions either. In the node degree case, the reasons are a bit different. As node connections accumulate over time, node degree values increase. When compared with node degree distributions calculated at individual high-propagation time steps, the cumulative node degree distribution shifts to the right and changes shape quite drastically. The lower degree range falls and the mid to upper degree ranges rise. The resulting cumulative distributions do not seem to have any useful physical meaning.

To accomplish integration over space and time, I devise a spatial/temporal approach very similar to the path length distribution approach (see Section 17.4.3.2). For the three degree types (out, in, and combined), I develop time series of node degrees achieved at each individual time step. I then derive the associated distributions from the time series. The result is node degree distributions that are integrated over space and time. These distributions, and corresponding time series, capture the node degree dynamics.

To test the robustness of this approach for node degree, I try another measure of per-time-step node degree: the node *mean degree* achieved at each time step. Here's the procedure: per-time-step values of node mean degree are calculated, a time series for the simulation run is developed, and the associated distribution is derived. The mean-degree measure also appropriately captures the node degree dynamics. The approach appears to be robust.

17.4.5.1.2 Per-Time-Step and Over-Time Cumulative Node Degree Distributions

As previously noted, this is one of the (few) analysis subactivities that did not yield useful results. Of course, I would not have known that unless I had developed and implemented the analyses, and then generated and scrutinized the results. The capability, therefore, is part of the software code. So, for completeness, I will briefly describe the development.

Individual-time-step node degree distributions as well as over-time cumulative node degree distributions (cumulative from time step one to the current time step) are developed and examined for useful information. Each distribution provides the number of nodes with degree x vs. x. For all of these node degree distributions, the analysis starting point is the generation of node degree arrays that provide the degree of each node at each time step. The node degree distributions are developed as follows:

- Sort the node degree values from smallest to largest.
- Define a degree size interval (≥ 1), partition the degree size domain into those intervals, and count the number of nodes in each interval.
- Generate a distribution with the ordered node degree intervals as abscissa and the number of nodes in each of those intervals as ordinate.
- Plot and display each distribution in both normal coordinates and log-log coordinates.

The resulting distributions can be scrutinized and analyzed.

17.4.5.2 Other Connectivity Considerations

Network critical connectivity/percolation analysis is performed. At any given time step, the network may achieve critical connectivity. When that occurs, several pertinent questions arise: What is the value of node mean degree? (The theoretical value for a directed random network at critical connectivity is node mean degree = 2.) Is the network sparsely connected at that time step? How many nodes are linked together? What is the fractional size of the resulting "giant cluster"? To answer these questions, the individual-time-step values of four network connectivity parameters are needed. (1) Node mean degree individual-time-step values are being calculated as part of node degree analysis (see the previous subsection) and will be available. (2) The network connection density at each time step must be calculated. (3) The number of nodes linked together at each time step (vector *NumLN_t_TS*) must be calculated. This can be done using propagation event data. (4) The fractional size of any candidate "giant cluster" at individual time steps (vector *CompSize_t_TS*) must be calculated. The fractional size of a candidate giant cluster equals the number of linked nodes in the cluster divided by the total number of nodes in the network.

Time series for each of the previous four network connectivity parameters are developed and plotted. From the time series, time steps of potential critical connectivity can be identified, and the values of the parameters at those time steps can be observed in order to draw conclusions regarding important network connectivity traits.

Please see the book's companion website for the MATLAB *analysis m-file,* which provides the detailed implementation of all of the analysis activities covered in Section 17.4.

17.5 GRAPHICS GENERATION

The *graphics generation* subset of the model software produces the graphs and diagrams associated with our analysis of the dynamics of propagation in complex ecological networks. The required graphics include:

- Network node-and-link flow diagrams
- Network flow value diagrams and input, output, and stock histories
- Network propagation event time series (event size vs. time) and distribution (number of events vs. size of events)
- Network propagation event time series frequency spectrum
- Path length time series and distributions (number of paths vs. path length)
- Indirect effects and direct effects indicators vs. time
- Indirect-path and direct-path time series (number of paths at each time step vs. time) and their distributions
- Network node degree time series and distributions
- Network critical connectivity graphics

Network flow graphics include node-and-link flow diagrams, flow value diagrams, and input/output/stock history diagrams. The MATLAB *gplot* function is used to generate the network node-and-link flow diagrams. An important advantage of *gplot* is that it preserves the spatial positioning of the network nodes. Illustrations of node-and-link flow at individual time steps as well as cumulative node-and-link flow over time are provided. For improved visual clarity, the nodes involved in propagation are color-coded (green for input nodes, red for propagating from-nodes, and dark blue for nonpropagating to-nodes). In addition to those graphics, cumulative flow value diagrams are generated to provide an "adjacency matrix" depiction of network flow. Network input, output, and stock histories are also plotted, and displayed on 3-D network grid bar charts.

The *network propagation event* graphics include propagation event time series and distributions. The distribution is plotted in both normal and log-log coordinates. On the log-log plot, we fit a straight line to the data. The slope and y-intercept of the straight line are used to create a power-law curve, which is then overlaid on the normal-coordinates plot of the network propagation event distribution for comparison purposes. A network propagation event frequency spectrum is also generated and an appropriate power-law overlay is added for comparison.

The *path length* graphics consist of time series and associated distributions. A path length time series and integrated path length distribution is produced using an approach analogous to the network "box covering" method. A time series and integrated distribution based on the mean path length at each individual time step is also produced. Distribution power-law overlays are generated for both cases. In addition, per-time-step path length distributions (for individual time steps as well as cumulative across time steps) are generated so that their behavior can be examined.

To illustrate ecological network *indirect effects* and direct effects, a variety of time series and, in some cases, their corresponding distributions are produced. These graphics can be used to explore possible dominance of indirect effects as well as other important behaviors. Direct effects ratio, indirect effects ratio, and indirect effects index time series are generated. Cumulative versions of these time series are also generated. In addition, various other indirect path quantity and direct path quantity graphs are plotted so that aspects of their dynamical behavior can be observed.

The *network connectivity* graphics consist of node degree three-dimensional depictions, appropriate node degree time series and distributions, and other connectivity graphs. Node degree grids (node degree values overlaid on the plane of the network grid) are plotted at various simulation-run time steps. These three-dimensional graphics facilitate explicit visualization of node degree dynamics over time. Node degree time series and distribution plots that capture the degree dynamics are also produced. Power-law overlays are added to these distributions to check for scale-invariant/fractal behavior. To test for network critical connectivity, graphs are generated that allow us to observe instances of possible criticality and the values of network connectivity parameters that occur at those instances.

See the book's companion website for the MATLAB *graphics m-file* that implements the generation of all of these graphics. The graphics m-file employs mostly standard MATLAB graphics capabilities, and provides detailed comments that describe how to use the MATLAB capabilities, including MATLAB *Plot Tools* to generate the required graphics.

17.6 A NOTE ABOUT RESULTS

This chapter has covered the ecological network dynamics model software design and development. All of the analysis activities and subactivities that are described here have been implemented in the software. One cannot know *a priori*, with certainty, whether every specified analysis subactivity will yield productive results, until it is implemented and the results are scrutinized. Not all, but the overwhelming majority of the analysis subactivities specified here in Chapter 17 have yielded meaningful and useful results. Those are the results discussed next, in Chapter 18.

CHAPTER 18

Ecological Network Dynamics Results

18.1 INTRODUCTION

In this chapter, we generate results, analyze and display the results, and reach conclusions. The material is presented in five categories:

- Operational Propagation Flow
- Network Propagation Events
- Path Length
- Indirect Effects
- Network Connectivity

The work of these five areas is discussed and described in detail in Sections 18.2–18.6, respectively. In the operational propagation flow category, comprehensive network node-and-link propagation flow results as well as flow history results are provided. Regarding network propagation events, the time series (event size vs. time), distribution (number of events vs. event size), and propagation event frequency spectrum (Fourier transform of the event time series) are generated and analyzed. Propagation path length behavior is both spatial and temporal (changes with time). To fully capture path length dynamics, therefore, I have devised an approach that integrates over both space and time. The results are described. To analyze network indirect effects, I have defined several indirect effects indicators (in Chapter 17) and generate both noncumulative and over-time cumulative time series for each of the indicators. Time series and distributions for indirect and direct path quantities are also generated and analyzed. Regarding network connectivity, node degree time series and distributions are produced and analyzed. (Node degree is the number of connections to/from a node.) Several other important network connectivity traits are also examined.

Here's a brief overview of the results outcomes. Operational propagation flow results indicate punctuated, local-to-global propagation behavior that continually changes with time. Ecological networks are ever-changing, "flickering" networks. Network propagation event time series and

Understanding Complex Ecosystem Dynamics
http://dx.doi.org/10.1016/B978-0-12-802031-9.00018-8

distribution results indicate power-law/fractal behavior, and the propagation event frequency spectrum results reflect "pink 1/f noise" behavior. Propagation path length results confirm that path length dynamics are punctuated and fractal. Network indirect effects analysis results show that indirect effects are indeed dominant in the model network. Regarding ecological network connectivity, results indicate several important connectivity traits: node degree exhibits power-law/scale-invariant behavior; network critical connectivity is achieved during high-propagation events; and ecological networks are sparsely connected networks. The full set of analysis activities and results provide comprehensive testing—and corroboration—of the ecosystem dynamics *characteristics hypothesis*.

18.2 OPERATIONAL PROPAGATION FLOW RESULTS

Network node-and-link propagation flow results are described and displayed in Section 18.2.1. Input and output flow history and related results are described and displayed in Section 18.2.2.

18.2.1 Network Node-and-Link Propagation Flow Results

Our modeling and analysis focus is network propagation. To set the context for the network node-and-link propagation flow results, we begin with the network propagation time series shown in Figure 18.1. Details of this time series (and related statistical results) are discussed later. For now, just note the ever-changing propagation behavior–from very small to very large propagation events. We'll examine the propagation flow for eight specific events. The red selection circles in Figure 18.1 highlight those events. (The second and third events are small events that are close together in time (time step 28

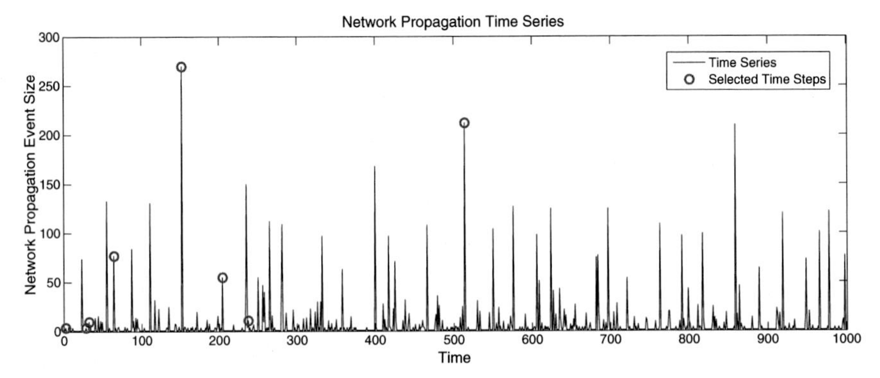

Figure 18.1 Network propagation time series.

and time step 32), and their selection circles may be difficult to see separately.)

Before proceeding, let's quickly review the propagation process. At each model simulation time step, a unit input is applied to a randomly selected input node. That input may or may not cause the node to propagate (per the node stock and propagation rules). If the node propagates, then that from-node propagates a unit of stock to each of several to-nodes. Propagation proceeds by neighborhood, compartment preference, and node preference, in that order. Figure 18.2 depicts a propagation from-node (in red) and its potential propagation to-nodes (in blue) for each of the three neighborhood levels.

Propagation is first attempted within the local neighborhood, then the extended local neighborhood, and finally the nonlocal neighborhood. Within each neighborhood, propagation is to preferred compartments (e.g., species groups) in rank order. Within each neighborhood and preferred compartment, propagation is to preferred nodes (e.g., species) according to their attachment preference probabilities. Propagation proceeds until the from-node flow quantity is exhausted. This first-stage propagation flow may cause one or more of the to-nodes to reach the propagation threshold (per the node stock and propagation rules). If so, each of those nodes propagates according to these same procedures, and so on.

For the following results (in this section and throughout the chapter), the simulation-run node propagation threshold variable is set to four and the node propagation flow quantity variable is set to four; that is, when a node's stock value reaches four, the node propagates four units of stock. Node inputs from the external environment, outputs to the environment, and internal network node-to-node propagation flows all have unit value (value = one). Given these settings and values, note that if a propagating node is also an output node, it

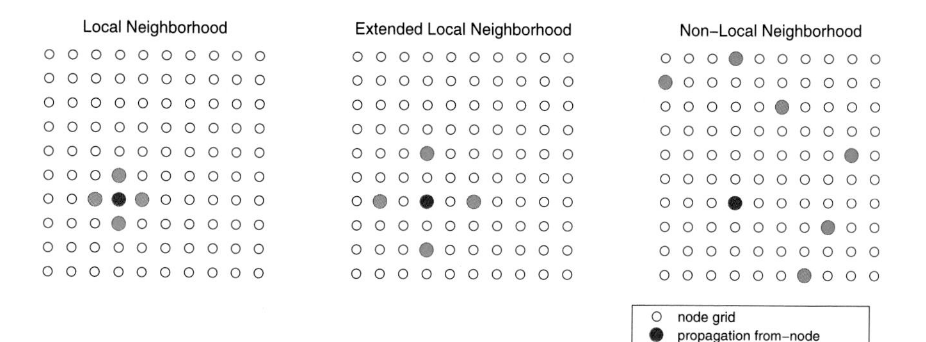

Figure 18.2 Review of propagation process.

outputs one unit of flow to the external environment and propagates three units of flow internally. Otherwise, four units of flow are propagated internally.

Network node-and-link propagation flow diagrams follow. In each diagram, the background is the network spatial grid consisting of 100 nodes. The solid lines represent propagation-flow links. Here is the node color code. The green node is the propagating input node. The red nodes are subsequent propagating nodes. The dark blue nodes are propagation to-nodes that do not propagate further. Any of these nodes can also be an output node.

See the Appendix, Code example 4 for the MATLAB code that creates node-and-link propagation flow diagrams.

Figure 18.3 displays the flow diagrams for four small network propagation events. The title caption above each diagram gives the time-step number and the event size (number of involved nodes). At time step 2, the input

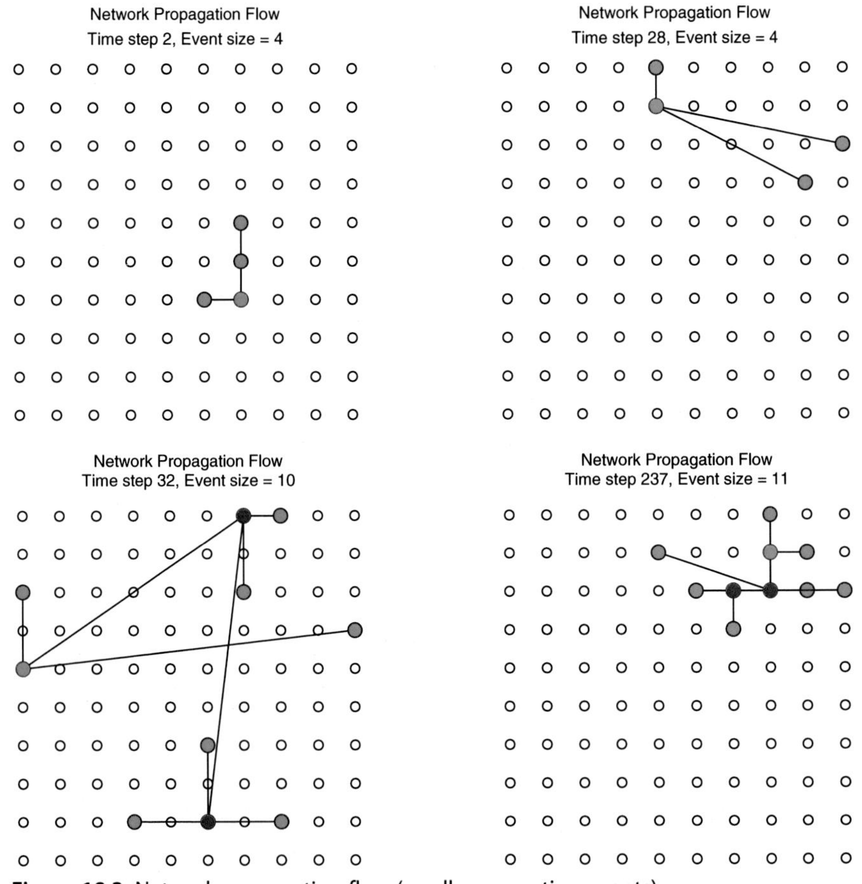

Figure 18.3 Network propagation flow (small propagation events).

node (which is also an output node) propagates one unit of stock to the external environment (not shown explicitly), one unit of stock to each of two nodes in the local neighborhood, and one unit of stock to a node in the extended local neighborhood. At time step 28, the input node (also an output node) propagates to one node in the local neighborhood and two nodes in the nonlocal neighborhood. At time step 32, we have a three-stage network propagation event involving local, extended local, and nonlocal neighborhoods. The event size is only 10, but it has a relatively large network span. The time step 237 event is also a three-stage event, but has more local processing. Count the colored nodes. The count does not equal the event size. The red propagating node on the right is involved more than once. It is first a propagation to-node, then a propagating from-node, and then a to-node again. This is our first glimpse of cycling in the model network.

Figure 18.4 displays flow diagrams for two midsize network propagation events (at time steps 64 and 204). The event sizes are 77 and 55, respectively. We see a mix of local, extended local, and nonlocal neighborhoods being utilized. We see relatively large network spans. Cycling is evident. An approximate "back of the envelope" expression for cycling index can be defined as follows:

$$\text{Cycling index} = \frac{\text{Number of cycling propagation flows}}{\text{Total number of propagation flows}}$$

$$\text{Cycling index} \cong \frac{(\text{Event size} - \text{Node count})}{\text{Event size}}$$

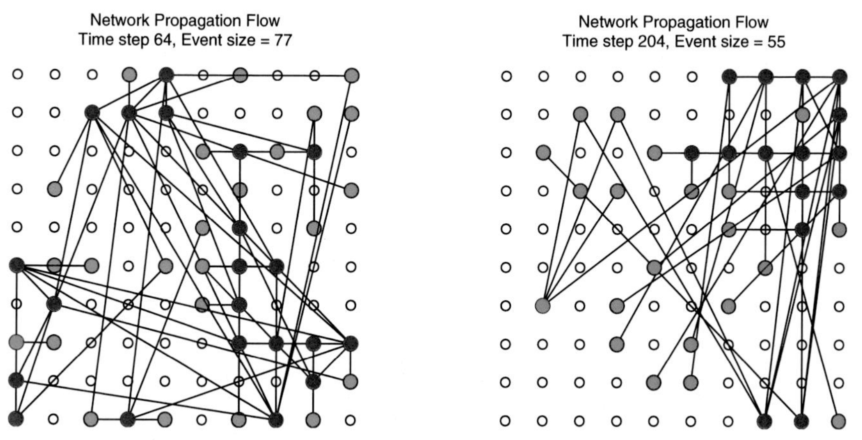

Figure 18.4 Network propagation flow (midsize propagation events).

Using this expression, at time step 64, the cycling index is approximately 0.4. At time step 204, the cycling index is about 0.3.

Two of the largest individual-time-step network propagation events are illustrated in Figures 18.5 and 18.6. Compared to the midsize events, we see greater network span and higher levels of local-to-global processing. A large fraction of all network nodes is involved. Significant cycling is occurring. The node-and-link flow diagram of Figure 18.5 has a cycling index of about 0.7, and the diagram of Figure 18.6 has a cycling index around 0.6.

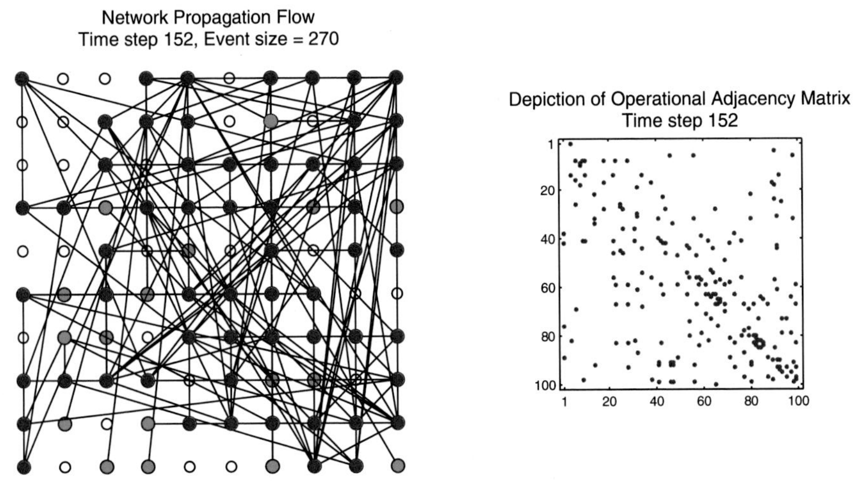

Figure 18.5 Network propagation flow (large event; time step 152).

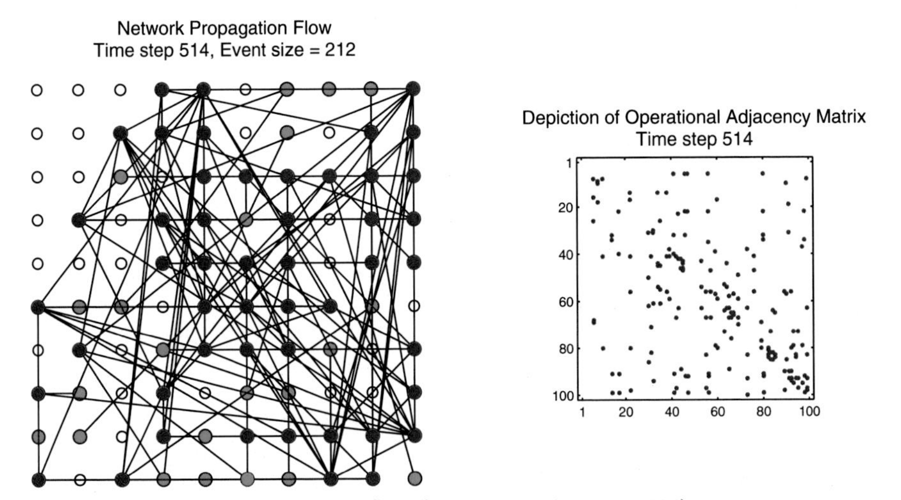

Figure 18.6 Network propagation flow (large event; time step 514).

For each of these two figures, I have added a depiction of the network operational adjacency matrix for the time step. For these 100×100 adjacency matrices, the markers (blue dots) indicate ones (i.e., connections) and white space indicates zeros (no connection). The matrices have ones distributed throughout with a particular concentration near the matrix major diagonal. I'll have more to say about this pattern shortly.

As we look back over Figures 18.1–18.6 (the time series and the flow diagrams for the eight representative time steps), several things become evident. Propagation behavior over time is punctuated. Most propagation events, and perhaps all of the midsize to large events, exhibit local-to-global propagation behavior. The network continually changes with time, and sometimes dramatically so. The view of the network operational dynamics is clear: we are dealing with ever-changing, "flickering" ecological networks.

It is also instructive to look at the network cumulative propagation flow over time to see the dense web of node interactions that develops and the participation of (very nearly) the entire network in propagation.

Figure 18.7 displays the cumulative propagation flow at time step 100 (i.e., the cumulative flow from time steps 1 through 100). The colored (light blue) nodes are the nodes participating in propagation. By time step 100, very few network nodes are not participating. Figure 18.8 displays the cumulative propagation flow at time step 1000 (the cumulative flow

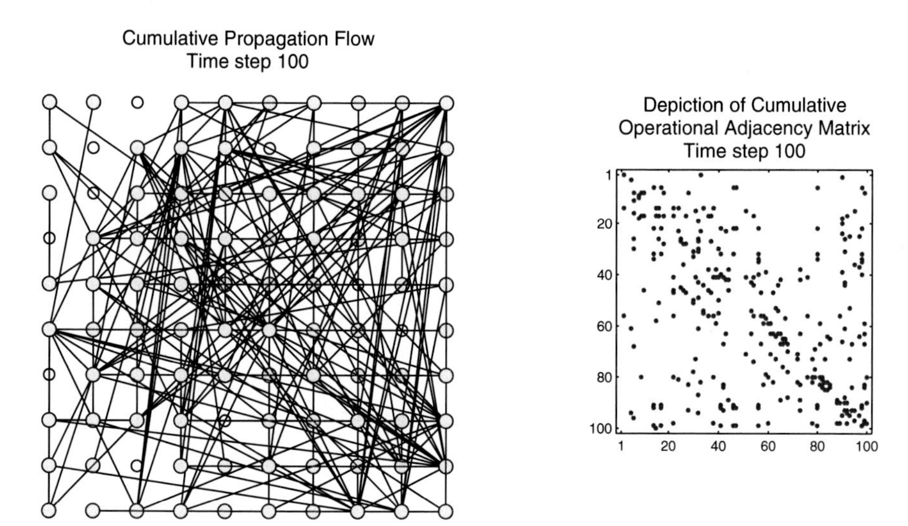

Figure 18.7 Cumulative propagation flow (time step 100).

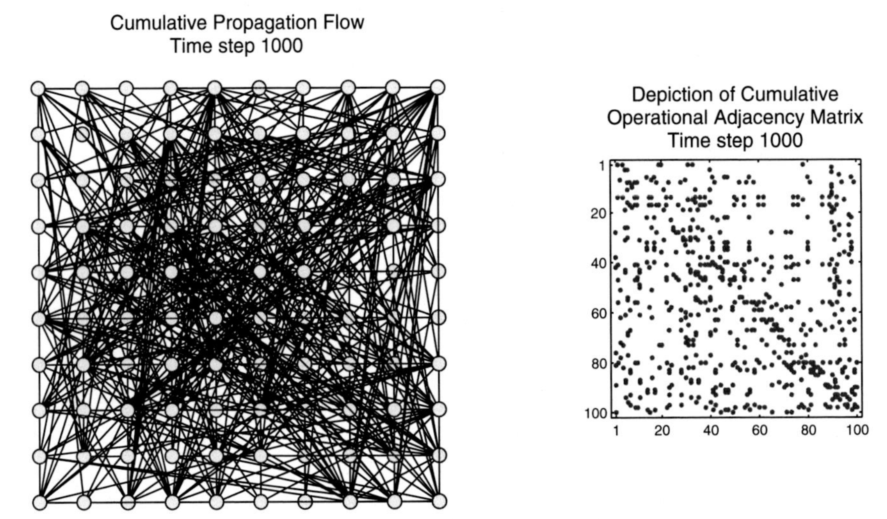

Figure 18.8 Cumulative propagation flow (time step 1000).

for the entire simulation run). Only *one* node (node 75—when counting top-to-bottom and left-to-right) does not participate in propagation. Here's why: as it turns out, node 75 is not an input node, not in the local or extended local neighborhoods of potential connecting nodes, and has low attachment preference strength. The network here could represent an ecological species network in which node 75 (i.e., species 75) is not participating in the community of species and is therefore in danger.

Figures 18.7 and 18.8 include a depiction of the network cumulative operational adjacency matrix. We see that the distribution pattern of ones and zeros for these cumulative matrices is similar to the distribution pattern we saw earlier for the individual-time-step matrices, only denser. All of these matrices have ones distributed throughout, with a particular concentration near the matrix major diagonal (again, more about this pattern shortly, at the end of the subsection).

The cumulative operational adjacency matrix depictions show where the network connections are, but do not show the intensity (flow value) of the connections. I have devised a companion view that provides the cumulative flow values of these connections. Figure 18.9 displays the cumulative flow value "adjacency matrix" (or *intensity matrix*) at time step 100. Figure 18.10 displays the cumulative intensity matrix at time step 1000.

The underlying distribution patterns in Figures 18.9 and 18.10, of course, match the patterns in Figures 18.7 and 18.8, respectively. We can

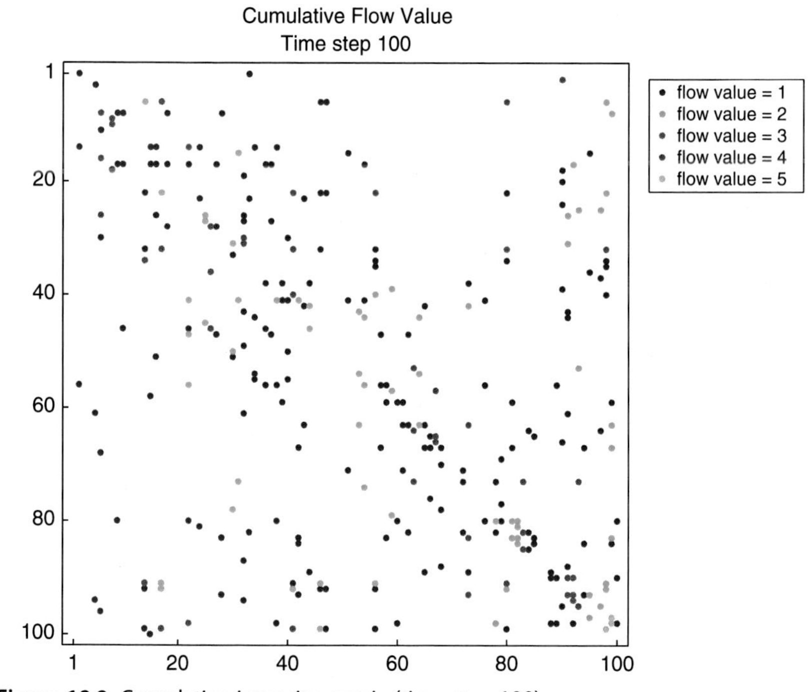

Figure 18.9 Cumulative intensity matrix (time step 100).

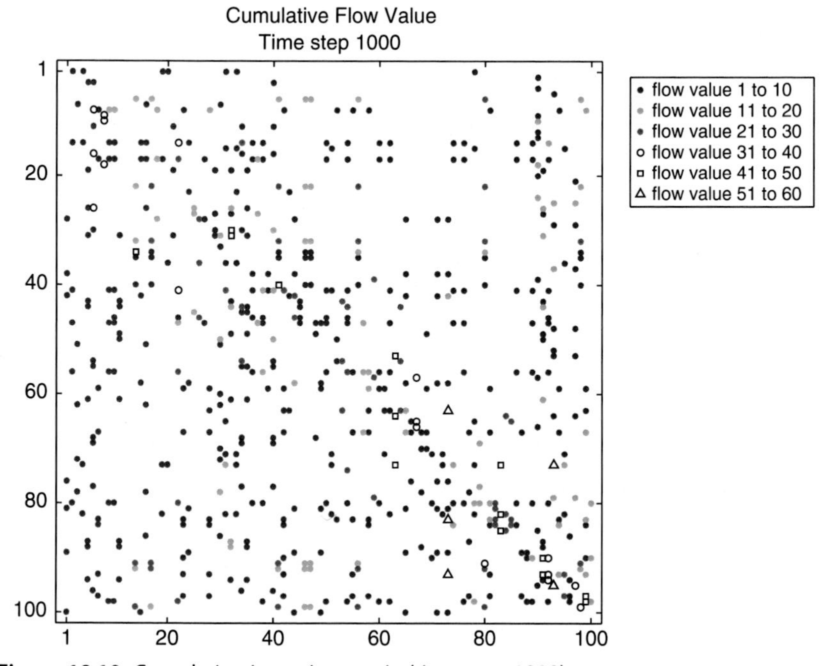

Figure 18.10 Cumulative intensity matrix (time step 1000).

make an important additional observation: the highest flow *intensities* also tend to concentrate near the matrix major diagonal.

What is the significance of the adjacency matrix and intensity matrix patterns we observe here in our results, and how does this behavior relate to the behavior of real-world complex systems? Herbert Simon has addressed such patterns and has offered insights and explanations. In Simon (1977), he said:

> *For real-world complex systems, if we construct an intensity matrix which reflects the intensity of interactions, "the large entries will be close to the diagonal, for these near-diagonal entries represent the interactions among elements that are close neighbors."*

Simon (1962) had explained that the reasons for this involve system architecture. He first explained that complex system architecture is very often *hierarchical*:

> *"Empirically, a large proportion of the complex systems we observe in nature exhibit hierarchic structure."*

Furthermore, these hierarchic complex systems are very often *nearly decomposable;* that is, at a given level of hierarchy, interactions among subsystems are weaker—often orders of magnitude weaker—than interactions within subsystems:

> *"In organic substances, intermolecular forces will generally be weaker than molecular forces, and molecular forces than nuclear forces."*

So, close-neighbor (within a subsystem) interactions occur with higher intensity, as reflected by larger values close to the major diagonal of the intensity matrix. Moreover, Simon suggested that these close-neighbor (within a subsystem) interactions occur with higher frequency:

> *"It is probably true that in social as in physical systems, the higher frequency dynamics are associated with the subsystems, the lower frequency dynamics with the larger systems."*

This is reflected by both a higher concentration of network connections close to the major diagonal of the adjacency matrix and higher flow values close to the major diagonal of the intensity matrix.

We have, of course, observed the effects described by Simon in our adjacency matrix and intensity matrix results in Figures 18.5–18.10. Let me emphasize two important points. First, the node neighborhood concept that I have implemented in my network dynamics model is very much consistent with Simon's insights and explanations regarding complex systems architecture. Second, the resulting behaviors of my network dynamics model are very much consistent with the behaviors of real-world complex systems.

18.2.2 Input/Output/Stock Histories

We can generate representations of cumulative input, output, and stock values for the nodes in our model network at each simulation time step. The representations are three-dimensional bar charts overlaid on the network node grid. Results for time steps 100, 400, 700, and 1000 are displayed in the following sets of diagrams.

Figure 18.11 displays the input value results. Recall that there is one unit-valued input to a randomly selected input node at each time step. In the underlying ecological network compartment model, five of the ten compartments are input compartments. There are, therefore, approximately

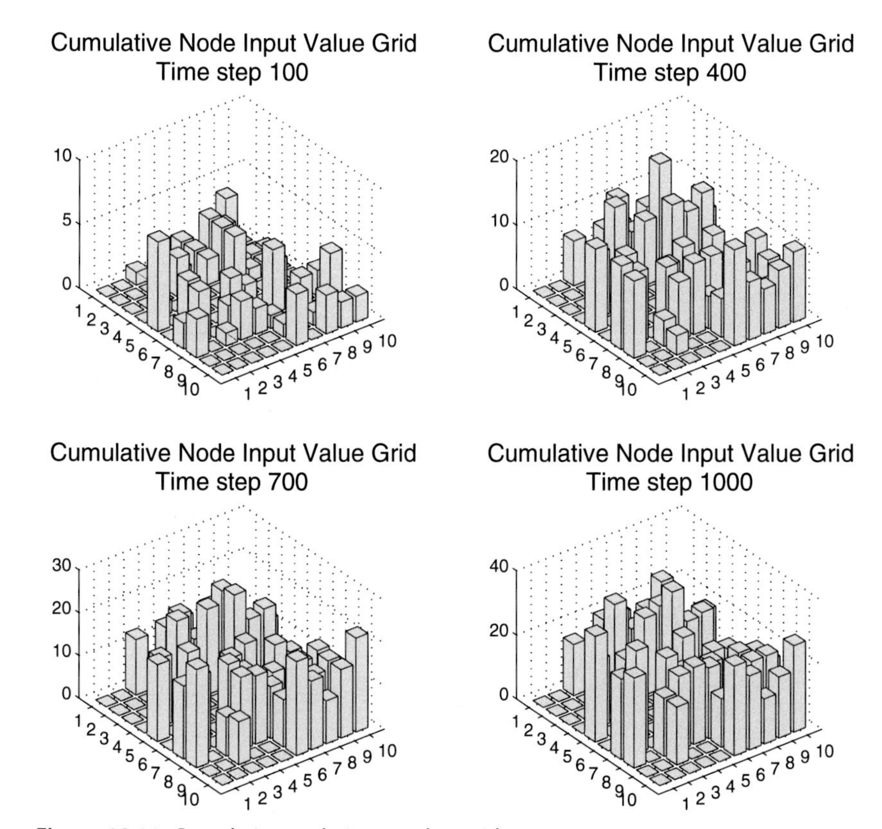

Figure 18.11 Cumulative node input value grid.

50 input nodes (~50% of the nodes) in our network node model. (I say "approximately" because I have randomly assigned nodes to compartments using a random number generator with a uniform distribution.) In the figure, note that some of the zero-valued grid elements are blocked from view by nonzero-valued elements. Also, because these are cumulative charts, the sequence of diagrams in Figure 18.11 shows monotonically increasing input values.

Figure 18.12 displays the cumulative output value results. At each time step, each propagating node that is also an output node sends one unit-valued output to the external environment. If the node propagates more than once at a time step (due to cycling), it outputs more than once at the time step. In the underlying ecological network compartment model, six of the ten compartments are output compartments, so there are approximately 60 output nodes in our network node model. Some nodes, of

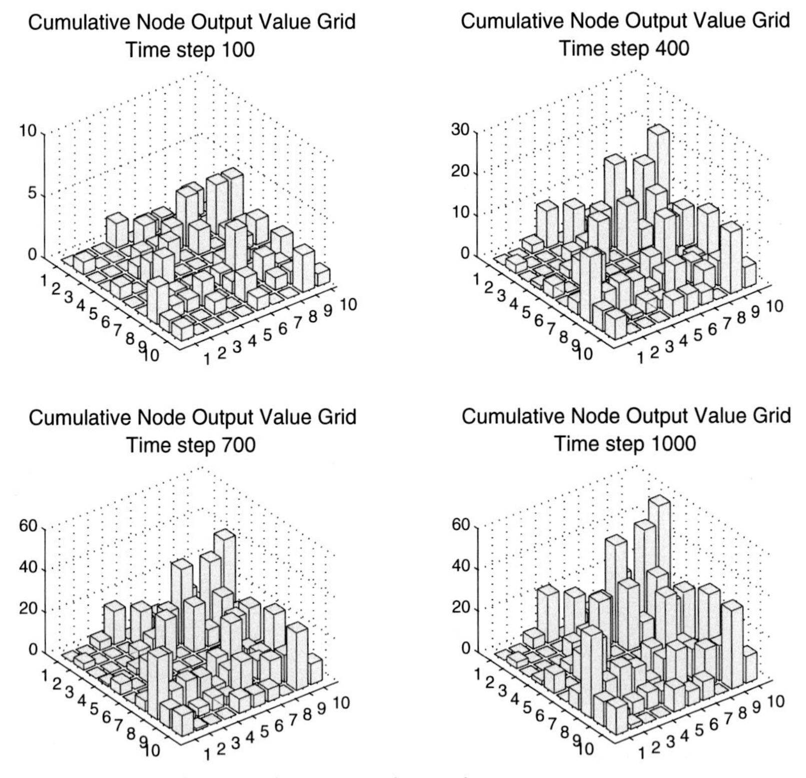

Figure 18.12 Cumulative node output value grid.

course, are both input nodes and output nodes. Again, because these are cumulative charts, the sequence of diagrams in Figure 18.12 shows monotonically increasing values.

Figure 18.13 provides the cumulative stock value of each of the network nodes. The node stock value initial conditions have been set using a uniform random assignment of zero to three before the start of the simulation run. Recall that a node propagation threshold of four and a node propagation flow quantity of four are used for the simulation run; that is, when a node's stock value reaches 4, the node propagates 4 units of stock. As shown in Figure 18.13, therefore, individual node stock values continually change but remain in the zero-to-three range.

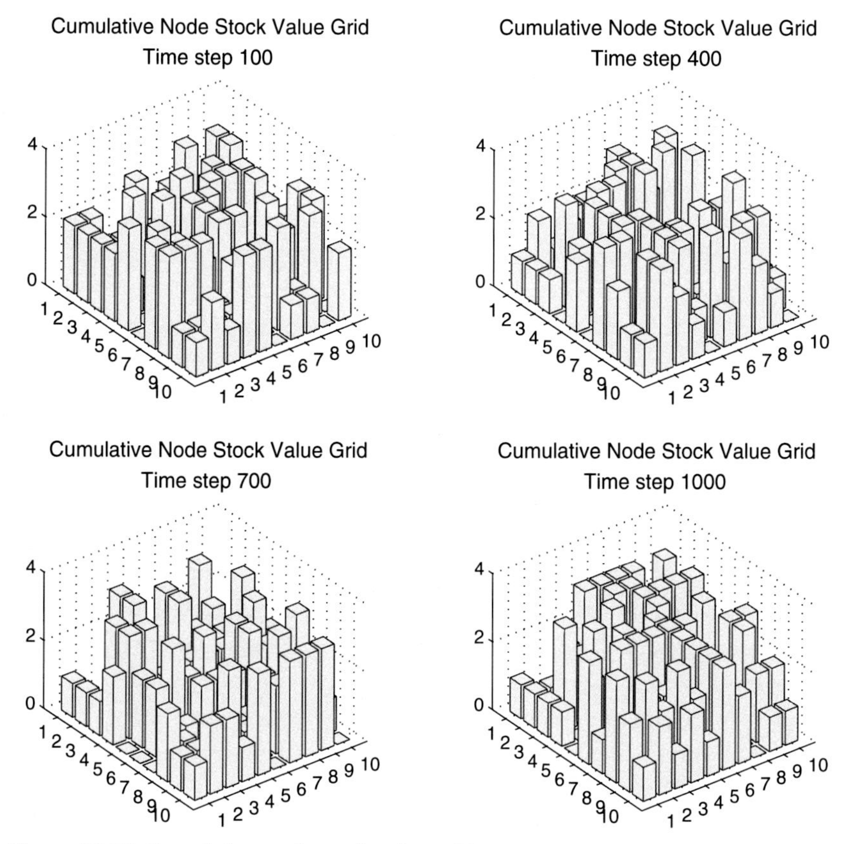

Figure 18.13 Cumulative node stock value grid.

18.3 NETWORK PROPAGATION EVENT RESULTS

Figure 18.14 displays the network propagation event time series and distribution. The time series plots network propagation event size vs. time. Network propagation event size at a time step is equal to the total number of node instances involved in propagation at that time step, that is, the number of node propagation instances (variable *nnpi*) plus one (for the initial input node instance). Because any given node can participate in propagation more than once at a time step (due to cycling), propagation event size can be larger than the number of physical nodes involved in propagation at the time step and sometimes larger than the total number of nodes in the network. The resulting time series clearly indicates punctuated and local-to-global dynamics, with many small events but also medium, large, and even a few very large propagation events interspersed. Smaller events are mostly local and larger events are global. The network propagation event distribution (number of events vs. size of events) shown in Figure 18.14 is derived from the time series.

See the Appendix, Code example 5 for the MATLAB code that generates the network propagation event distribution.

Let's take a closer look at the distribution with the help of Figure 18.15. At the upper left-hand corner of the figure, we again see the distribution plotted in normal coordinates. The curve is definitely "long-tailed," and perhaps it is a power-law curve.

Let's see if power-law behavior can be confirmed. First, note that a power-law curve has the form

$$y = Cx^{-\lambda}$$

where C is a constant and λ is the *power* or *scaling exponent*. Taking the logarithm of both sides of the power-law equation, we obtain

$$\log y = \log C - \lambda \log x$$

In log-log space this, of course, is the equation of a straight line with *y-intercept* equal to $\log C$ and *slope* equal to the negative of the scaling exponent. So here is the determining question: does our network propagation event distribution approximate a straight line in log-log coordinates? The log-log plot of the propagation event distribution data points is shown at the upper right-hand corner of Figure 18.15. We fit a straight line to the data points and observe that the equation of the log-log straight line is

$$y = -1.5x + 3.5$$

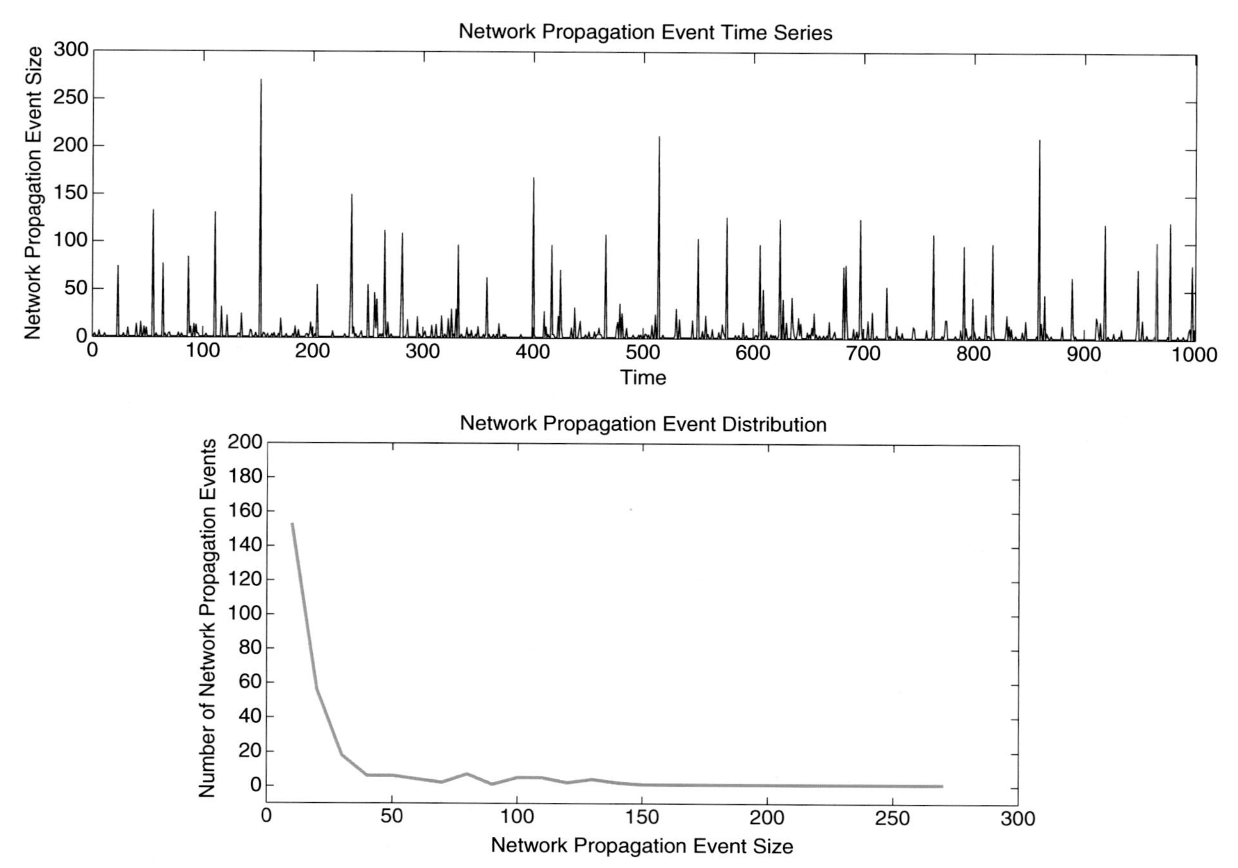

Figure 18.14 Network propagation event time series and distribution.

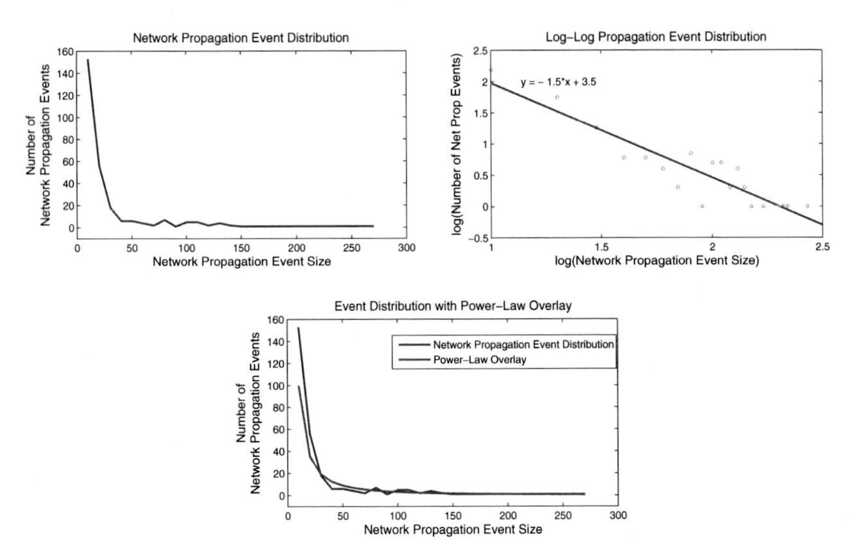

Figure 18.15 Network propagation event distribution set.

Now let's do one more plot in normal coordinates (lower half of Figure 18.15). We plot the original propagation event distribution (in blue) and overlay the power-law distribution calculated from the log-log plot (in red). The equation of the power-law overlay is

$$y = C\,x^{-\lambda}$$
$$y = \left[10^{(\text{loglog y-intercept})}\right] * \left[x^{(\text{loglog slope})}\right]$$
$$y = 10^{3.5}x^{-1.5}$$

As shown in Figure 18.15, the fit is quite good! Power-law dynamics are confirmed. Network propagation events exhibit power-law/process-fractal behavior.

Next questions: What does the network propagation event frequency spectrum look like? Does the "noise-like"[1] network propagation event time series have a "pink 1/f noise" frequency spectrum? Recall, from Chapter 15, that the power-law "pink 1/f noise" spectrum is a process-fractal frequency spectrum.

The network propagation event frequency spectrum is obtained by taking the discrete Fourier transform of the discrete network propagation event time series. The result is shown in the top diagram of Figure 18.16.

[1] The network propagation event time series appears to be "noise-like," but it is certainly not noise. The time series reflects complex ecological system fluctuations.

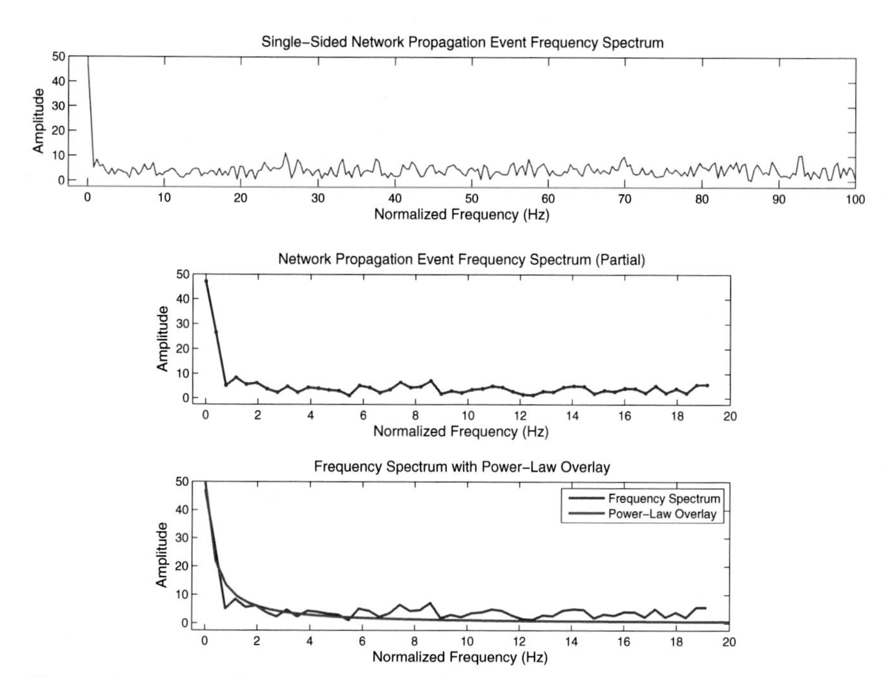

Figure 18.16 Network propagation event frequency spectrum.

See the Appendix, Code example 6 for the MATLAB code that generates the network propagation event frequency spectrum.

Note that the normalized frequency abscissa in Figure 18.16 is somewhat arbitrary. The frequency range depends on the signal sampling frequency, which depends on the time duration of a time step (sampling frequency in Hz = 1/time-step duration). Because my model is intended to apply to a broad set of real ecological networks, I have not specified any particular time-step duration.

Because the lower frequencies are of most interest, the middle diagram of Figure 18.16 focuses on the first 20% or so of the waveform. The spectrum appears to follow a power-law curve, but I cannot confirm this with a straight-line log-log plot. The accuracy of the Fourier transform data that I was able to obtain is not high enough for that.[2] On the bottom diagram

[2] The accuracy of these discrete Fourier transform computations is limited by the *relatively* short length of the signal time series (1000 time steps for my model). Much improved Fourier transform accuracy would result from signal lengths on the order of 10,000 or even 25,000 time steps. For my model, however, such values are way beyond the capacity of my computer. Perhaps a reader with a high-capacity computer (random access memory is the primary limiting factor) could check the 25,000 time-step case.

of Figure 18.16, however, I do overlay a power-law curve on the frequency spectrum for comparison. The power-law curve parameters are chosen for approximate "best fit." The equation of the power-law overlay is

$$y = 50(x + 1)^{-1.2}$$

In a power-law equation, when the independent variable is zero, the dependent variable goes to infinity. To avoid that situation, I have shifted the independent variable by one. As shown in the bottom diagram of Figure 18.16, the frequency spectrum fit to a power-law curve is pretty good. This result[3] strongly suggests that the spectrum reflects "pink 1/f noise" and network propagation event power-law/process-fractal behavior.

Overall, our propagation event results indicate that network propagation event behavior is punctuated, local-to-global, and fractal.

18.4 PATH LENGTH RESULTS

Path length is usually viewed as a spatial concept. My analysis results, however, show that propagation path length behavior changes (in dramatic punctuated fashion) with time, so that temporal considerations are also very important. To fully capture path length dynamics, therefore, I have devised an approach that integrates over both space and time. The approach is a spatial/temporal analog of the conventional "box covering method" for network path length analysis (as discussed in Chapter 17). My approach uses the path length values achieved at each time step as a measure of path length dynamic behavior. Figure 18.17 shows a corresponding path length time series and space/time integrated distribution for the simulation run. The figure uses the maximum path length achieved at each time step.

The time series is clearly punctuated and exhibits convincing evidence of local-to-global propagation. Path length values vary from very small (local propagation) to very large (global propagation). The integrated path length distribution is "long-tailed" and suggests a power-law curve. Note that the path length distribution is a bit "ragged." This is due to the small interval size on the abscissa (it's equal to 1) and, therefore, the relatively small number of samples per interval. The simulation run consists of 1000 time steps, but less than 300 are propagation time steps. These < 300 time steps yield path lengths from 1 to 22. In the distribution, therefore, we have a relatively large

[3] Notice that our results here are quite similar to the results of the heart rate dynamics analysis that we discussed in Chapter 15.

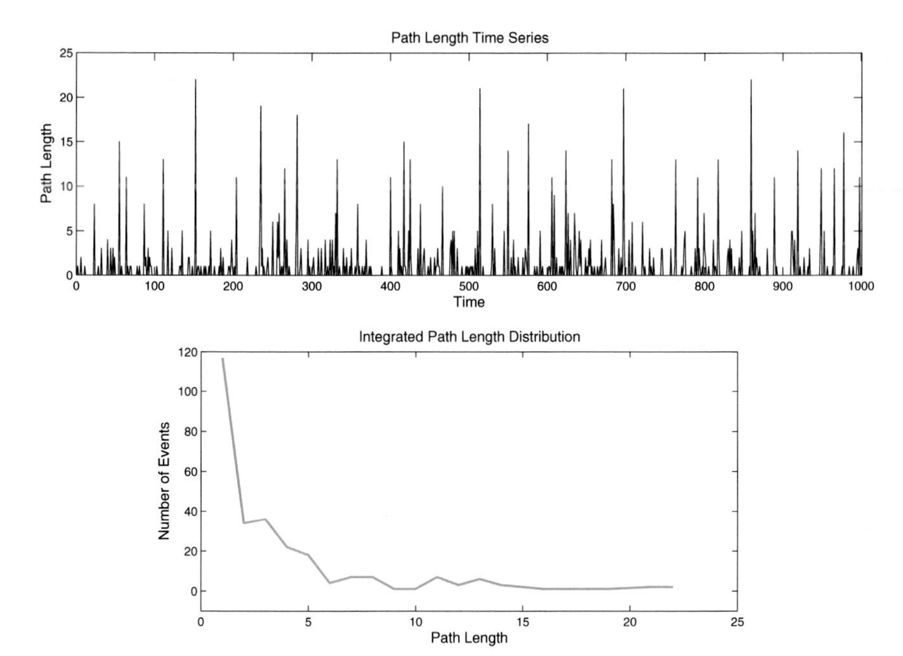

Figure 18.17 Path length time series and integrated distribution.

number of abscissa intervals (22) and a relatively small number of sample sets (less than 300). The distribution data set is large enough to yield definitive behavior results, but not large enough to yield a "continuous-looking" curve connecting the data points. Of course, more time steps would be expected to reduce or eliminate the raggedness. In the literature, I've seen cellular automata models (e.g., the sandpile model and a model for binary genetic networks) that use 25,000 time steps. For my complex ecological network model, however, I am limited to 1000 time steps by the capabilities of my computer.

We take a closer look at the path length distribution via Figure 18.18. The upper-left diagram in the figure is a repeat of the path length distribution plotted in normal coordinates.

Let's see if power-law behavior can be confirmed. Let's see if the log-log plot of the path length distribution data points approximates a straight line. That plot is shown in the upper right of Figure 18.18. We fit a straight line to the data points and observe that the equation of the log-log straight line is

$$y = -1.5x + 2.1$$

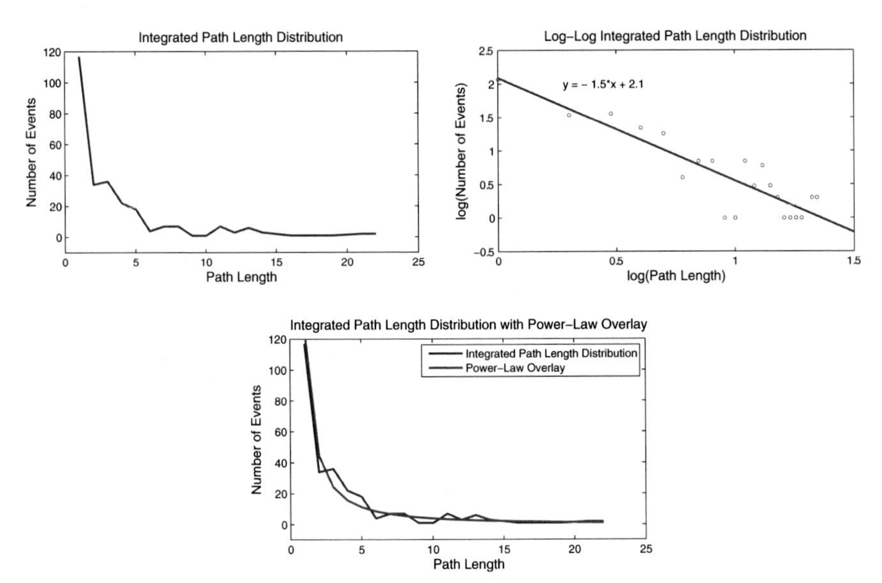

Figure 18.18 Integrated path length distribution set.

The graph in the lower half of Figure 18.18 plots the original normal-coordinates path length distribution (in blue) and overlays the power-law distribution calculated from the log-log plot (in red). The equation of the power-law overlay is

$$y = C\,x^{-\lambda}$$
$$y = \left[10^{(\text{loglog y-intercept})}\right] * \left[x^{(\text{loglog slope})}\right]$$
$$y = 10^{2.1}x^{-1.5}$$

As shown in the figure, the fit is good. Power-law dynamics are confirmed. Path length exhibits fractal behavior.

In Chapter 15, we mathematically derived a distribution-dependent condition for dominance of indirect effects: if the path length distribution scaling exponent is greater than 1 and less than 1.7, then indirect effects are dominant. Our data results here (i.e., scaling exponent = 1.5) satisfy that condition. According to the mathematics, indirect effects should dominate in this simulation run. In the next section, indirect effects results are presented. We will see if our indirect effects data results agree with the mathematical prediction. (Hint: they do.)

To obtain these path length time series and distribution results (Figures 18.17 and 18.18), I used the maximum path length achieved at each time step. Are our observations and conclusions about path length dynamics sensitive to the choice of path length "measure" at each time step? Let's pick another measure and see. Let's try the *mean* path length achieved at each time step as the measure.[4] Figure 18.19 provides the resulting mean path length time series and integrated distribution set.

The time series (top diagram) is clearly punctuated and exhibits convincing evidence of local-to-global propagation (for the same reasons given for Figure 18.17). The integrated path length distribution (mid-left diagram of Figure 18.19) is "long-tailed" and suggests a power-law curve. In the log-log plot of the distribution data (mid-right diagram), we fit a straight line to the data points and observe that the equation of the log-log straight line is

$$y = -2.1x + 1.9$$

The bottom diagram of Figure 18.19 plots the normal-coordinates path length distribution from the mid-left diagram (in blue) and overlays the corresponding curve-fit power-law distribution (in red). The difference between the data plot (blue) and the curve-fit overlay (red) at mean path length equal to one is due to "cutoff effects." For the data plot, I considered only time steps with propagation (which have a minimum mean path length of one). This "cuts off" the data distribution at mean path length = 1 and alters the low end (low abscissa values) of the curve-fit overlay. The equation of the power-law overlay is

$$y = C\,x^{-\lambda}$$
$$y = \left[10^{(\text{loglog y-intercept})}\right] * \left[x^{(\text{loglog slope})}\right]$$
$$y = 10^{1.9}x^{-2.1}$$

Path length power-law/fractal behavior is corroborated. Important additional points are suggested by the results shown in Figure 18.19. Our path length analysis is not particularly sensitive[5] to the "measure" of path length dynamics we use, as long as that measure reflects the time-varying nature of the dynamics. It appears that per-time-step averages (means) do reflect and generally preserve those dynamics. It appears that it may be valid to use such mean values in this and other similar analyses. (We will try this again as one of our approaches in the node degree analysis of Section 18.6.)

[4] The mean path length we use here is an average at each time step and not an average over time. We know that averages over time lose the instantaneous dynamics of a network. We will see if per-time-step means preserve instantaneous dynamics.

[5] Although the specific distribution parameters differ somewhat, the analysis general conclusions (punctuated, local-to-global, and fractal behavior) are the same.

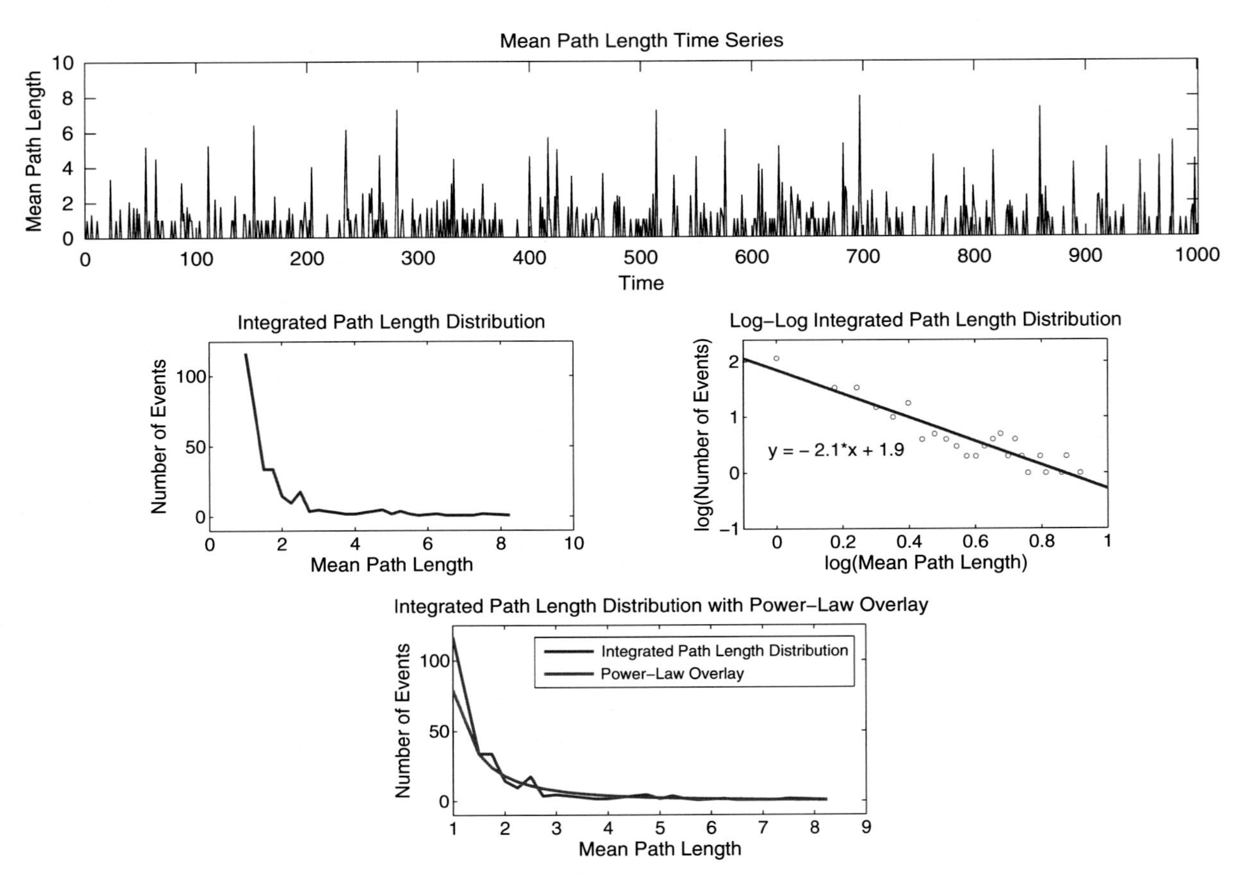

Figure 18.19 Mean path length time series and integrated distribution set.

Overall, our results here indicate that path length exhibits punctuated dynamics, local-to-global propagation, and fractal behavior.

18.5 INDIRECT EFFECTS RESULTS

I have defined several ecological network direct effects and indirect effects indicators. The *direct effects ratio* equals the number of direct paths (path length = 1) divided by the total number of paths. The *indirect effects ratio* equals the number of indirect paths (path length > 1) divided by the total number of paths. The *indirect effects index* equals the number of indirect paths divided by the number of direct paths. These indicators are calculated for the given simulation-run network at each point in time (each time step).

Figure 18.20 provides the individual-time-step time series for each of the indicators, that is, the individual-time-step values vs. time. Perhaps the first thing you notice about the figure is that all of these time series are highly punctuated. There is not a hint of gradual or continuous behavior. Next, you may notice that the direct effects ratio time series (the top diagram in Figure 18.20) looks particularly "dense" with values of 1. Why is that? Recall, from the network propagation event time series and distribution, that most propagation events are small. In small one-stage propagation events, all paths are direct paths. These events yield all the values of 1 in the direct effects time series. Notice also that there are quite a few values of two-thirds and one-half in that time series. Due to the propagation granularity in the model, small two-stage events often have twice as many direct paths as indirect paths, and small three-stage events often have an equal number of direct and indirect paths. These effects yield the two-thirds and one-half values, respectively. The appearance of the direct effects ratio time series is simply a reflection of the large number of small propagation events. The indirect effects ratio time series (the middle diagram in Figure 18.20), on the other hand, is a reflection of the high percentage of indirect paths that occur in midsize to large propagation events. For these events, 50-90% or more of the paths are indirect paths. The indirect effects index time series (the bottom diagram in Figure 18.20) shows that the number of indirect paths can be 5-10 or more times the number of direct paths at these time steps. For the larger propagation events, therefore, indirect effects far exceed direct effects.

Earlier in this chapter, in the node-and-link propagation flow diagrams, we saw that the midsize to large propagation events have broad network span. From a single (local) input node, there is propagation to an extensive (global) portion of the network. The path between any two given nodes is

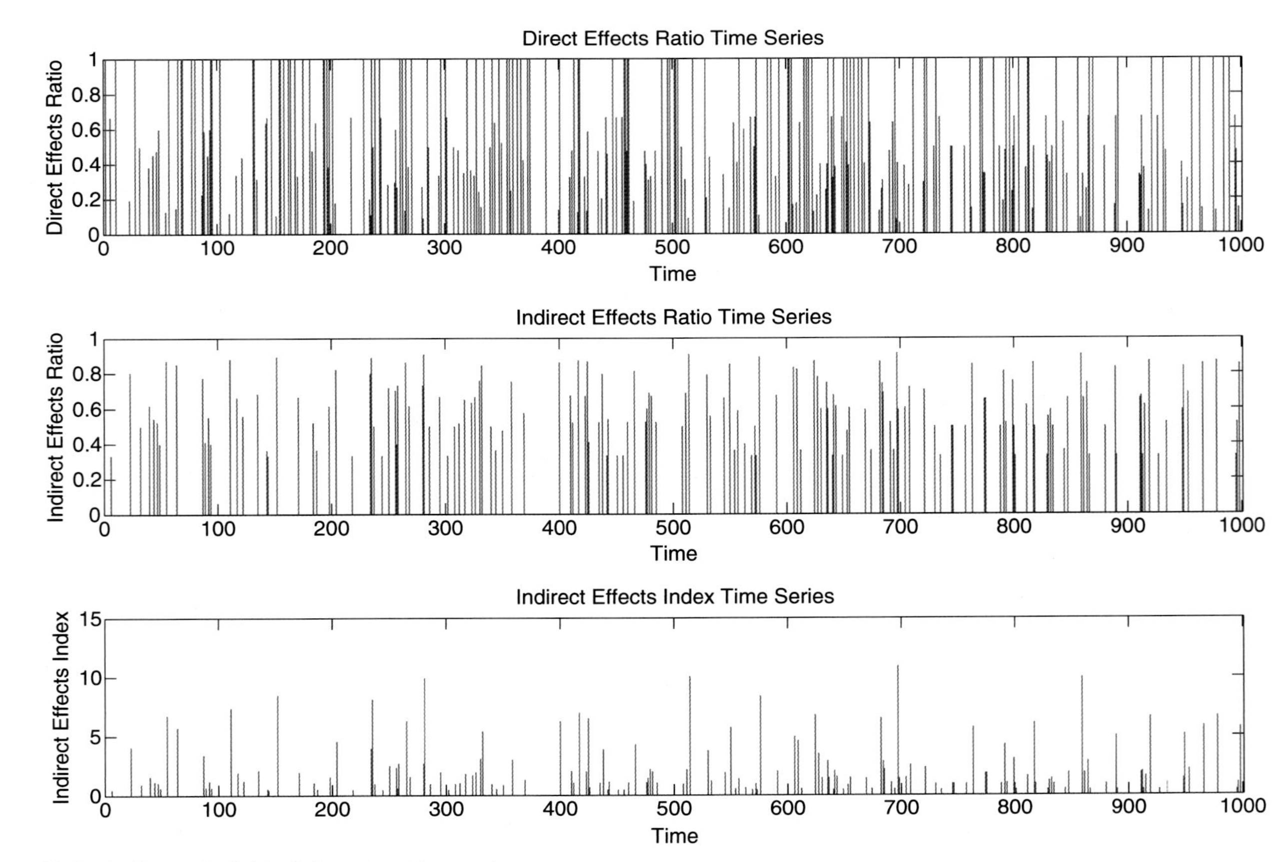

Figure 18.20 Indicator individual-time-step time series.

very often an indirect path. Clearly, indirect effects are enablers of local-to-global processing.

We see that indirect effects are dominant at high-propagation time steps. Are indirect effects dominant overall? To answer that question, we can look at the over-time cumulative indicator time series in Figure 18.21. At any given time (time step), each of these time series shows the cumulative value of the indicator up to that point in time.

The first thing you may notice in Figure 18.21 (compared with Figure 18.20) is that these cumulative averages over time "smooth" the instantaneous dynamics. The cumulative direct effects ratio settles in at a value of about 0.17. The cumulative indirect effects ratio reaches and maintains a value of about 0.83. Cumulative direct paths are $< 20\%$ of total paths and cumulative indirect paths are $> 80\%$ of total paths. The cumulative number of indirect paths is almost 5 times (actually 4.8 times) the number of direct paths. The 4.8 result is confirmed by the indirect effects index time series (bottom diagram of Figure 18.21). The next figure (Figure 18.22) is also interesting.

Figure 18.22 shows the cumulative quantity of indirect paths vs. time and the cumulative quantity of direct paths vs. time. Both plots are monotonically increasing (actually nondecreasing) as expected, but note the final values. At the end of the simulation run, there are about 6000 direct paths and almost 30,000 indirect paths (that's the 4.8 factor again). *Indirect effects are definitely dominant* in this simulation-run network.

Next, let's turn to path quantity time series and distributions, and test them for fractal behavior. Figure 18.23 provides the indirect path results. The time series (upper diagram in the figure) shows that there are very small (or zero) quantities of indirect paths at time steps corresponding to small propagation events and very large quantities of indirect paths at time steps corresponding to midsize and large propagation events. For the largest propagation events, the quantity exceeds 2000. The lower-left diagram in Figure 18.23 displays the indirect path quantity distribution. The lower-right diagram repeats that distribution (in blue) and adds a power-law overlay (in red) for comparison. The equation of the power-law overlay is

$$y = 10^6 x^{-2}$$

The fit is good. Indirect path quantity is fractal.

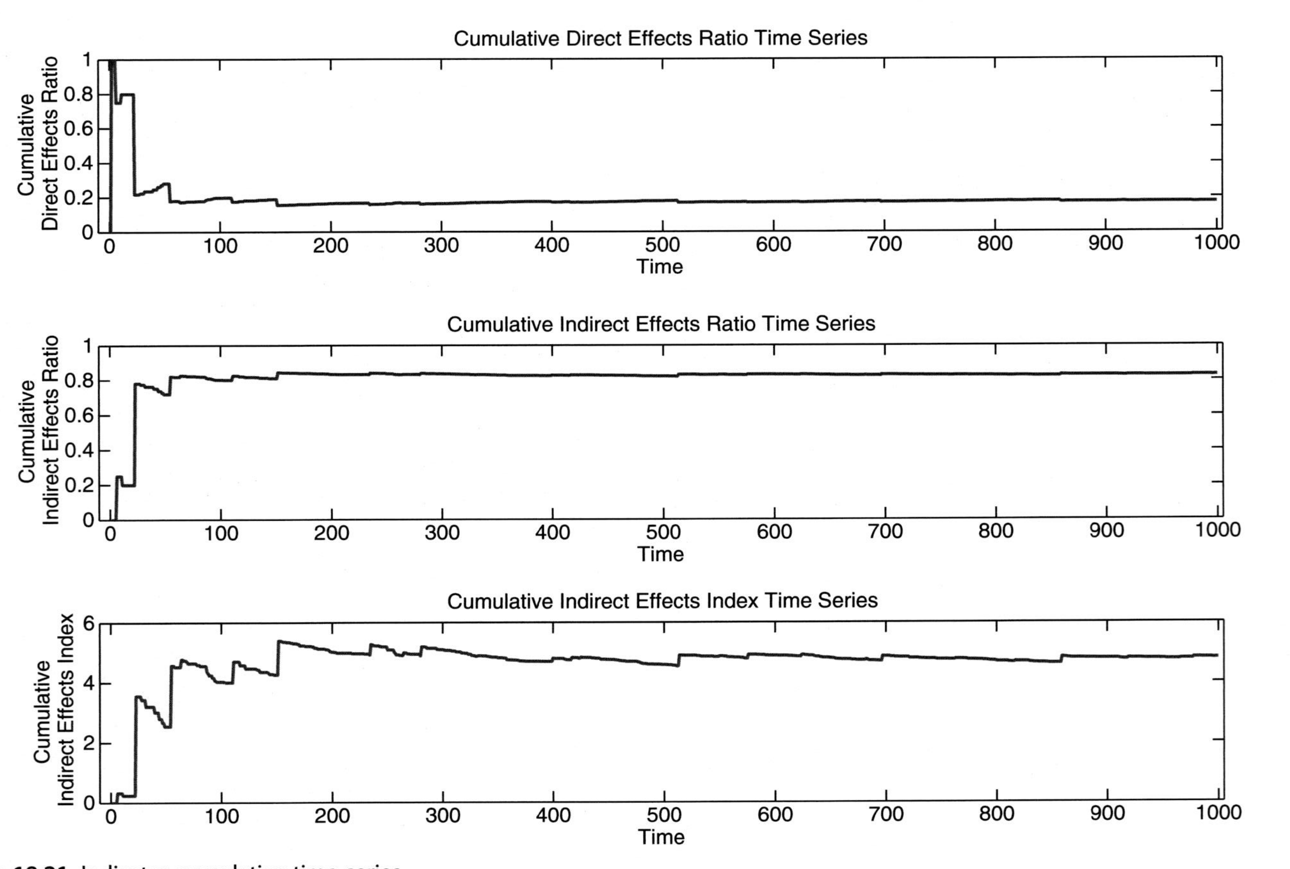

Figure 18.21 Indicator cumulative time series.

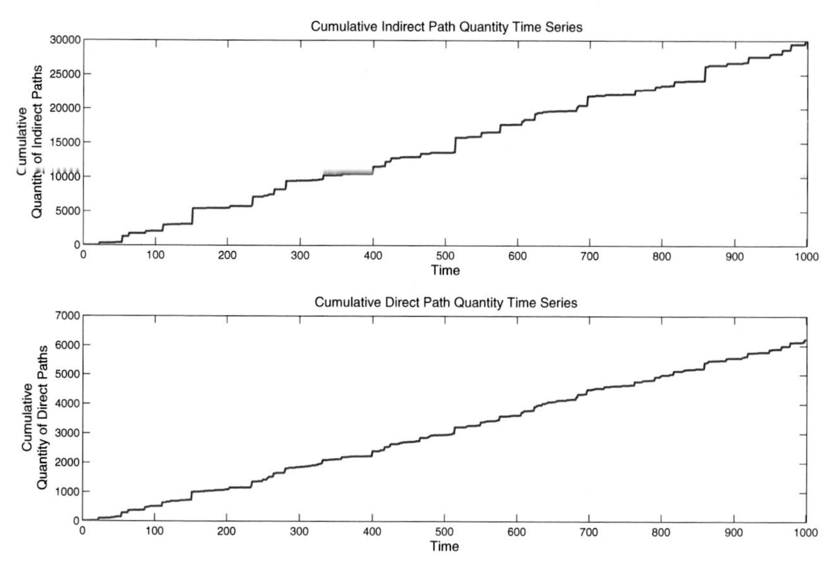

Figure 18.22 Cumulative path quantity time series.

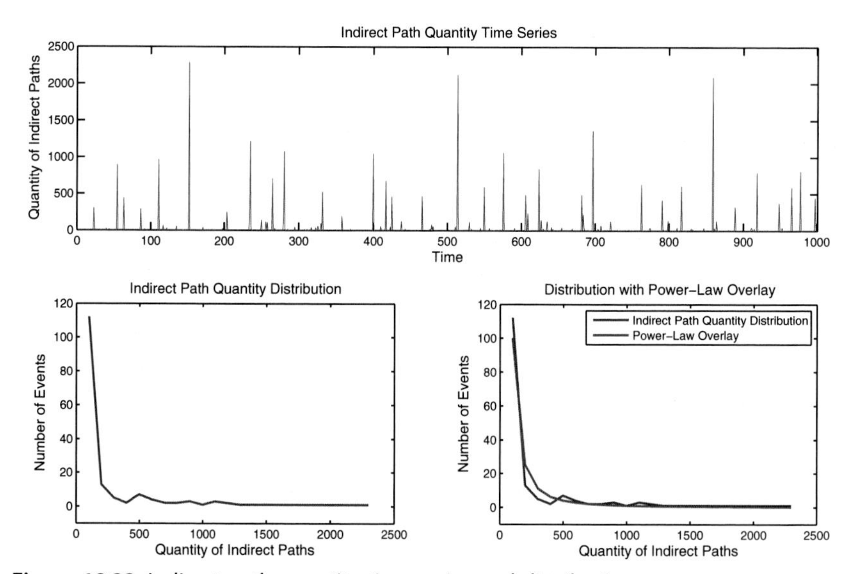

Figure 18.23 Indirect path quantity time series and distribution.

The analogous set of results for direct paths is provided in Figure 18.24. The equation of the power-law overlay (shown in red on the lower-right diagram of Figure 18.24) is

$$y = 1.5\left(10^4\right)x^{-2}$$

Direct path quantity is also fractal.

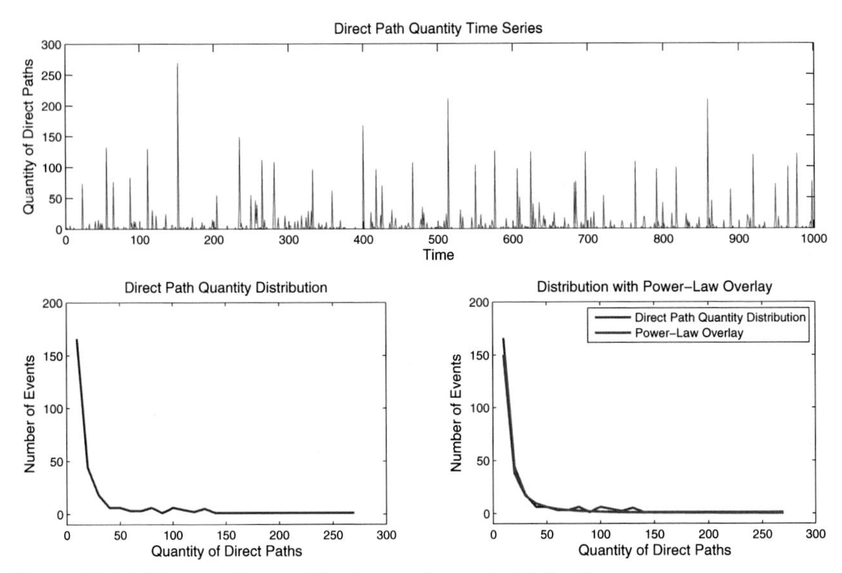

Figure 18.24 Direct path quantity time series and distribution.

In summary, the results here show that indirect effects are dominant and that they are punctuated, enablers of local-to-global propagation, and they exhibit fractal behavior.

18.6 NETWORK CONNECTIVITY RESULTS

Along with network propagation events, path length, and indirect effects, our results show that network connectivity exhibits punctuated, fluctuating dynamics. Node degree (the number of connections to/from a node) is an important indicator of network connectivity. Let's visually observe node degree behavior at several representative time steps. As shown on Figure 18.25, I have selected three time steps (32, 64, and 152) that correspond to small, midsize, and large propagation events, respectively.

To observe node degree behavior at these time steps, I have devised three-dimensional node degree grids (node degree overlays on the network grid). The node "in-degree" results are displayed in Figure 18.26, and the "combined-degree" (in-degree plus out-degree) results are displayed in Figure 18.27.

See the Appendix for the MATLAB code that creates the three-dimensional node degree grids. Code example 7 displays the code that

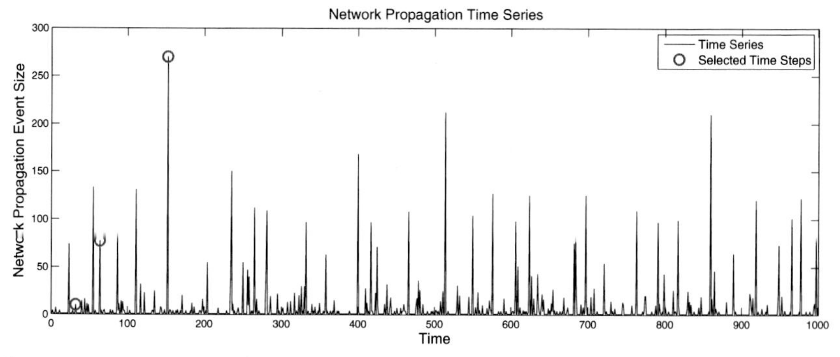

Figure 18.25 Network propagation time series with selections.

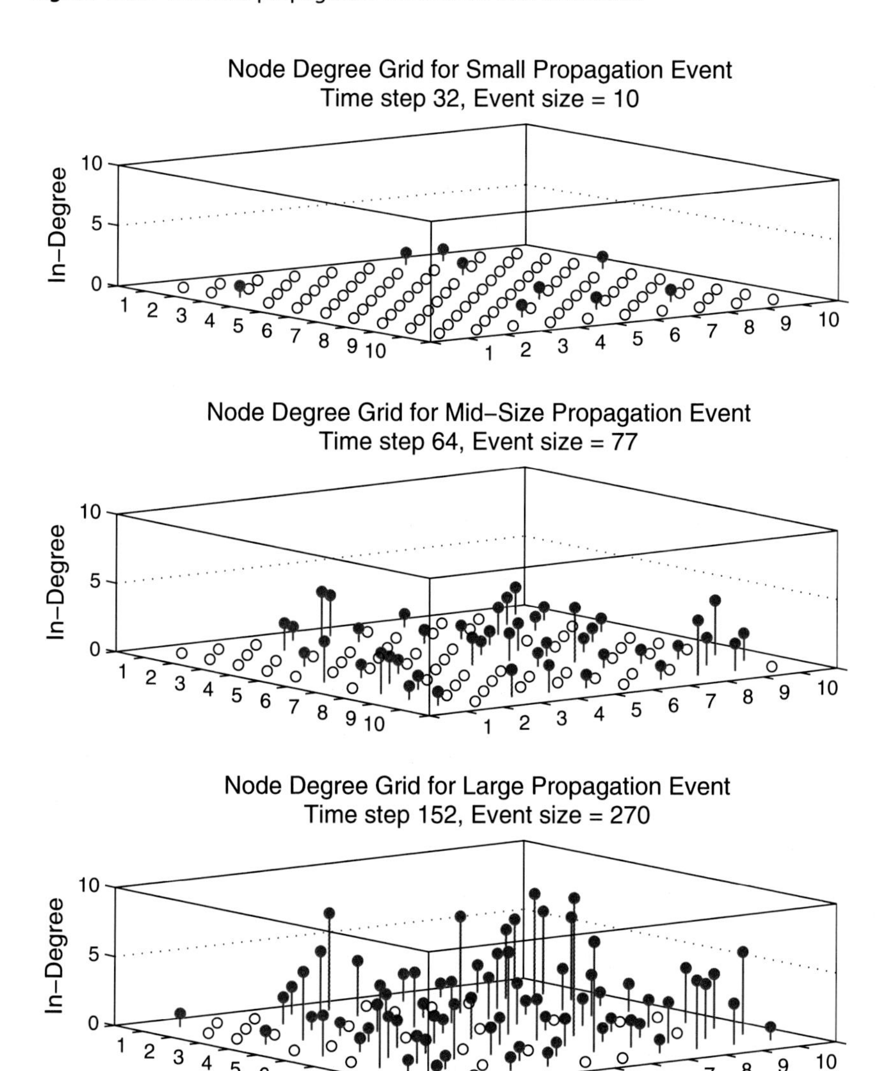

Figure 18.26 Node in-degree grid at representative time steps.

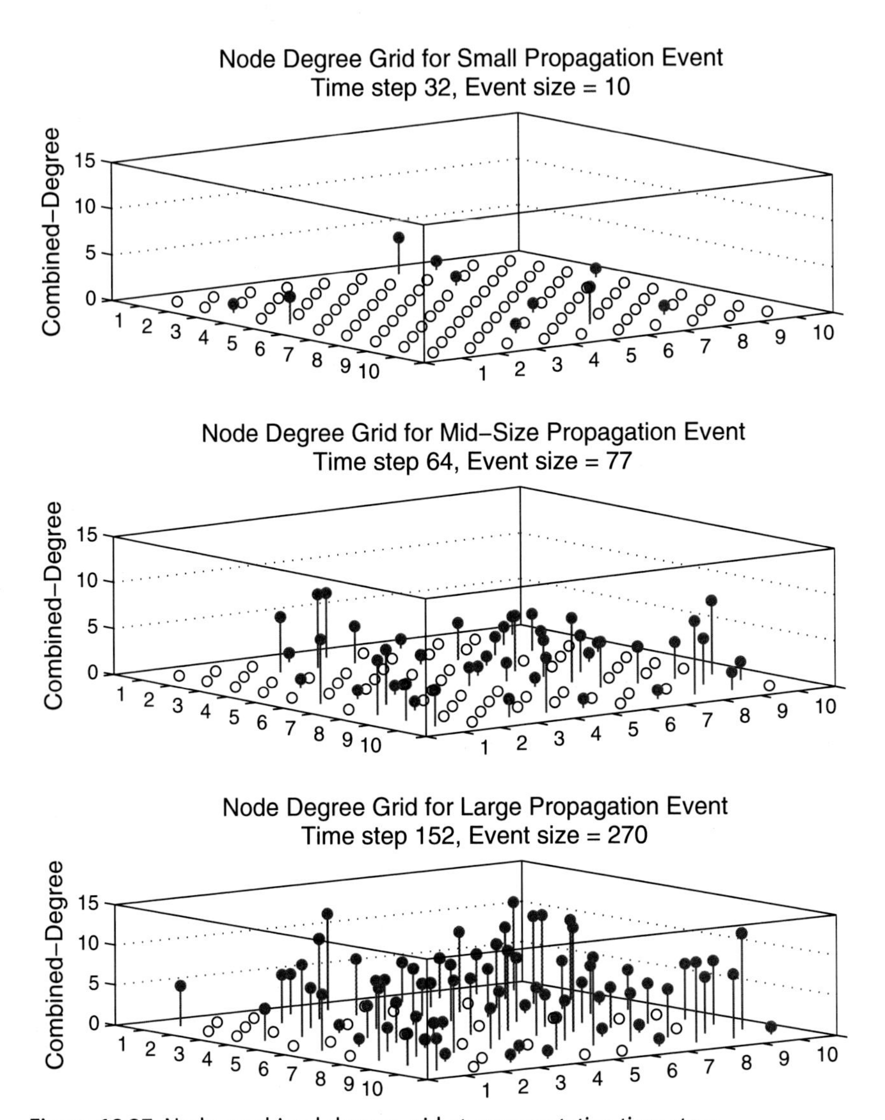

Figure 18.27 Node combined-degree grid at representative time steps.

generates individual-time-step node degree *vector arrays* that provide the degree values of each node at each time step. Code example 8 displays the code that generates the node degree *grid arrays* that overlay these degree values on the node grid to obtain a spatial view at each time step.

Figures 18.26 and 18.27 clearly illustrate the fluctuating dynamics of node degree over time. You can see that the instantaneous dynamics are ever-changing.

Before proceeding, note that there are some model "granularity effects" at play here. Because individual node inputs always occur in increments of 1, these effects are minimized for the in-degree case. For the out-degree case (not shown in the figures), on the other hand, granularity effects are maximized. Individual node outputs to other nodes occur in quanta of 3 or 4. Out-degree data, therefore, contain the most "jumps." The combined-degree case lies between the in-degree and out-degree cases. These granularity effects show up especially in some of the distribution results, for example, the combined-degree distribution of Figure 18.29 (that we will discuss shortly).

Next, let's focus on node degree time series and distributions. We need to be able to capture the fluctuating dynamics (depicted in Figures 18.26 and 18.27) in an integrated fashion over space and time. To accomplish that, a spatial/temporal approach very similar to the path length approach (discussed and applied earlier) is employed. The results are displayed in the time series and distribution sets of Figures 18.28 and 18.29 for in-degree and combined-degree, respectively. For these results, the measure of node degree dynamics used is the maximum node degree achieved at each individual time step.

See the Appendix, Code example 9 for the MATLAB code that generates the node in-degree integrated distribution.

Figure 18.28 displays the node in-degree results. The time series is clearly punctuated. The integrated node degree distribution is "long-tailed" and suggests a power-law curve. On the log-log plot, we fit a straight line to the data points and observe that the equation of the log-log straight line is

$$y = -2.2x + 2.4$$

Finally, we plot the original normal-coordinates integrated node degree distribution (in blue) and overlay the power-law distribution calculated from the log-log plot (in red). The equation of the power-law overlay is

$$\begin{aligned} y &= C\,x^{-\lambda} \\ y &= \left[10^{(\text{loglog y-intercept})}\right] * \left[x^{(\text{loglog slope})}\right] \\ y &= 10^{2.4}x^{-2.2} \end{aligned}$$

As shown in Figure 18.28, the fit is good given that the largest value of node degree here is 10, so that we are working with only about 10 data points. Power-law dynamics are corroborated. Node in-degree exhibits scale-invariant fractal behavior.

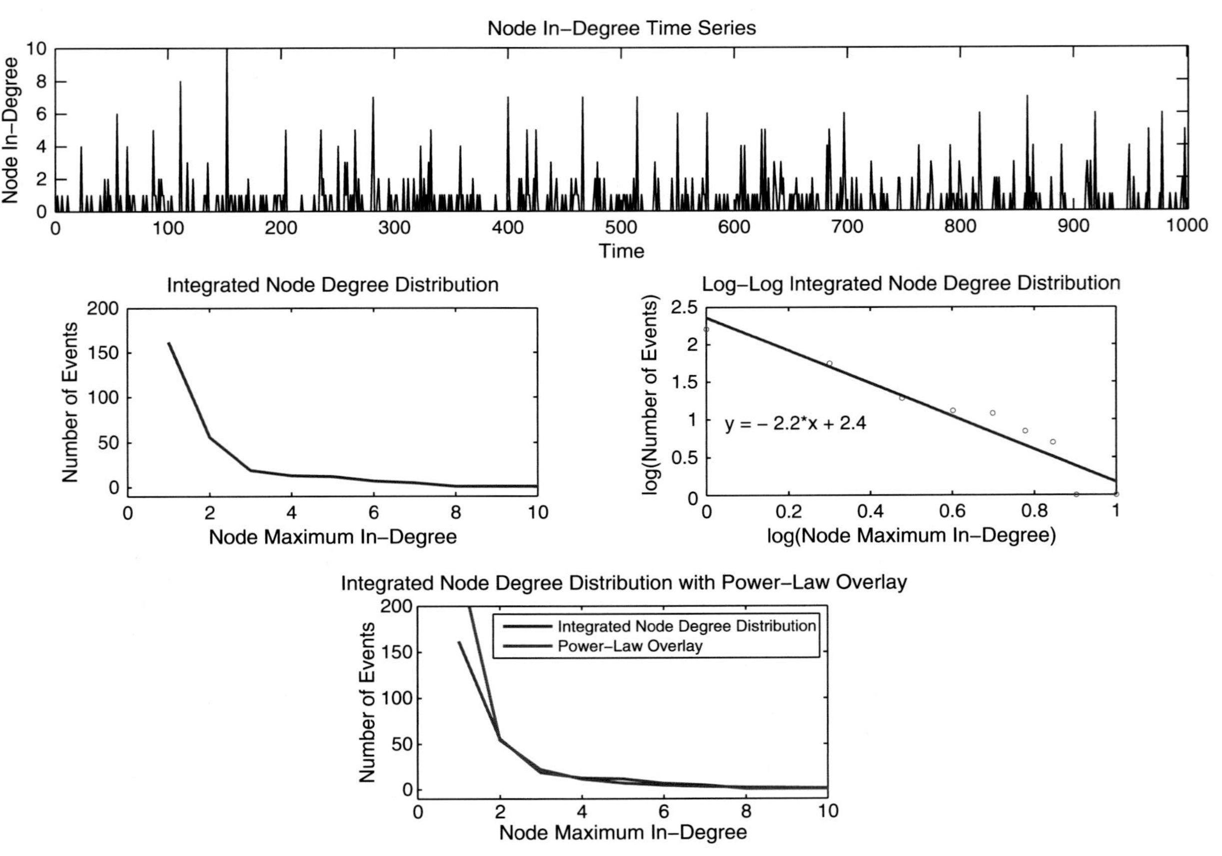

Figure 18.28 Node in-degree time series and distribution set.

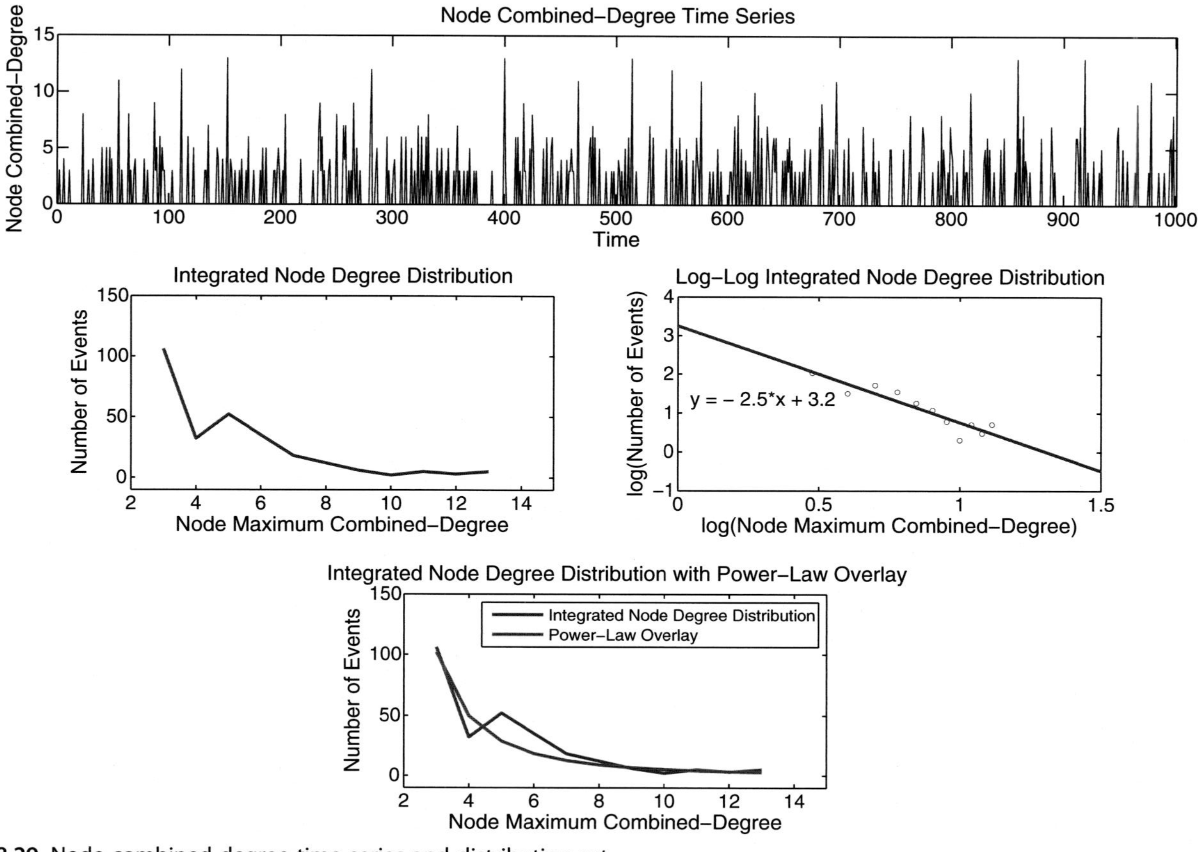

Figure 18.29 Node combined-degree time series and distribution set.

Consider the node combined-degree results in Figure 18.29 for a moment. These combined-degree results are similar to the in-degree results. The granularity effects mentioned earlier are reflected in the "jumps" in the distribution and in the deviation from the power-law overlay curve in that region. Still, node combined-degree power-law dynamics and scale-invariant fractal behavior are corroborated.

Another measure that may capture the node degree dynamics is the node *mean degree* calculated at each individual time step. These mean values are averages calculated at a point in time. They are not averages over time. The node-mean-degree time series and distribution set is shown in Figure 18.30. The time series is punctuated and the degree distribution follows a "long-tailed" power law. Even though we are dealing here with both granularity effects and relatively few data points, node-mean-degree power-law dynamics and scale-invariant fractal behavior are corroborated.

Let us now examine some other important network connectivity traits. To do this, the set of time series provided in Figure 18.31 is used. The first time series is the node mean degree time series. The second is the network

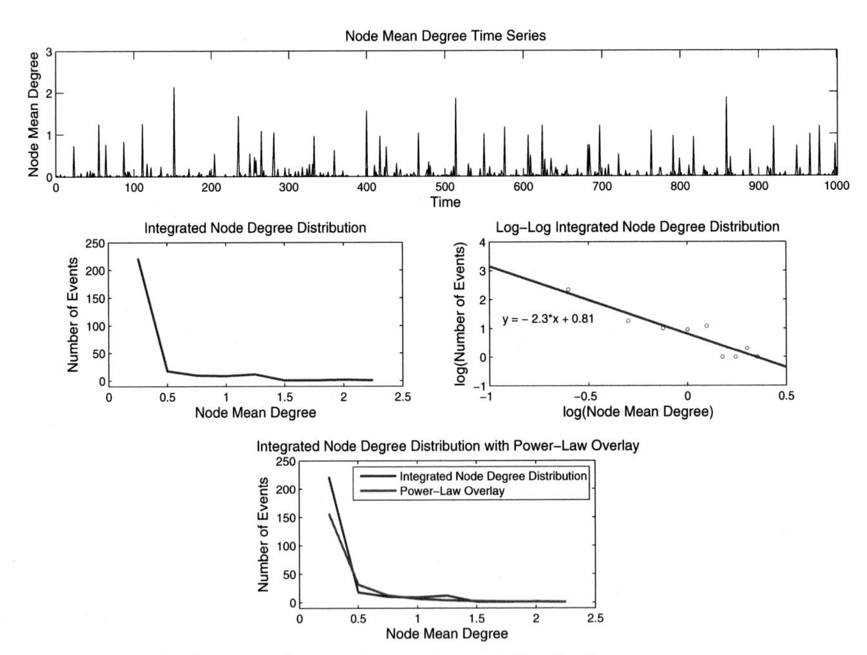

Figure 18.30 Node mean degree time series and distribution set.

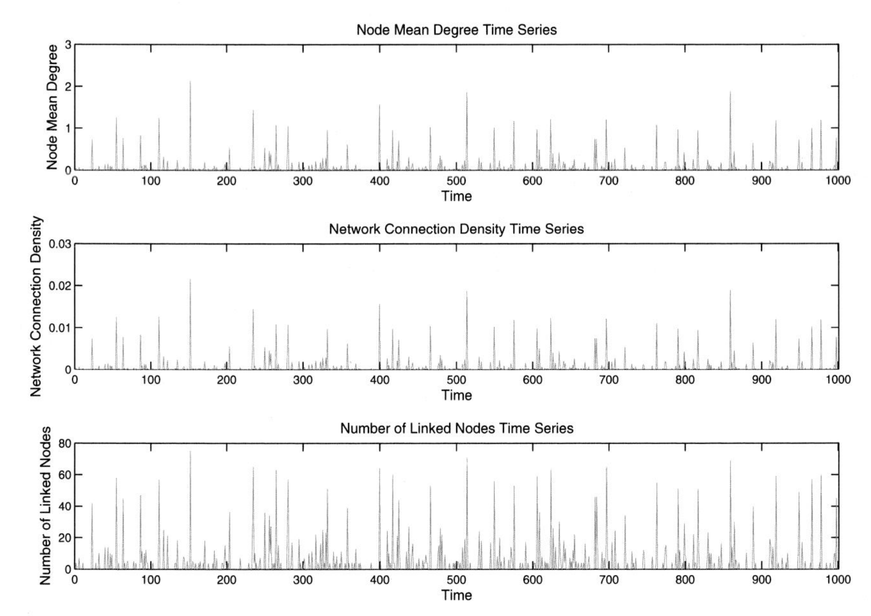

Figure 18.31 Network connectivity time series set.

connection density time series. Network connection density (also called connectance) is given by

$$\text{Network density} = \frac{\langle k \rangle}{n-1}$$

where

$\langle k \rangle$ = node mean degree

n = number of nodes in the network

Because node mean degree and network connection density differ only by a multiplicative factor, these two time series have the same form. The third time series provides the number of linked nodes at each time step.

Let's look at the time series values at the time steps that correspond to the largest propagation events in the simulation run: time steps 152, 514, and 859. At these time steps, the number of linked nodes ranges from 69 to 75, out of 100 nodes in the network. This is a very significant fraction of the total network. Although our model does not support a full network critical connectivity/percolation analysis,[6] the results here strongly suggest that

[6] A full network critical connectivity/percolation analysis requires consideration of multiple clusters at a simulation time step. If the multiple clusters should coalesce into a "giant cluster" that covers a major fraction of the network, then critical connectivity/percolation is achieved. Our model has one input and, therefore, develops just one cluster per time step. We can, however, observe that even a single cluster can approach a giant cluster at high-propagation time steps.

network critical connectivity is achieved at these three time steps. We know that, for a directed random network at critical connectivity, the theoretical value of node mean degree is 2. For our network, we see from Figure 18.31 that node mean degree is approximately 2 at all three of the "critical" time steps. We also see that network connection density is approximately 0.02 at these time steps.

So, at the highest propagation, highest density time steps, node mean degree is approximately 2 and network connection density is approximately 0.02. This is very *sparse* connectivity! As Stuart Kauffman (1995) has said in a somewhat similar context (Boolean networks), this result should "blow your socks off." Kauffman says further, "If the network is 'sparsely connected,' then the system exhibits stunning order. . . . Our intuitions about the requirements for order have, I contend, been wrong for millennia. We do not need careful construction; we do not require crafting. We require only that extremely complex webs of interacting elements are sparsely coupled."

In his *How Complex Are Complex Systems?* paper, Herbert Simon (1977) also talks about sparseness. "It will be convenient to represent the interconnectedness of a system by an incidence matrix [adjacency matrix], a matrix of zeros and ones, the $(i, j)^{th}$ element being 1 if the ith element interacts with the jth, and 0 otherwise. There are a number of different reasons why we might expect most real-world systems to have rather sparse incidence matrices." The "different reasons" involve the *hierarchical architecture* and the associated *nearly decomposable* property of complex systems, discussed earlier in this chapter.

Our network connectivity results here can be summarized as follows. Ecological network connectivity exhibits punctuated, fluctuating dynamics. Node degree is fractal. It appears that network critical connectivity is achieved at high-propagation time steps. All of this occurs in sparsely connected networks.

PART VI

Pulling It All Together

This part begins with a brief summary of the work covered in the book. We then take a broader and more interpretative look at the work with respect to the perspective taken, what we have found, what it means, and its potential influence on work in this area going forward. We see that, although our focus has been ecological systems, there are implications for all complex systems.

CHAPTER 19

An Increased Understanding of Complex System Dynamics

In this chapter, we pull everything together. The chapter begins (Section 19.1) with a brief summary of the key aspects of the work covered in the book. I then take a broader and more interpretative look at the work with respect to the perspective taken, what we have found, what it means, and its potential influence on work in this area going forward. This portion of the chapter (Sections **19.2–19.6**, respectively) covers: the substantial advantages of the systems and engineering perspective; the mechanism of complexity; the characteristics of complex ecosystem dynamics; complex system universal behavior; and a going-forward integrated view of complex system dynamics. We see that, although our focus has been ecological systems, there are implications for all complex systems.

19.1 SUMMARY OF THE WORK

Let's start with a brief summary of the work covered in this book. Part I of the book provides a comprehensive look at the systems and engineering perspective that is foundational to this effort. Consistent with that perspective, the work presented in the subsequent parts of the book is an effective blend of synthesis and analysis that encompasses the systems triad, network thinking, and complex systems theory. The result is a fresh and innovative view of complex system behavioral dynamics.

Early on, in Part II of the book, I *synthesize* a functional framework to set the context and direction for the work. In systems engineering, I have found that such a framework is essential for specifying and guiding the design and development of artificial (human-made) systems. In systems ecology, I believe that such a framework is equally essential for understanding natural systems (i.e., ecosystems). The framework recognizes that ecological systems consistently take the form of networks. The implementation architecture of ecosystems is the network. Next, in Part III, we conduct an extensive review of the pertinent extant complex systems theory in the areas of networks, nonlinear dynamics, cellular automata, and roughness (fractals). That serves

Understanding Complex Ecosystem Dynamics
http://dx.doi.org/10.1016/B978-0-12-802031-9.00019-X

as a valuable and necessary resource for the work of the remainder of the book. Then, in Part IV, we *synthesize* a comprehensive view of the propagation dynamics of ecological system networks. This view, in turn, is the basis for the central hypothesis of the book: ecological networks are ever-changing, "flickering" networks with propagation dynamics that are punctuated, fractal, local-to-global, and enabled by indirect effects. In Part V of the book, *analysis* takes the spotlight. I define, design, and develop an ecological network dynamics model to analyze and fully test the hypothesis. I make every effort to produce a realistic and effective ecological network propagation model. For example, the model is linked to an underlying existing and verified ecological network compartment model; the network nodes are spatially distributed and von Neumann node propagation neighborhoods are implemented; the preferential attachment behavior that is observed in many real-world networks is an integral part of the model. The resulting model software development effort is very substantial and includes not only a comprehensive implementation of the ecosystem propagation process, but also a full complement of analysis capabilities and graphics capabilities. We then generate results and analyze the dynamics of operational propagation flow, network propagation events, propagation path length, indirect effects, and network connectivity. Our hypothesis is fully tested—and corroborated.

Let's now take a broader and more interpretative look at the work with respect to the perspective taken, what we have found, what it means, and its potential influence on work in this area going forward. The objective of this book has been to increase our understanding of complex ecosystem dynamics. I think we have achieved that, and more. We have seen universality in system dynamical behavior that seems to apply, not only across ecological systems, but also to other types of complex systems—both natural systems and, in some cases, even artificial (human-made) systems. In my view, the material in this book can serve to increase our understanding of ecological systems specifically, and many other types of complex systems more generally.

I'll describe the broader significance of the work in five areas:

- An Advantageous Perspective
- The Mechanism of Complexity
- The Characteristics of Complex Ecosystem Dynamics
- Complex System Universal Behavior
- A Going-Forward Integrated View of Complex System Dynamics

These five important areas are addressed in Sections 19.2–19.6, respectively.

19.2 AN ADVANTAGEOUS PERSPECTIVE

The systems and engineering perspective is extremely beneficial for understanding complex systems. I am very well positioned in that regard. I had a 34-year professional career as a systems engineer at Bell Laboratories. Throughout my career, I worked to solve systems problems for Bell Labs and for its clients across the United States and around the world. During that time, I acquired a very significant set of skills and perspectives with respect to the practice of systems engineering and the systems approach. From that experience, I have developed "A Systems Engineering Skills Framework," which is described in Part I of the book. At Bell Labs, I focused on *building* human-made systems, but this same set of skills and perspectives can be applied very beneficially to *understanding* natural systems—specifically, natural ecological systems. That, of course, requires very substantial ecological system study and research. To that end, I have earned a PhD in systems ecology at the University of Georgia's Eugene P. Odum School of Ecology.

Other specifics of my background also play an important role in my current work. My early degrees (bachelor's and master's) were in electrical engineering. My master's degree work and early Bell Labs work included the areas of communication theory, signal processing theory, information theory, control theory, and network theory. My later work at Bell Labs was in software systems and technology. All of these disciplines and their associated methods and tools contribute beneficially to my current complex system dynamics work.

The combined set of experiences, approaches, methods, and tools employed in this book, I believe, have not before been applied to ecological systems or other complex natural systems. I have begun with substantial electrical engineering and systems engineering knowledge and experience; combined that with PhD-level study of ecology; and then further supplemented those resources with additional study of complex systems theory in the areas of networks, nonlinear dynamics, cellular automata, and roughness (fractals).

The modeling perspective used in this work is new and innovative in the field of ecology. My approach is to model complex ecological networks as discrete dynamic systems that emulate relevant features of real-world ecological network behavior. Traditionally, much ecological network modeling and analysis assumes a static network structure, steady state network stocks and flows, and (sometimes) linear system behavior. I, on the other hand, do not make these *a priori* assumptions about network dynamics. My objective is to model and analyze ecological networks in order to *determine* the nature of their dynamics. I let the dynamics develop as the model simulation runs.

19.3 THE MECHANISM OF COMPLEXITY

Stephen Wolfram has articulated an innovative and bold hypothesis (described in Chapter 11) on the underlying mechanism of highly complex behavior in systems. The mechanism is *simple rules, repeated over and over.* Wolfram (2002) says it is the mechanism "that allows nature seemingly so effortlessly to produce so much that appears to us so complex. . . . It takes only very simple rules to produce highly complex behavior." This hypothesis "implies a radical rethinking of how processes in nature and elsewhere work." It seems to apply "to systems throughout the natural world and elsewhere."

Benoit Mandelbrot was also aware, by the 1980s, of the mechanism of simple programs/simple rules repeated over and over. Mandelbrot (1982) talks about biological form and simple mechanisms in his classic book on fractal geometry. "Biological form being often very complicated, it may seem that the programs that encode this form must be very lengthy. . . . However, the complications in question are often highly repetitive in their structure." The generating rule can be systematic and simple. "The key is that the rule is applied again and again, in successive loops."

In Chapter 7, *Evolution and Universal Development Concepts*, we made the case that the evolution process model can be considered a universal development model. This model is essentially a "simple program" that operates with "simple rules." The program and rules are repeated again and again. The evolution/development process seems consistent with the hypothesized *mechanism of complexity.*

As I explained in Chapter 16, my ecological network dynamics model employs the mechanism of simple rules repeated over and over. For propagation, there is a simple node stock threshold and simple node-to-node flow patterns. The propagating from-node "looks" closer for available to-nodes before looking farther. When there are multiple available to-node candidates within a given neighborhood, the from-node looks for the more preferred to-node. Simply stated, for propagation, closer is better and more preferred is better. Using these simple rules, my network model generates highly complex dynamical behavior that matches reality.

19.3.1 Probabilistic Context

The node-to-node interactions that take place in complex system networks follow simple rules repeated over and over, and, in my view, they often occur in a probabilistic context (Chapter 14). William Drury and Melanie Mitchell support that view. With respect to natural systems, William Holland Drury Jr. (1998) says, "A first principle is that chance and change are the rule." Both "chance and change are ubiquitous. . . . Nature works on the basis of one-on-one species interactions, variability, and chance." Melanie Mitchell (2009) says, in complex systems, "it appears that . . . intrinsic random and probabilistic elements are needed in order for a comparatively small population of simple components . . . to explore an enormously larger space of possibilities."

My ecological network dynamics model (Chapters 16–18) reflects the probabilistic nature of real-world ecological networks. The model is probabilistic in many key respects, that is, in the spatial distribution of nodes (and, therefore, their compartments); in the establishment of model network initial conditions; in the application of environmental inputs to network input nodes; in the node-to-compartment attachment preferences; and in the node-to-node attachment preferences. The probabilistic nature of the model (and the simulation runs) authentically reflects the "chance" aspects of real ecosystems. Each simulation run is different due to its probabilistic nature, but all runs that I have made demonstrate the same general behaviors.

19.3.2 *Optimal vs. Adequate*

Is the complex behavior, which results from the simple-rules mechanism, optimal or just "good enough"? It is good enough (Chapter 14).

Biological system development does not require "maximum fitness" outcomes, but rather "good enough" outcomes. These are outcomes that work but are not optimum solutions. Stephen Wolfram (2002) has said, "In the past, the idea of optimization for some sophisticated purpose seemed to be the only conceivable explanation for the level of complexity that is seen in many biological systems." But no, it seems optimization is not the reason for the extreme complexity of living systems. It turns out that complexity is relatively "easy" to obtain. The simple-rules mechanism very frequently produces behavior of great complexity in nature.

Why not optimality? There is the issue of feasibility. Finding an optimum solution requires a comprehensive search through a "design" space (the space of all possible solutions). That search is often not practical and, for high-functioning natural systems, very likely not possible. So, perhaps a natural system "tries" (at random) a simple program with simple rules. If it works satisfactorily, it is used. A "best" simple program cannot, in any practical sense, be determined *a priori*. One has to choose a program, run it, and see what happens. If it "works," then it may be good enough. If not, "selection" will find one that is. Herbert Simon (1996) has pointed out that, even for artificial systems, an optimization approach for finding the "best" solution is often not realistically feasible. One turns "to procedures that find good enough answers to questions whose best answers are unknowable."

19.4 THE CHARACTERISTICS OF COMPLEX ECOSYSTEM DYNAMICS

In the previous section, we argue that the mechanism of simple rules repeated over and over appears to generate the extremely complex behavior we see in systems everywhere. The next question, then, is: How can the resulting extremely complex system behavior be characterized? To answer that question, we have synthesized a characteristics hypothesis (Chapter 15). Our modeling and analysis results (Chapters 16–18) fully and comprehensively test and corroborate the hypothesized characteristics, which are:

- Flickering Networks
- Punctuated Dynamics
- Fractal Behavior
- Local-to-Global Interaction
- Indirect Effects

Fractal behavior is perhaps the dominant characteristic of complex network dynamics. It is central to our understanding of these dynamics. The other hypothesized characteristics are related, in various ways, to fractal behavior. In addition, we have investigated the network connectivity traits associated with the full set of dynamics characteristics.

I will summarize all that we have found, in five analysis categories, in the following five paragraphs.

We have analyzed *operational propagation flow* for a range of propagation events from very small to very large. Several points are evident. Flow behavior over time is punctuated. The smaller propagation events are mostly local. Many of the midsize events and all observed large events exhibit broad network span and local-to-global propagation behavior. Cycling is evident in the flow diagrams. The network continually changes with time, and sometimes dramatically so. The view of the network operational dynamics is clear: we are dealing with ever-changing, "flickering" ecological networks.

Our *network propagation event* analysis results complement the propagation flow results, and go further. Network propagation event time series indicate punctuated and local-to-global dynamics, with many small events but also medium, large, and even a few very large propagation events. The corresponding event distributions confirm power-law dynamics. Network propagation events, therefore, exhibit process fractal behavior. The event frequency spectrum strongly suggests "*pink 1/f noise*" and associated process fractal behavior as well.

To capture the network time-varying *path length* dynamics, I devised an analysis approach that integrates over both space and time. Examination of the resulting time series and distribution confirms that path length exhibits punctuated, local-to-global, and fractal characteristics. We checked the path length distribution against a mathematically derived condition for dominance of indirect effects. The path length distribution satisfies that condition. Accordingly, indirect effects should dominate direct effects in our model network. They do.

To analyze ecological network *indirect effects* dynamics, I defined several indirect effects indicators, calculated their values at each simulation-run individual time step, and plotted the resulting collection of time series. We have seen that the time series are highly punctuated, with no sign of gradual or continuous behavior. We have seen, at time steps of larger propagation events, that indirect effects far exceed direct effects. The number of indirect paths can be 5-10 or more times the number of direct paths. Given the large number of indirect paths and the global span of network propagation in these cases, we can conclude that indirect effects are enablers of local-to-global processing. In addition to their dominance at high-propagation time steps, are indirect effects dominant overall? To answer that question, I developed indicator over-time cumulative time series. Results show that cumulative direct paths are <20% of total paths and cumulative indirect paths are >80% of total paths. Indirect effects are dominant overall. We have also generated path quantity time series and distributions. Those results show that indirect path quantity is fractal. In summary, for the model network, indirect effects are dominant—and they are punctuated, they are enablers of local-to-global propagation, and they exhibit fractal behavior.

Network connectivity also exhibits punctuated, fluctuating dynamics. Node degree (the number of connections to/from a node) is an important indicator of network connectivity. Using three-dimensional node degree grids, we have observed dramatically changing node degree behavior over time. Those fluctuating dynamics must be captured in our node degree time series/distribution analyses. To do that, I used an integrated spatial/temporal approach very similar to the path length approach (applied earlier). The resulting time series is punctuated and the power-law distribution indicates node degree fractal dynamics. Other important network connectivity traits were also examined. Our analysis results strongly suggest that network critical connectivity is achieved at time steps that correspond to the largest propagation events. At these high-propagation time steps, node mean degree is approximately 2 and network connection density is approximately 0.02. This is very *sparse* connectivity (as expected; see Chapter 18). In summary, our overall results indicate that ecological network connectivity exhibits punctuated, fluctuating dynamics. Node degree is fractal. Network critical connectivity is likely achieved at high-propagation time steps. All of this occurs in sparsely connected networks.

19.4.1 Model Robustness

Our network dynamics model that has produced all of the previous results has proven robust and reliable. The general behaviors of the model seem to be independent of network and analysis specific details. The model behaviors in all of the simulation runs I have made are insensitive to changes in specific network parameters and settings (e.g., node propagation threshold, node propagation flow quantity). Model results are not sensitive to variations in initial conditions. Analysis results are robust with respect to choice of measure of the instantaneous dynamics of a given network propagation parameter. It seems that any reasonable measure will do; for example, when determining path length dynamics or node degree dynamics, even per-time-step mean measures can appropriately capture the instantaneous dynamics.

19.4.2 Descriptive vs. Predictive Results

Our complex network dynamics results are essentially *descriptive*, not *predictive*. That is the best one can do for highly complex systems.

A goal of traditional science, for the past several centuries, has been to predict system behavior over time, preferably via concise and elegant mathematical formulations. It turns out that goal can be realized only for relatively simple systems. For highly complex systems, mathematical predictive success is limited. As we have seen, however, mathematics can be very useful in a descriptive sense—describing, explaining, and characterizing the behavior of complex systems. For example, we can describe and evaluate the probabilities of occurrence of system events, although specific times of occurrence and event sizes (from small to extreme) cannot be predicted with certainty.

Complex systems are largely *unknowable* in specific predictive terms. As discussed in Chapter 11, unknowability can involve irreducibility, undecidability, and intractability.[1] I'll very briefly summarize here. Most great historical triumphs of theoretical science have found some reduced description (usually a mathematical formula) of a system's behavior that seems able to predict behavior over time. But highly complex systems have no reduced description. Rather, they exhibit *computational irreducibility*. For such systems, one cannot describe their behavior except by computing all (or almost all) steps of the system evolution in time. We have an irreducible amount of computation. Irreducibility can yield undecidability and intractability. We are unable to determine/predict the ultimate outcome of system behavior. The best one can do is to let the system run and see what happens.

The mechanism of complexity is driven by simple rules. Per Wolfram (2002), models with simple rules can "capture the essential features ... of systems with very complex behavior .... Given these models the only way to find out what they do will usually be just to run them." That is what I have done in my work.

19.5 COMPLEX SYSTEM UNIVERSAL BEHAVIOR

The phenomenon of complexity is universal. It can be found everywhere in nature and elsewhere—in ecological, biological, physical, and even some artificial systems. The hypothesized mechanism of complexity seems universal. That mechanism is simple rules, repeated over and over. It can be found in all sorts of processes and systems, independent of the details of those specific systems. In Chapter 7, for example, we have suggested and demonstrated that the preeminently fundamental evolution process is a universal development process that operates in many system development scenarios. The evolution process mechanism is simple rules repeated again and again. The evolution/development process has extremely broad applicability, and applies across spatial and temporal scales and across system domains (natural and artificial) and system types. The process and the process mechanism appear to be universal.

The complex system fractal behavior characteristic and the related characteristics (flickering networks, punctuated dynamics, local-to-global interaction, and indirect effects) that I have hypothesized, as well as the associated network connectivity traits, also seem to be universal.

[1] These ideas are from Stephen Wolfram (2002).

Let's focus for a few moments on the complex system characteristics hypothesis. First, consider the universal aspects of fractal behavior. It seems that most things in the natural world—both structures and processes—are fractal. Process fractal behavior is the fundamental form of behavior that is observed over and over again in complex system dynamics. Accordingly, "normal" (small, gradual) events occur most often, but "extreme" (large, abrupt) events also occur. These complex system events have punctuated time series and power-law probability distributions. Recall the process fractal self-similarity and cross-similarity discussions from Chapter 15. Process fractals have the property of scale-invariance, that is, self-similarity across scales. Their event distributions indicate self-similarity across event "size" (spatial) scales. Their time series indicate self-similarity across time scales. This is a clear reflection of universal behavior. It is independent of spatial scale and time scale. One can observe the same form of behavior at all scales (within the fractal validity range of a given process). Going even further, we discussed fractal behavior similarity with respect to *very different* processes that occur at *vastly different* spatial frames and time frames. I used the term *cross*-similarity for this phenomenon. We saw that the dynamics of short-term operational processes that occur in smaller spaces can match the dynamics of long-term developmental processes that occur in much larger spaces. We saw equivalence in system dynamics—*universality*—across process, space, and time. The point is that fractal behavior is very widespread in nature. It is a universal concept. It occurs across a wide spectrum of natural system structures and processes and across a wide spectrum of parameters within any given system. As Benoit Mandelbrot (2004) has said, "fractals . . . are present everywhere."

The complex network connectivity traits associated with the behavioral dynamics characteristics also appear universal. Consider the interaction patterns among nodes in complex networks. In the results presented in Chapter 18, we observed that ecological network *operational adjacency matrices* show node-to-node connections distributed throughout the network, but with a particular concentration near the matrix major diagonal. Furthermore, ecological network *intensity matrices* have a similar underlying distribution pattern and show that the highest flow intensities also tend to concentrate near the matrix major diagonal. Herbert Simon (1977) has explained that these effects are seen in many real-world complex systems and seem universal. In Chapter 18, we also saw that complex ecological networks are very *sparsely connected*. Stuart Kauffman (1995) suggests that such connectivity sparseness is a universal property that is associated with order and complexity in networks. Herbert Simon (1977) agrees that sparse connectivity is an expected property of most real-world complex systems.

The behavioral dynamics and the suggestions of universality we are seeing are not limited to ecological systems. It seems that our findings may also apply to a wide range of complex systems across other disciplines and subject areas as well. We have larger emerging universal principles that may be true for complex systems in general. In my view, the material in this book can serve to increase our understanding of ecological systems specifically, and many other types of complex systems more generally.

19.6 A GOING-FORWARD INTEGRATED VIEW OF COMPLEX SYSTEM DYNAMICS

At the outset of the book, I explained that my systems work is always a blend of synthesis and analysis, although at some stages of the work it may be appropriate to emphasize one over the other. Accordingly, in the earlier portions of the book, I mostly emphasized synthesis. In the later (most recent) chapters, I have focused mostly on analysis. I would like to conclude the book with one more flourish of synthesis.

I think we have the necessary ingredients for proposing a new going-forward integrated perspective on complex system dynamics. We know that complexity is widespread and universal. We have two highly relevant, and apparently universal, hypotheses for understanding complex system behavior: the complexity *mechanism* hypothesis and the dynamics *characteristics* hypothesis.[2] Stephen Wolfram (2002) has provided very substantial corroborating evidence for the repeated-simple-rules mechanism hypothesis. Benoit Mandelbrot (2010a) has supported this concept and has said that "bottomless wonders spring from simple rules . . . repeated without end." The wonders he was talking about are fractals. That brings us to the fractal-based dynamics characteristics. We have spent all of Part V of this book comprehensively testing and corroborating the characteristics hypothesis.

We can combine the two hypotheses and formulate an integrated dynamics hierarchy in the context of the *systems triad*. The systems triad is a major element of the systems approach. We first discussed the triad in Part I and have revisited and relied upon it several times since then in the book. Figure 19.1 presents the systems triad view of a going-forward integrated

[2] Merriam-Webster online provides definitions for the terms: *Mechanisms* are the fundamental processes involved in or responsible for an action, reaction, or other natural phenomenon. *Characteristics* are distinguishing traits, qualities, or properties. (Note that in my use of the term *mechanism*, I specifically am not referring to the Cartesian/Newtonian mechanism worldview and philosophy.)

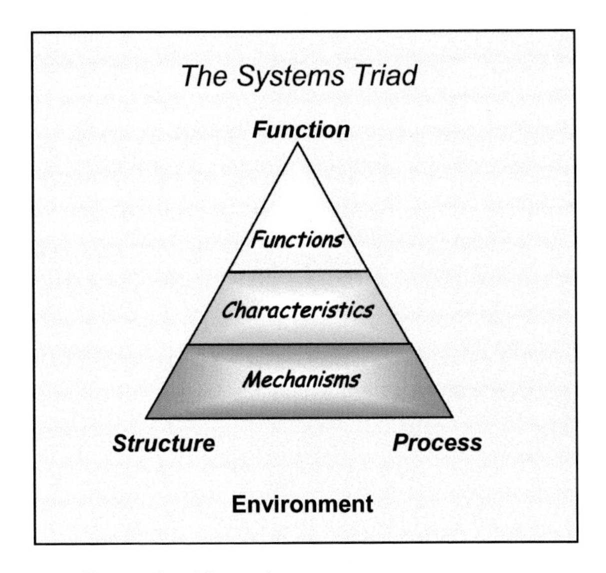

Dynamics Hierarchy:

- Simple-rules *mechanisms* produce process-over-structure flow.
- Outcomes of process-over-structure flow (events) exhibit fractal-based behavioral *characteristics*.
- One or more fractal-based event sets (composite event sets) deliver system *functions*.

Figure 19.1 An integrated complex system dynamics hierarchy.

perspective on complex system dynamics. As indicated in the figure, process-over-structure flow is generated by repeated-simple-rules mechanisms—the *mechanism* hypothesis. Outcomes of process-over-structure flow (i.e., events) have fractal-based behavioral characteristics—the *characteristics* hypothesis. Sets of events (composite event sets) deliver system *functions*. The systems triad brings together and integrates the two component hypotheses with a strong systems grounding.

Here is something interesting to think about. David Bohm's work on *implicate order* seems consistent with the Figure 19.1 view of complex system dynamics. Bohm (1983) defines implicate order as *enfolded* order. (According to Bohm, *implicate* is "from a Latin root meaning 'to enfold' or 'to fold inward.'") Bohm proposes that implicate order "be taken as fundamental" – and that "what is primary, independently existent, and universal has to be expressed in terms of the implicate order." He says further that the "laws" of implicate order "are only vaguely known" and "may even be ultimately unknowable in their totality." Bohm's work suggests the following to me. Perhaps the order of highly complex systems can be considered to be enfolded implicate order. The simple order of simple rules repeated over and over is enfolded into the implicate order of developed complex systems. This implicate order is not well understood and perhaps is not fully knowable to us. Accordingly, per Figure 19.1, implicate order outcome *characteristics* can be identified; however, specific outcomes are "not fully knowable to us" and cannot be predicted (as we discussed in Section 19.4 of this chapter). In addition, with respect to the joint occurrence of complexity and fractals, Bohm and Peat (2000) say that the implicate order of complexity "is in a close relationship to that of fractals in the sense that, in both, there is a kind of whole generated from certain basic principles." The "basics" are simple rules repeated over and over. The "whole" is characterized by fractal behavior.

In my opinion, the work of Wolfram and the work of Mandelbrot come together very nicely. The extreme complexity observed by Wolfram in his cellular automata explicit experimentation and the bottomless wonders observed by Mandelbrot in his fractal investigations have the same generating mechanism. The same fundamental simple-rules mechanism produces complexity and fractals. Complexity is widespread in nature. Fractals are widespread in nature. Why? It is because they have a widespread common source. When we add to those findings my new work on the behavioral characteristics of complex system dynamics, I think we get a compelling total picture.

In my view, the integrated perspective of Figure 19.1 encapsulates and unifies emerging complex system principles and could prove extremely valuable for guiding and setting the context for ongoing work in complex system dynamics.

Herbert Simon (1996) has said that a central task of natural science is "to show that the wonderful is not incomprehensible." When a certain level of comprehension is achieved, "a new wonder arises at how complexity was woven out of simplicity." The work described in this book is quite consistent with those thoughts.

We began the book, in the Preface, with Einstein's advice to "look deep, deep into nature, and then you will understand everything better." I think that now I really do understand everything better. I hope that you do too.

APPENDIX

Selections from the Dynamics Model Programming Code

Complex system dynamics modeling is an important part of my work and an important part of this book. I want readers to have access to my complex ecosystem dynamics model and the complete MATLAB[1] programming code. The reader is invited to explore, run, and experiment with the model software. One can, for example, change model parameters and even change the code, and then generate the associated results. The model provides a tool for enhancing understanding of complex system network dynamics. Readers can test further the ideas in this book, as well as develop and test their own ideas.

The full complex ecosystem dynamics model code is available on the book's companion website. The programming code is heavily commented to explain and describe the software.

Here, in the Appendix, I provide selected excerpts of the MATLAB code that implements the model software. I refer to this material as I discuss the model in the main body of the book. These excerpts provide the reader with easy access to examples of the software without having to navigate through the online full set of MATLAB m-files while reading the book. There are nine code excerpts ranging from code that establishes the model network structure and relationships, to code that describes and implements propagation process flow, to code that describes and implements ecosystem dynamics analysis activities.

The complete list of excerpts from the dynamics model programming code follows:

Code example 1: Assigning nodes to compartments
Code example 2: Creating the adjacency-type matrix (AAT)
Code example 3: Propagation process "pseudocode"
Code example 4: Creating node-and-link propagation flow diagrams
Code example 5: Developing the network propagation event distribution

[1] MATLAB release R2009a, The MathWorks, Inc., February 12, 2009.

Code example 6: Developing the network propagation event frequency spectrum
Code example 7: Generating node degree vector arrays
Code example 8: Generating node degree grid arrays
Code example 9: Developing a node degree distribution

CODE EXAMPLE 1

Assigning Nodes to Compartments

```
% NCG = node-compartment grid matrix (10x10)
% in which the elements identify a node's
% home compartment number.

NCG = randi(10,10);
NCG =
   9  2  7  8  5  3  8  9  4  1
  10 10  1  1  4  7  3  3  9  1
   2 10  9  3  8  7  6  9  6  6
  10  5 10  1  8  2  7  3  6  8
   7  9  7  1  2  2  9 10 10 10
   1  2  8  9  5  5 10  4  3  2
   3  5  8  7  5 10  6  2  8  6
   6 10  4  4  7  4  2  3  8  5
  10  8  7 10  8  6  2  7  4  1
  10 10  2  1  8  3  3  5  6  4

% QNinC = quantity of nodes in each compartment
% (10x1)

  QNinC' = 9 11 10 8 8 9 10 12 8 15
```

CODE EXAMPLE 2

Creating the Adjacency-Type Matrix (AAT)

```
%% Create a node adjacency-type
%% "adjacency matrix" (AAT)

% AA = network node basic adjacency matrix for
% "candidate" connections (100x100)

% AAT = adjacency adjacency-type matrix
% (100x100) (element values are 1, 2, 3)

% AAT provides the adjacency type of "to" nodes
% type 1 = local neighbor
% type 2 = extended local neighbor
% type 3 = non-local neighbor
```

```
AAT = zeros(100,100);  % preallocation
for j = 1:100  % columns ("from" nodes)
    N2temp = find(AA(:,j)==1);  % "to" nodes
    for k = 1:sum(AA(:,j))  % rows ("to" nodes)
        % grid comparison
        if logical(abs(N2temp(k) - j) == 1) ||...
               logical(abs(N2temp(k) - j) == 10)
            AAT(N2temp(k),j) = 1;
        elseif logical(abs(N2temp(k) - j) == 2) ||...
                   logical(abs(N2temp(k) - j) == 20)
            AAT(N2temp(k),j) = 2;
        else
            AAT(N2temp(k),j) = 3;
        end
    end
end
```

CODE EXAMPLE 3

Propagation Process "Pseudocode"

```
High-level view of the program flow for the overall
propagation process
This is a narrative "pseudocode-like" rendering of the flow

% Time step loop follows
for i = 1:NumTS
% NumTS = number of model simulation time steps
perform input node processing
if input node stock < thd, continue
% thd = node stock threshold for propagation
prop-event-node-set = input node
    % Stage loop follows
    for j = 1:100
    NumCStages = j - 1
    % NumCStages = current number of completed prop stages
    if prop-event-node-set is empty, break
    perform propagating node processing
        % Node propagation event loop follows
        for k = 1:size of prop-event-node-set
        available flow F = npfq
        % npfq = node propagation flow quantity
        from-node = prop-event-node-set(k)
        if from-node is an output node
        perform output node processing
        F = F - 1
            % Neighborhood selection loop follows
            for u = 1:3
```

```
            candidate to-node-set = to nodes with adjacency type u
            if to-node-set is empty, continue
            find to-node-set compartment numbers
            find ranks of to-node-set compartment numbers
                % Compartment selection loop follows
                for v = 1:QNCC(from-node)
                % QNCC = quantity of node connecting compartments
                revise to-node-set
                % include only the nodes in compartment with rank v
                if revised to-node-set is empty, continue
                    % Node selection loop follows
                    for w = 1:size of revised to-node-set
                    find revised to-node-set attachment pref strengths
                    find attachment pref probability set
                    determine selected-node
                    % use mnrnd(1, probability set)
                    % propagate to selected-node
                    perform node propagation instance processing
                    F = F - 1
                    if F = 0, break
                    revise to-node-set by removing selected-node
                    % proceed until propagate to all nodes in the
                    % set or until flow is exhausted
                    end
                    if F = 0, break
                    end
                if F = 0, break
                end
        end
    check for new node propagation events [>= thd]
    populate prop-event-node-set
    end
NumCStages_TS(i) = NumCStages
% NumCStages_TS = number of propagation stages in each
% completed time step
end
```

CODE EXAMPLE 4

Creating Node-and-Link Propagation Flow Diagrams

```
%% Propagation flow at individual time steps
% AAO_t_TS = per-time-step operational node adjacency
% multidimensional array (100x100xNumTS)
% NumTS = number of model simulation time steps

% tsn = time step number
```

```
% establish axes
axis square
axis ([1 21 1 21])
axis off
hold on

% plot node grid
xx= NCoord(:,1);
yy= NCoord(:,2);
plot(xx,yy,'o')
hold on

% plot propagation flow
gplot(AAO_t_TS(:,:,tsn),NCoord,'-ok')
hold on

% color propagation event nodes dark blue:
% RGB 11, 132, 199

% plot and color propagating nodes (red)
% PNCoord = propagating node coordinates
% PNL_t_TS = per-time-step multidimensional array
% of individual time step (stage-cumulative)
% propagating-node-logical matrices
% (10x10xNumTS) (matrix values are 0, 1)

PNCoord = NCoord;
PNCoord(find(PNL_t_TS(:,:,tsn) == 0),:) = [ ];
xx= PNCoord(:,1);
yy= PNCoord(:,2);
plot(xx,yy,'o','MarkerFaceColor', ...
    [.824 .004 .216],'MarkerSize',12)
hold on

% plot and color input node (green)
% INCoord = input node coordinates
% INSL_t_TS = per-time-step
% input-node-selection-logical multidimensional
% array (10x10xNumTS) (matrix values are 0, 1)

INCoord = NCoord;
INCoord(find(INSL_t_TS(:,:,tsn) == 0),:) = [ ];
xx= INCoord(:,1);
yy= INCoord(:,2);
plot(xx,yy,'o','MarkerFaceColor', ...
    [0 1 .392],'MarkerSize',12)
hold off
```

CODE EXAMPLE 5

Developing the Network Propagation Event Distribution

```
%% Develop network propagation event distribution

% Develop a network propagation event distribution
% (number of events vs. size of events) as follows:
%     Define a size interval (>= 1) and partition the
%        NetPESize_TS domain into intervals.
%     Count the number of events in each interval.
%     Generate a distribution with the ordered event
%        size intervals as abscissa and the number of
%        events in each of those size intervals as
%        ordinate.

% nsi = network propagation event size interval
% (scalar)

% nintervals = number of intervals (scalar)
% (temporary variable)

% NetPESize_TS = per-time-step
% network-propagation-event-size vector (1xNumTS)

% NetPEDistr = network propagation event distribution
% matrix (#size intervals x 2) (column 1 contains the
% ordered event size intervals and column 2 contains
% the number of events in each of those size
% intervals)

% set and initialize variables
nsi = 10;

nintervals = ceil(max(NetPESize_TS) / nsi);
% number of intervals rounded up

NetPEDistr = zeros(nintervals, 2);

% populate distribution matrix
for j = 1: nintervals
    NetPEDistr(j, 1) = j * nsi;
    NetPEDistr(j, 2) = sum((j - 1) * nsi < ...
        NetPESize_TS & NetPESize_TS <= j * nsi);
end
```

CODE EXAMPLE 6

Developing the Network Propagation Event Frequency Spectrum

```
%% Develop network propagation event
```

```
%% frequency spectrum

% Take the discrete Fourier transform (DFT) of the
% event time series to obtain its frequency spectrum.
% Use the MATLAB fast Fourier function (fft) to
% accomplish that. The procedure follows.

% NetPESize_TS = per-time-step
% network-propagation-event-size vector (for this
% frequency spectrum procedure, remove
% non-propagation events)

% Lsig = length of signal time series (scalar)

% NFFT = next power of 2 up from Lsig (scalar)

% PEdft = propagation event discrete
% Fourier transform vector (1 x NFFT)

% PEssAmp = propagation event single-sided amplitude
% vector (1 x NFFT/2+1)

% normfreq = vector of normalized frequency values
% (1 x NFFT/2+1)

% numpoints = number of frequency spectrum points
% for partial spectrum (scalar)
% (the maximum number of points available = NFFT/2+1)

% freq = vector of normalized frequency values
% for partial spectrum (1 x numpoints)

% amp = vector of amplitude values for
% partial spectrum (1 x numpoints)

Lsig = size(NetPESize_TS, 2);
NFFT = 2^nextpow2(Lsig);
PEdft = fft(NetPESize_TS, NFFT)/Lsig;
PEssAmp = 2*abs(PEdft(1:NFFT/2+1));
normfreq = 100*linspace(0,1,NFFT/2+1);
numpoints = 50;
freq = normfreq(1:numpoints);
amp = PEssAmp(1:numpoints);
```

CODE EXAMPLE 7

Generating Node Degree Vector Arrays

```
%% Develop individual-time-step node degree vector
%% arrays

% Develop multidimensional arrays that provide the
% out-degree,in-degree, and combined-degree (both out
```

```
% and in) of each node at each individual time step
% vs. time. Arrays are (1x100xNumTS).

% NodeODeg_t_TS = node-out-degree multidimensional
% array (1x100xNumTS)

% NodeIDeg_t_TS = node-in-degree multidimensional
% array (1x100xNumTS)

% NodeCDeg_t_TS = node-combined-degree
% multidimensional array (1x100xNumTS)

% Each of the above three variables provides a node
% degree vector (out, in, or combined) for each
% individual time step per time step.

% NumTS = number of model simulation time steps

% AAO_t_TS = per-time-step operational node adjacency
% multidimensional array (100x100x NumTS)

% preallocation and initialization
NodeODeg_t_TS = zeros(1,100, NumTS);
NodeIDeg_t_TS = zeros(1,100, NumTS);
NodeCDeg_t_TS = zeros(1,100, NumTS);

% calculate the vector values
% External connections (inputs and outputs) with the
% environment are not counted in these node degree
% calculations
for i = 1: NumTS
   NodeODeg_t_TS(1, :, i) = sum(AAO_t_TS(:, :, i));
   NodeIDeg_t_TS(1, :, i) = ...
       sum(transpose(AAO_t_TS(:, :, i)));
   NodeCDeg_t_TS(1, :, i) = NodeODeg_t_TS(1, :, i)...
       + NodeIDeg_t_TS(1, :, i);
end
```

CODE EXAMPLE 8

Generating Node Degree Grid Arrays

```
%% Develop individual-time-step node degree
%% grid arrays

% Develop multidimensional arrays that provide
% out-degree, in-degree, and combined-degree node
% grids at each individual time step vs. time.
% The arrays are (10x10xNumTS).

% NodeODegGrid_t_TS = node-out-degree-grid
```

```
% multidimensional array (10x10xNumTS)

% NodeIDegGrid_t_TS = node-in-degree-grid
% multidimensional array (10x10xNumTS)

% NodeCDegGrid_t_TS = node-combined-degree-grid
% multidimensional array (10x10xNumTS)

% Each of the above three variables provides a
% node degree grid (out, in, or combined) for each
% individual time step per time step.

% preallocation and initialization
NodeODegGrid_t_TS = zeros(10,10, NumTS);
NodeIDegGrid_t_TS = zeros(10,10, NumTS);
NodeCDegGrid_t_TS = zeros(10,10, NumTS);

% calculate the grid values
% External connections (inputs and outputs) with
% the environment are not counted in these
% node degree calculations
for i=1:NumTS  % time steps
    for j=1:10  % columns
        for k=1:10  % rows
            NodeODegGrid_t_TS(k, j, i) = ...
                NodeODeg_t_TS(1, k+10*(j-1), i);
            NodeIDegGrid_t_TS(k, j, i) = ...
                NodeIDeg_t_TS(1, k+10*(j-1), i);
            NodeCDegGrid_t_TS(k, j, i) = ...
                NodeCDeg_t_TS(1, k+10*(j-1), i);
        end
    end
end
```

CODE EXAMPLE 9

Developing a Node Degree Distribution

```
%% Develop a node in-degree distribution for
% the entire simulation run

% Note that our approach here has both spatial
% and temporal dimensions.
% We investigate individual-time-step spatial events
% and integrate them over time.

% Create a two-column node degree distribution
% matrix. Column 1 contains the node degree
% intervals and column 2 contains the number of
% events in each of those intervals.
```

```
% MaxIDDistr = maximum in-degree distribution matrix
% (#size intervals x 2)

% mdsi = maximum node degree size interval (scalar)

% nmdintervals = number of maximum node degree
% intervals (scalar) (temporary variable)

% IDmax_t_TS = vector (1x NumTS) that provides the
% maximum in-degree in each time step per time step


% set and initialize out-degree variables
mdsi = 1;

nmdintervals = ceil(max(IDmax_t_TS) / mdsi);
% number of intervals rounded up

MaxIDDistr = zeros(nmdintervals, 2);


% populate in-degree distribution matrix
for j = 1: nmdintervals
    MaxIDDistr(j, 1) = j * mdsi;
    MaxIDDistr(j, 2) = sum((j - 1) * mdsi < ...
        IDmax_t_TS & ...
        IDmax_t_TS <= j * mdsi);
end
```

REFERENCES

Albert, R., Barabási, A.-L., 2002. Statistical mechanics of complex networks. Rev. Mod. Phys. 74 (1), 47–97.

Albert, R., et al., 2000. Error and attack tolerance of complex networks. Nature 406, 378–382.

Alessi, S., 2000. Thoughts on the extent of human systems within systems engineering. INCOSE Insight 3(1).

Alexander, C., 1964. Notes on the Synthesis of Form. Harvard University Press, Cambridge, MA.

Alexander, C., 1979. The Timeless Way of Building. Oxford University Press, New York, NY.

Alexander, C., et al., 1977. A Pattern Language: Towns, Buildings, Construction. Oxford University Press, New York, NY.

Arora, J.S., 1989. Introduction to Optimum Design. McGraw-Hill, New York, NY.

Bacon, F., 1902. Novum Organum. American Home Library Company, New York, NY [first published in1620].

Bak, P., 1996. How Nature Works—The Science of Self-Organized Criticality. Copernicus (Springer-Verlag), New York, NY.

Bak, P., Sneppen, K., 1993. Punctuated equilibrium and criticality in a simple model of evolution. Phys. Rev. Lett. 71 (24), 4083–4086.

Bak, P., et al., 1987. Self-organized criticality: an explanation of 1/f noise. Phys. Rev. Lett. 59 (4), 381–384.

Barabási, A.-L., 2002. Linked: The New Science of Networks. Perseus Publishing, Cambridge, MA.

Barabási, A.-L., Albert, R., 1999. Emergence of scaling in random networks. Science 286 (5439), 509–512.

Barenboim, D., 2006. In the beginning was sound. British Broadcasting Corporation (BBC) Reith Lectures.

Barnes, J., 2000. Aristotle: A Very Short Introduction. Oxford University Press, New York, NY.

Barnsley, M.F., 1993. Fractals Everywhere, second ed. Morgan Kaufmann (an imprint of Elsevier), San Diego, CA.

Bar-Yam, Y., 2003. Dynamics of Complex Systems. Westview Press, Boulder, CO.

Bar-Yam, Y., 2004. Making Things Work—Solving Complex Problems in a Complex World. NECSI Knowledge Press, Cambridge, MA.

Bar-Yam, Y., Epstein, I.R., 2004. Response of complex networks to stimuli. Proc. Natl. Acad. Sci. U. S. A. 101 (13), 4341–4345.

Benton, M.J., 1993. The Fossil Record 2. Chapman and Hall, London. http://www.fossilrecord.net/fossilrecord/summaries.html.

Benton, M.J., 1995. Diversification and extinction in the history of life. Science 268 (5207), 52–58.

Bloomer, C.M., 1976. Principles of Visual Perception. Van Nostrand Reinhold, New York, NY.

Bohm, D., 1983. Wholeness and the Implicate Order. Ark Paperbacks, London.

Bohm, D., Peat, F.D., 2000. Science, Order, and Creativity, second ed. Routledge (Taylor & Francis Group), New York, NY.

Bolles, E.B. (Ed.), 1999. Galileo's Commandment: 2,500 Years of Great Science Writing. Henry Holt and Company, New York, NY.

Bonacich, P., 2003. Cellular automata for the network researcher. J. Math. Soc. 27, 263–278.

Box, G.E.P., Draper, N.R., 1987. Empirical Model-Building and Response Surfaces. Wiley, Hoboken, NJ.

Broder, A., et al., 2000. Graph structure in the web. Comput. Netw. 33, 309–320.

Buede, D.M., 2000. The Engineering Design of Systems: Models and Methods. Wiley, Hoboken, NJ.

Butterfield, H., 1960. The Origins of Modern Science. Macmillan, New York, NY.

Calaprice, A. (Ed.), 1996. The Quotable Einstein. Princeton University Press, Princeton, NJ.

Callaway, D.S., Newman, M.E.J., Strogatz, S.H., Watts, D.J., 2000. Network robustness and fragility: percolation on random graphs. Phys. Rev. Lett. 85 (25), 5468–5471.

Cancho, R.F., Solé, R.V., 2001. Optimization in complex networks. Santa Fe Institute, Working Paper 01-11-068.

Capra, F., 1996. The Web of Life. Anchor Books Doubleday, New York, NY.

Capra, F., 2002. The Hidden Connections. Doubleday, New York, NY.

Capra, F., 2007. The Science of Leonardo. Doubleday, New York, NY.

Casaday, G., 1996. Rationale in practice: templates for capturing and applying design experience. In: Moran, T.P., Carroll, J.M. (Eds.), Chapter 12, Design Rationale. CRC Press, Taylor & Francis Group.

Checkland, P.B., 1981. Systems Thinking, Systems Practice. Wiley, New York, NY.

Csermely, P., 2006. Weak Links—Stabilizers of Complex Systems from Proteins to Social Networks. Springer, Berlin.

Curtis Jr., C.P., Greenslet, F. (Eds.), 1962. The Practical Cogitator or The Thinker's Anthology. third ed. Houghton Mifflin Company, Boston, MA.

Dawkins, R., 1996. The Blind Watchmaker. Norton, New York, NY.

Derényi, I., et al., 2004. Topological phase transitions of random networks. Phys. A 334, 583–590.

Dorigo, M., Gambardella, L.M., 1997. Ant colonies for the travelling salesman problem. BioSystems 43, 73–81.

Drury Jr., W.H., 1998. Chance and Change: Ecology for Conservationists. University of California Press, Berkeley, CA (edited by J. G.T. Anderson).

Dutton, D., 2009. The Art Instinct: Beauty, Pleasure, and Human Evolution. Bloomsbury Press, New York, NY.

Echenique, M. 1963. Models: a discussion. University of Cambridge, working paper number 6, Cambridge, UK.

Einstein, A., 1950. Out of My Later Years. The Philosophical Library, New York, NY.

Ellson, J., et al., 2004. Graphviz and dynagraph—static and dynamic graph drawing tools. In: Junger, M., Mutzel, P. (Eds.), Graph Drawing Software. Springer-Verlag, Berlin, Heidelberg, pp. 127–148.

Emerson, R.W., 1983. Experience. In: Emerson: Essays and Lectures. The Library of America, New York, NY.

Erdös, P., Rényi, A., 1960. On the evolution of random graphs. Publ. Math. Inst. Hung. Acad. Sci. 5, 17–61.

Evans, J.R., 1991. Creative Thinking in the Decision and Management Sciences. South-Western Publishing, Cincinnati, OH.

Feigenbaum, M.J., 1978. Quantitative universality for a class of nonlinear transformations. J. Stat. Phys. 19 (1), 25–52.

Flake, G.W., Pennock, D.M., 2010. Chapter 6: self-organization, self-regulation, and self-similarity on the fractal web. In: Lesmoir-Gordon, N. (Ed.), The Colours of Infinity: The Beauty and Power of Fractals. Springer-Verlag, London.

Flood, R.L., Carson, E.R., 1993. Dealing with Complexity: An Introduction to the Theory and Application of Systems Science. Plenum Press, New York, NY.

Fractal Geometry, Yale University. Available from: http://classes.yale.edu/fractals/.
Gharajedaghi, J., 1999. Systems Thinking—Managing Chaos and Complexity. Butterworth-Heinemann, Boston, MA.
Gladwell, M., 2000. The Tipping Point. Little Brown.
Goguen, J., 1998. Tossing algebraic flowers down the great divide. In: Calude, C. (Ed.), People and Ideas in Theoretical Computer Science. Springer, Berlin, Heidelberg, pp. 93–129.
Goldberger, A.L., 1991. Is the normal heartbeat chaotic or homeostatic? Am. Physiol. Soc. 6 (2), 87–91.
Goldberger, A.L., 1996. Non-linear dynamics for clinicians: chaos theory, fractals, and complexity at the bedside. Lancet 347, 1312–1314.
Goldberger, A.L., Rigney, D.R., West, B.J., 1990. Chaos and fractals in human physiology. Sci. Am. 262 (2), 42–49.
Gould, S.J., 2003. The Hedgehog, the Fox, and the Magister's Pox. Harmony Books, New York, NY.
Gould, S.J., Eldredge, N., 1977. Punctuated equilibria: the tempo and mode of evolution reconsidered. Paleobiology 3 (2), 115–151.
Granovetter, M., 1973. The strength of weak ties. Am. J. Sociol. 78, 1360–1380.
Gribbin, J., 2005. Deep Simplicity: Bringing Order to Chaos and Complexity. Random House, New York, NY.
Grudin, R., 1990. The Grace of Great Things: Creativity and Innovation. Ticknor & Fields, New York, NY.
Halley, J.M., 1996. Ecology, evolution and 1/f-noise. Trends Ecol. Evol. 11 (1), 33–37.
Harris, D., 2001. Supporting human communication in network-based SE. Syst. Eng. J. INCOSE 4 (3), 213–221.
Hazelrigg, G.A., 1996. Systems Engineering: An Approach to Information-Based Design. Prentice Hall, Upper Saddle River, NJ.
Holland, J.H., 1996. Hidden Order—How Adaptation Builds Complexity. Helix Books, New York, NY.
International Council on Systems Engineering. Available from: http://www.incose.org/.
Ivanov, P.C., et al., 1996. Scaling behavior of heartbeat intervals obtained by wavelet–based time-series analysis. Nature 383 (6598), 323–327.
Jacob, F., 1977. Evolution and tinkering. Science 196 (4295), 1161–1166.
Jeong, H., et al., 2000. The large-scale organization of metabolic networks. Nature 407, 651–654.
Ji, L.-J., Peng, K., Nisbett, R.E., 2000. Culture, control, and perception of relationships in the environment. J. Pers. Soc. Psychol. 78 (5), 943–955.
Jørgensen, S.E., Svirezhev, Y.M., 2004. Towards a Thermodynamic Theory for Ecological Systems, first ed. Pergamon Press, Oxford.
Kauffman, S.A., 1993. The Origins of Order: Self Organization and Selection in Evolution. Oxford University Press, New York, NY.
Kauffman, S.A., 1995. At Home in the Universe—The Search for Laws of Self-Organization and Complexity. Oxford University Press, New York, NY.
Kauffman, S.A., 2002. Investigations. Oxford University Press, New York, NY.
Kazanci, C., 2007. EcoNet: a new software for ecological modelling, simulation and network analysis. Ecol. Model. 208 (1), 3–8.
Klein, N., 2007. The Shock Doctrine: The Rise of Disaster Capitalism. Metropolitan Books/ Henry Holt, New York, NY.
Klir, G.J., 1985. Complexity: some general observations. Syst. Res. 2 (2), 131–140.
Klir, G.J., 1991. Facets of Systems Science. Plenum Press, New York, NY.
Krugman, P., 1996. The Self-Organizing Economy. Blackwell Publishers, Cambridge, MA.
Kuhn, T.S., 1996. The Structure of Scientific Revolutions, third ed. The University of Chicago Press, Chicago, IL.

Lack, D., 1954. The Natural Regulation of Animal Numbers. Clarendon Press, Oxford, UK.
Larman, C., 1998. Applying UML and Patterns. Prentice Hall, Upper Saddle River, NJ.
Laszlo, E., 1972. The Systems View of the World. G. Braziller, New York, NY.
Lesmoir-Gordon, N. (Ed.), 2010. The Colours of Infinity: The Beauty and Power of Fractals. Springer-Verlag, London.
Lipsitz, L.A., Mietus, J., Moody, G.B., Goldberger, A.L., 1990. Spectral characteristics of heart rate variability before and during postural tilt. Circulation 81 (6), 1803–1810.
Mandelbrot, B.B., 1982. The Fractal Geometry of Nature. W.H. Freeman and Company, San Francisco, CA.
Mandelbrot, B.B., 2004. Interview with the edge foundation. December 20, 2004, http://www.edge.org/3rd_culture/mandelbrot04/mandelbrot04_index.html.
Mandelbrot, B.B., 2010a. Fractals and the art of roughness. Talk presented at the TED2010 conference.
Mandelbrot, B.B., 2010b. Chapter 3: a geometry able to include mountains and clouds. In: Lesmoir-Gordon, N. (Ed.), The Colours of Infinity: The Beauty and Power of Fractals. Springer-Verlag, London.
Mandelbrot, B.B., 2012. The Fractalist: Memoir of a Scientific Maverick. Pantheon Books, New York, NY.
Manrubia, S.C., Paczuski, M., 1998. A simple model of large-scale organization in evolution. Int. J. Modern Phys. C 9, 1025–1032.
Marsh, T., Pfleiderer, P., 2012. "Black Swans" and the financial crisis. Rev. Pac. Basin Fin. Mark. Polic. 15 (02), 1250008-1–1250008-12.
MATLAB release R2009a, The MathWorks, Inc., February 12, 2009.
Maxwell, J.C., 1873. Molecules. Nature 8 (204), 437–441.
May, R.M., 1974. Biological populations with non-overlapping generations: stable points, stable cycles, and chaos. Science 186 (4164), 645–647.
May, R.M., 1976. Simple mathematical models with very complicated dynamics. Nature 261 (5560), 459–467.
Meiss, J.D., 2012. IFS software. Macintosh application, Version 2.2.1, October 22, 2012, http://amath.colorado.edu/faculty/jdm/ifs.htm.
Merriam-Webster Online. Available from: http://www.merriam-webster.com/.
Miettinen, K., 2008. Chapter in: Vermaas, P.E. et al., (Eds.), Philosophy and Design: From Engineering to Architecture, Part II. Springer, The Netherlands.
Milgram, S., 1967. The small world problem. Psychol. Today 1 (1), 61–67.
Mitchell, M., 2009. Complexity: A Guided Tour. Oxford University Press, New York, NY.
Molofsky, J., Bever, J.D., 2004. A new kind of ecology? Bioscience 54 (5), 440–446.
Montoya, J.M., Solé, R.V., 2000. Small world patterns in food webs. Santa Fe Institute, Working Paper 00-10-059.
Mowbray, T.J., Zahavi, R., 1995. The Essential CORBA: Systems Integration Using Distributed Objects. Wiley, New York, NY.
Müller, F., 1996. Emergent properties of ecosystems—consequences of self-organizing processes? Senckenberg. Marit. 27, 151–168.
Naeem, S., Golley, F.B., et al., 1999. Biodiversity and ecosystem functioning: maintaining natural life support processes. Ecol. Soc. Am. 4, 1–12, Issues in Ecology.
National Human Genome Research Institute. Available from: http://www.genome.gov/.
Newman, M.E.J., 2003. The structure and function of complex networks. SIAM Rev. 45 (2), 167–256.
Newman, M.E.J., Watts, D.J., 1999. Scaling and percolation in the small-world network model. Phys. Rev. E 60 (6), 7332–7342.
Next Generation Education Project of MDRC (originally the Manpower Demonstration Research Corporation). Available from: http://www.mdrc.org/.

Nicolis, G., Prigogine, I., 1989. Exploring Complexity: An Introduction. W. H. Freeman and Company, New York, NY.
Nisbett, R.E., 2003. The Geography of Thought: How Asians and Westerners Think Differently . . . and Why. Free Press, New York, NY.
O'Connor, J., McDermott, I., 1997. The Art of Systems Thinking: Essential Skills for Creativity and Problem Solving. Thorsons, London.
Odom, H.T., 1971. Environment, Power, and Society. Wiley-Interscience, New York, NY.
Odum, E.P., Barrett, G.W., 2005. Fundamentals of Ecology, fifth ed. Thomson Brooks/Cole, Belmont, CA.
Oppenheimer, J.R., 1954. A science in change. In: Science and the Common Understanding. Simon and Schuster, New York, NY.
Pajek—Program for Large Network Analysis. Available from: http://vlado.fmf.uni-lj.si/pub/networks/pajek/.
Patten, B.C., Odum, E.P., 1981. The cybernetic nature of ecosystems. Am. Nat. 118, 886–895.
Popper, K.R., 1983. Realism and the Aim of Science. Rowman and Littlefield, Totowa, NJ.
Popper, K.R., 1999. "Heroic Science" from replies to my critics. In: Bolles, E.B. (Ed.), Galileo's Commandment. Henry Holt and Company, New York, NY.
Rechtin, E., Maier, M.W., 1997. The Art of Systems Architecting. CRC Press, New York, NY.
Reilly, N.B., 1993. Successful Systems Engineering for Engineers and Managers. Van Nostrand Reinhold, New York, NY.
Richardson Jr., R.D., 1986. Henry Thoreau: A Life of the Mind. University of California Press, Berkeley, CA.
Rood, W., 2010. Chapter 5: fractal limits. In: Lesmoir-Gordon, N. (Ed.), The Colours of Infinity: The Beauty and Power of Fractals. Springer-Verlag, London.
Roughgarden, J., 1979. Theory of Population Genetics and Evolutionary Ecology: An Introduction. Macmillan Publishing Company, New York, NY.
Rumbaugh, J.R., et al., 1990. Object-Oriented Modeling and Design. Prentice Hall, Upper Saddle River, NJ.
Schiff, J.L., 2008. Cellular Automata: A Discrete View of the World. Wiley-Interscience, New York, NY.
Schneider, E.D., Kay, J.J., 1995. Order from disorder: the thermodynamics of complexity in biology. In: Murphy, M.P., O'Neill, L.A.J. (Eds.), What is Life: The Next Fifty Years. Reflections on the Future of Biology. Cambridge University Press, Cambridge, UK, pp. 161–172.
Schön, D.A., 1983. The Reflective Practitioner. Basic Books, New York, NY.
Schön, D.A., 1990. Educating the Reflective Practitioner. Jossey-Bass Publishers, San Francisco, CA.
Shannon, C.E., 1948. A mathematical theory of communication. Bell Syst. Techn. J. 27, 379–423, 623–656.
Shishikura, M., 1998. The Hausdorff dimension of the boundary of the Mandelbrot set and Julia sets. Ann. Math. 147 (2), 225–267 (Second Series).
Simon, H.A., 1955. On a class of Skew distribution functions. Biometrika 42 (3/4), 425–440.
Simon, H.A., 1962. The architecture of complexity. Proc. Am. Philos. Soc. 106 (6), 467–482.
Simon, H.A., 1977. How complex are complex systems? In: Proceedings of the 1976 Biennial Meeting of the Philosophy of Science Association. American Philosophical Society, pp. 507–522, Volume Two: Symposia and Invited Papers.
Simon, H.A., 1983. Discovery, invention, and development: human creative thinking. Proc. Natl. Acad. Sci. U. S. A. 80 (14), 4569–4571, Part 2: Physical Sciences.
Simon, H.A., 1996. The Sciences of the Artificial, third ed. MIT Press, Cambridge, MA.

Solé, R.V., Bascompte, J., 2006. Self-Organization in Complex Ecosystems. Princeton University Press, Princeton, NJ.

Solé, R.V., Goodwin, B., 2000. Signs of Life: How Complexity Pervades Biology. Basic Books (Perseus Books Group), New York, NY.

Solé, R.V., Montoya, J.M., 2001. Complexity and fragility in ecological networks. Proc. R. Soc. Lond. B 268 (1480), 2039–2045.

Solé, R.V., Valverde, S., 2007. Spontaneous emergence of modularity in cellular networks. Santa Fe Institute, Working Paper 07-06-013.

Solé, R.V., Manrubia, S.C., Pérez-Mercader, J., Benton, M., Bak, P., 1998. Long-range correlations in the fossil record and the fractal nature of macroevolution. Adv. Compl. Syst. 1 (2–3), 255–266.

Solé, R.V., et al., 1999. Criticality and scaling in evolutionary ecology. Trends Ecol. Evol. 14 (4), 156–160.

Solé, R.V., et al., 2003. Selection, tinkering, and emergence in complex networks: crossing the land of tinkering. Complexity 8 (1), 20–33.

Song, C., Havlin, S., Makse, H.A., 2005. Self-similarity of complex networks. Nature 433 (7024), 392–395.

Song, C., et al., 2007. How to calculate the fractal dimension of a complex network: the box covering algorithm. J. Stat. Mech. Theory Exp.. 2007(03).

Stewart, I., 2010. Chapter 1: the nature of fractal geometry. In: Lesmoir-Gordon, N. (Ed.), The Colours of Infinity: The Beauty and Power of Fractals. Springer-Verlag, London.

Strogatz, S.H., 2001. Exploring complex networks. Nature 410, 268–276.

Sullivan, L.H., 1896. The Tall Office Building Artistically Considered. Lippincott's Magazine.

Taleb, N., 2007. The Black Swan. Random House, New York, NY.

The MathWorks website. Available from: http://www.mathworks.com.

Thompson, D.W., 1992. On Growth and Form. Dover Publications, New York, NY (revised edition; first edition was in 1917).

Thoreau, H.D., 1927. The Heart of Thoreau's Journals. Houghton Mifflin Company, Boston, MA (edited by O. Shepard).

TIME Magazine, The I.Q. Gene?, 154 (11), September 13, 1999.

Ulanowicz, R.E., 1997. Ecology, the Ascendent Perspective. Columbia University Press, New York, NY.

Valverde, S., Cancho, R.F., Solé, R.V., 2002. Scale-free networks from optimal design. Europhys. Lett. 60, 512–517.

Vicsek, T., 2002. Complexity: the bigger picture. Nature 418, 131.

Vishwanathan, G.M., et al., 1999. Optimizing the success of random searches. Nature 401, 911–914.

Warfield, J.N., 1976. Societal Systems Planning, Policy and Complexity. John Wiley & Sons, New York, NY.

Watts, D.J., 2003. Six Degrees: The Science of a Connected Age. W. W. Norton, New York, NY.

Watts, D.J., Strogatz, S.H., 1998. Collective dynamics of 'small-world' networks. Nature 393, 440–442.

Weaver, W., 1948. Science and complexity. Am. Sci. 36, 536–544.

Weinberg, G.M., 1975. An Introduction to General Systems Thinking. Wiley, New York, NY.

Weinberg, G.M., Weinberg, D., 1988. General Principles of Systems Design. Dorset House, New York, NY.

Wesson, R.G., 1991. Beyond Natural Selection. MIT Press, Cambridge, MA.

White Jr., K.P., 1998. Systems design engineering. Syst. Eng. J. INCOSE 1 (4), 285–302.

Whitehead, A.N., 1959. Whitehead lectures in: Science and the Modern World: Lowell Lectures, pp. 10–14 and 19–22. New American Library, New York, NY.
Wiener, N., 1948. Cybernetics, or Communication and Control in the Animal and the Machine. MIT Press, Cambridge, MA.
Wilson, E.O., 1999. Consilience: The Unity of Knowledge. Vintage Books, New York, NY.
Wolfram, S., 2002. A New Kind of Science. Wolfram Media, Inc., Champaign, IL.
Yackinous, W.S., 2008. Dissertation Prospectus—Emerging Principles of Ecological Network Dynamics. University of Georgia.
Yackinous, W.S., 2010. Emerging principles of ecological network dynamics: innovative synthesis, modeling, analysis, and results. Doctor of Philosophy Dissertation, University of Georgia.
Yang, X.-S., 2006. An Introduction to Computational Engineering with Matlab. Cambridge International Science Publishing Limited, Cambridge, UK.

INDEX

Note: Page numbers followed by *f* indicate figures, *t* indicate tables and *b* indicate boxes.

D

E

I

J

L

T

Printed in the United States
By Bookmasters